THE
AVIAN MIGRANT

THE
AVIAN MIGRANT

The Biology of Bird Migration

JOHN H. RAPPOLE

COLUMBIA UNIVERSITY PRESS

NEW YORK

COLUMBIA UNIVERSITY PRESS

PUBLISHERS SINCE 1893

NEW YORK   CHICHESTER, WEST SUSSEX

cup.columbia.edu

Library of Congress Cataloging-in-Publication Data
Rappole, John H.
The avian migrant: the biology of bird migration / John H. Rappole.
pages cm
Includes bibliographical references and index.
ISBN 978-0-231-14678-4 (cloth : alk. paper)
ISBN 978-0-231-51863-5 (e-book)
1. Birds—Migration. 2. Migratory birds. I. Title.

QL698.9.R37 2013
598.156'8—dc23

2012036925

COVER IMAGE: © ALASKA STOCK/CORBIS
COVER DESIGN: CHANG JAE LEE

References to Internet Web sites (URLs) were accurate at the time of writing.
Neither the author nor Columbia University Press is responsible for URLs that may have expired
or changed since the manuscript was prepared.

Dr. Mario Alberto Ramos Olmos (1949–2006)

Great Ornithologist and Student of Migration

Extraordinary Conservation Biologist

Dear Friend, Sorely Missed

CONTENTS

PREFACE

But how to begin a beginning? A strange sound interrupted my tossing. I went to the window, the cold air against my eyes. At first I saw only starlight. Then they were there. Up in the March blackness, two entwined skeins of snow and blue geese honking north, an undulating W-shaped configuration across the deep sky, white bellies glowing eerily with the reflected light from town, necks stretched northward. Then another flock pulled by who knows what out of the south to breed and remake itself. A new season. Answer: begin by following spring as they did—darkly, with neck stuck out.

S o BEGINS William Least Heat Moon's *Blue Highways*, with the avian semi-annual journey through the blue standing in not only for the physical investigation of byways but also for metaphysical exploration. It seems to be an appropriate metaphorical beginning for this book as well because while several fine reviews of the voluminous literature on bird migration have already been published (e.g., Keast and Morton 1980; Gwinner 1990; Hagan and Johnston 1992; Alerstam and Hedenström 1998a; Berthold et al. 2003; Greenberg and Marra 2005; Newton 2008), none has taken the particular road I intend to travel. In this treatment, I will use the structure of the annual cycle as a means for examining how the avian migrant differs from its resident counterparts. The basic thesis is that these differences represent milestones along the evolutionary pathway that shape a migrant lifestyle through the process of natural selection. I am hopeful that this slight shift in emphasis, from a cataloging and description of the various aspects of bird migration to a synthesis and consideration of the larger picture into which the migrant fits, will result in a deeper understanding. Perforce, my treatment is selective rather than exhaustive, and topics considered important by others may not be addressed.

Pronouns are always a concern in a book of this kind, especially when the author has spent much of his life investigating many aspects of the topic in the field. Starting every sentence with "I" is not only annoying for the reader but also misleading in my own case, as more than 90 percent of my work has been done with the help of colleagues whose contributions were very large indeed—not to mention the fact that all inquirers, regardless of subject matter, stand on the shoulders of giants. Therefore, throughout this work, when referring to my published research and perceptions on major issues, I use the royal "we." This practice, however, is not meant to incriminate my collaborators. I take full responsibility for the opinions expressed.

The approach taken in this treatment is to provide "reflective inquiry" (*sensu* Reese 2012) on what a migrant is and how it came to be. In pursuing this end, I will strive to summarize what is known, what we think we know, and what we still don't understand about topics I consider to be critical. The reader will be referred to other sources for greater detail and will be exposed to competing hypotheses.

Ideas, of course, are the currency of the research biologist's profession, and it has been a concern of mine throughout to honor that concept. Therefore, my policy for literature citation in general is as follows:

- Cite the first-published or most influential paper that presents an important idea.
- Cite the most recent reviews and research of which I am aware.
- Cite my own work where it applies.
- Cite alternative explanations as pertinent.

Continental and epochal biases in literature citation characterize nearly all published scientific work and are, perhaps, not surprising given that a person's thought is formed largely in a particular intellectual milieu in space and time. I originally approached Barbara Helm, formerly of Germany's Max Planck Institute for Ornithology and currently lecturer at the University of Glasgow, to serve as co-author for this work, not only because of her brilliant and original work in collaboration with Eberhard Gwinner but also as a result of a conscious effort to address this problem; that is, my own intellectual consciousness was formed in the 1970s in the New World, whereas Barbara's was formed in the European tradition in the early 2000s. Unfortunately, work-level and other concerns forced her withdrawal, but she reviewed large portions of the manuscript in an attempt to help me address these issues. Nevertheless, biases of these kinds are, to some extent, unavoidable, and I apologize in advance where I have allowed them to creep in and persist.

One aspect of the continental divide is in the use of bird names: Old World biologists tend to follow different authorities from New World scientists. The procedure followed in this book represents an uneasy compromise. I use the taxonomic sequence and terminology of the American Ornithologists' Union (2012) *Check-list of North American Birds* wherever possible and Gill and Wright (2006) for everything else. This procedure results in some odd choices because the two references disagree substantially in some places, but it seemed best for my purposes.

The taxonomic focus for my research has been on passerines, and there is no doubt that a tendency to use examples from this group disproportionately exists in my treatment of the hypotheses and literature of migration. I mention this slant not only as an apology to colleagues who work with one of the many other orders in class Aves but also as a theoretical caveat. These other avian groups have ancient evolutionary histories and ecologies that are quite different from songbirds (e.g., several are pelagic). Certainly, there is room for future syntheses involving greater incorporation of what is known about migration among orders other than Passeriformes than I have attempted, and I look forward to reading the results of such efforts.

Much of what I will summarize and analyze concerning the adaptations of migrants in the following chapters is based on field observations. A generation ago, the great theoretical ecologist Robert MacArthur (1972) suggested that the era of descriptive biology was over: it was time for synthesis. In many ways, he was right. It was time for ecology to move beyond the descriptive work of field biologists with their dusty specimen collections and pedantic ways and begin to try and make some effort at understanding the larger picture using rigorous theoretical reasoning supported by modeling, mathematics, and statistics. An unintended consequence of this change in focus was the severe weakening of the authority of field biology as an important scientific endeavor relative to other branches of biological science. Yet this view ignores three contributions made by field biology studies to the understanding of the natural world that are of fundamental importance: (1) they provide the original data on which explanatory models, hypotheses, and theories are based; (2) they test the assumptions for these theoretical efforts; and (3) they track change. Of course, the latter two of these contributions are not always appreciated by the scientific establishment since they often involve presentation of negative data—that is, data that invalidate rather than support a popular hypothesis. A recent example of the tendency to denigrate field biology contributions is provided by the reception given to E. O. Wilson's (2012) book, *The Social Conquest of Earth,* where Wilson summarizes field data on human and social insect behavior that cannot be explained readily on the basis of individual or kin selection, and concludes that some form of group selection must be involved. This view is at odds with current theory and the majority of workers in the discipline of evolutionary biology (e.g., Dawkins 2012, Mithen 2012). Nevertheless, it is based on a variety of excellent data from field studies for which, at present, there appears to be no better explanation.

Despite the obvious continued need for field studies, particularly in the case of migratory birds, there are serious issues concerning the formulation of hypothetical explanations based on field data. Indeed, these explanations have been criticized, often correctly, as the telling of "just so" stories, as in "How the leopard got its spots" (Kipling 1912), because the investigator essentially accepts an explanation for observations without testing among the alternatives. Admittedly, it is often difficult to construct an experiment that will provide the data necessary to evaluate competing hypothetical explanations in the natural world. For instance, whereas it is relatively easy to assess the effects of nitrogen on a particular crop by assigning different nitrogen treatments to a set of test plots, all of which are treated in exactly the same way except for the amounts of nitrogen given to them, it is a good deal less simple to test how a female Northern Pintail (*Anas acuta*) selects a mate, even under relatively controlled conditions (Sorenson and Derrickson 1994). Nevertheless, without some form of testing among alternatives, no hypothesis can be considered as properly vetted. A way out of this dilemma is through the presentation of testable predictions, at least in some cases. When one has gathered the data necessary to formulate a hypothesis or several alternative hypotheses, one can then put forward a set of testable predictions. These predictions allow for direct or indirect testing among the various alternatives, as long as the assumptions on which the study is based are valid. Where possible, I use this approach in an effort to make clear what is known as well as what remains to be done to understand the avian migrant, stimulate future research, and help resolve important issues.

A critical aspect of my examination of the avian migrant is consideration of what physiologic and behavioral characteristics of migration are precisely programmed (i.e., in the migrant's genome) versus what remains flexible within well-defined genetic parameters (i.e., "reaction norms") (van Noordwijk 1989; Schlichting and Pigliucci 1998; van Noordwijk et al. 2006). Although this move "beyond nature and nurture" (Pigliucci 2001; Piersma and van Gils 2011) has profoundly influenced thinking in field biology, some remain skeptical about the entire area of reaction norm investigation (Zink 2002). As I understand their criticism, concepts such as reaction norms lack meaning because they have no physical aspect that can be tested in a laboratory. In this, they are quite right: We are at present only beginning to understand how the same genetic structure can result in different behavioral or physiologic manifestations in different environments. In some ways, this debate is similar to earlier controversies. In fact, Darwin's (1859) theory of evolution by adaptation through natural selection has never lacked for critics since its first publication. Although often motivated by challenges posed to sacred beliefs, many of these critics had valid observations concerning the theory's deficiencies. The most obvious weakness, at least for the first few decades after its proposal, was the lack of a plausible mechanism allowing the passing of information concerning the effects of natural selection from one generation to the next. Darwin's "gemmules" proffered a rather cumbersome but functional theoretical contraption to explain how natural selection might serve to use one generation to shape the next, just as Ptolemy's spheres worked for awhile to describe elliptical orbits for the planets. The work of Mendel, Hardy, Weinberg, Fisher, Watson, Crick, Mayr, E. O. Wilson, Dawkins, and many others built the foundation for the modern theory, placing it on much firmer functional ground. Nevertheless, the well-tested and studied principles of variation, heredity, differential reproduction, and mutation do not yet provide a complete explanation for how adaptation of populations occurs in the real world in general (Fodor and Piattelli-Palmarini 2009; Lewontin 2010) and in migratory bird behavior in particular (Helm et al. 2005; van Noordwijk et al. 2006; Piersma and van Gils 2011). Yet although the process may not be fully understood, the basic theory of adaptation through natural selection as proposed by Darwin remains the most powerful explanatory hypothesis we have for how organisms are shaped. Freeman Dyson (2003) wrote in response to critics of some of Isaac Newton's most profound thinking, "Newton was no fool." The same can be said for Darwin and for van Noordwijk as well.

Adaptations discussed for avian migrants in this book obviously are derived from natural selection, although some details of the mechanism remain to be clarified. Bird migration is one of the fields of biology where the concept of reaction norms appears to have its most obvious application. It may be that reaction norms are no more than another creaky contrivance, like gemmules, formulated to explain what at present cannot be explained by our modern theory of evolution. Nevertheless, they will serve my purposes well enough for the present, and I will use the concept to the best of my ability to emphasize those aspects of avian migration that make clear the fact that we do not yet have a complete understanding of how genes and environment interact to control the physiology and behavior of individuals.

The approach taken throughout the text has been to examine what is known versus what is hypothesized concerning the biology of the avian migrant. On occasion, this has resulted in detailed dissection of specific studies, a practice often referred to pejoratively (especially by those whose ideas have been subjected to such treatment) as setting up a "straw man" simply to knock it down. Although I sympathize wholeheartedly with the unhappy sensation caused by such treatment (having experienced it often enough myself), I disagree with the characterization. Those responsible for the hypotheses being parsed are giants in the field—straw is not the medium of which they or their ideas are constructed. Also, I am well aware of the fact that if my critique is flawed, I will receive a thunderous negative response in a variety of excruciatingly public venues. So be it. Progress in understanding is not achieved by blind acceptance or repetition. Failure to examine critically the assumptions on which important ideas are based or to consider alternative explanations are both serious disservices to science. Francis Bacon said, "If we begin with certainties, we will end in doubt, but if we begin with doubts and bear them patiently, we may end in certainty." This sentiment serves as the guiding principle for this book in which critical interpretation of ideas rather than a summary of facts or accepted explanations is the focus. Inevitably, this approach will bring me into some level of disagreement with a fair number of my colleagues and predecessors in the field of migratory bird studies. For these differences, I express to them my sincere apologies. In no way do I wish to minimize the importance of their contributions. I am well aware of the immense amount of work and commitment involved in the conduct of field and laboratory research, and I deeply honor their contributions. Nevertheless, progress in understanding requires synthesis, and synthesis requires judgment. As I have said at the beginning of this preface, I am sticking my neck out and flying into the dark in the hope of building a clearer understanding of the avian migrant.

ACKNOWLEDGMENTS

T HE ANCIENT Greeks knew something about sources of inspiration, usually citing one or more muses. For this work, I think Clio, goddess of historical accounts, might do. Lots of us certainly believe our best thoughts derive from the divine, although, as many long-suffering graduate assistants can attest, sources for many of the ideas of the major professor, maestro, or professor-doctor are often much more prosaic and closer to hand. I confess this to be the case for this work, which owes a great deal to my students and colleagues, past and present.

Not so long ago works of this kind could only be accomplished while situated at one of the world's great libraries. By the late 1970s, this was no longer the case. Although access to the services of an excellent library was still essential, computer searches and interlibrary loans made access to necessary references much easier than it had been. When I set about trying to find and read everything ever written on migrants in the neotropics with a grant from the U.S. Fish and Wildlife Service and the World Wildlife Fund in 1979, the task was far easier than it would have been a decade earlier. Nevertheless, I stressed the system at the University of Georgia Library and still fell short of my goal in the volume published on that research (Rappole et al. 1983). Today, the job is much simpler than it was in 1979 or at any previous time in history. With the online tools made available by a great research institution like the Smithsonian, most of the major ideas ever written on a topic can be obtained. So, although it is true that the Internet has perhaps McLuhanized and trivialized knowledge in some ways, it also allows complete immersion in the thought of the most outstanding workers in a person's field. I can attest that this is not always an uplifting feeling. One can feel completely beaten at the end of a day's work by the sheer intellectual weight of one's colleagues and predecessors. Another unfortunately common occurrence, for me at least, is to have finally synthesized a concept on paper only to have a key piece of information turn up that necessitates a complete revision.

In any event, the ideas presented in this book owe a great debt to those who came before me: a debt I feel has to be recognized specifically. The more work I have done, the clearer it has become to me that these thinkers laid the foundation on which we migrant workers all stand. My listing here is shamelessly idiosyncratic and incomplete, for which I apologize, but any such list must be, as it is a personal thing. So for me, this work owes the greatest intellectual debt to the following: Charles Darwin, W.W. Cooke, Ernst Mayr, David Lack, George G. Williams, Robert MacArthur, Stephen Fretwell, John Wiens, John Terborgh, Eberhard Gwinner, and Arie van Noordwijk. Of course, there are many others. Too often, I have written something, thinking that I had created the idea, only to find it written by somebody else long before me—which I probably had read and conveniently forgotten the source. I cannot say that I am happy when such errors are pointed out to me, but I do appreciate it and attempt to correct them. For those errors in proper sourcing for key concepts that remain, I apologize. They are a product of ignorance, not intent.

Several people gave invaluable aid with specific aspects of the project including Polly Lasker, National Zoological Park librarian, who provided me with dozens of references; Danny Bystrak of the U.S. Bird Banding Lab, who pulled together banding data for me on the Swainson's Thrush; Peter Jones, who helped me with extensive information and publications on his excellent work on inter-African movements of European migrants; and Clive Finlayson, director of the Gibraltar Museum, who generously made a preprint available to me of his book on the ornithogeography of the Palearctic (Finlayson 2011). Barbara Helm began as my co-author. The directions taken and work levels involved ultimately doomed the collaboration, but she performed thorough reviews of major parts of the manuscript, especially chapters 1 to 4. Darren Fa, Gene Morton, Dave King, Wylie Barrow, Peter Jones, Alan Pine, Kevin Winker, Richard Chandler, and Jorge Vega Rivera reviewed portions of the manuscript related to their areas of expertise and provided excellent advice. Jim Berry, director of the Roger Tory Peterson Institute in Jamestown, New York, allowed me use of a beautiful office at the institute during most of the period when the actual writing of the manuscript was done. My supervisors at the Smithsonian Conservation Biology Institute (Front Royal, Virginia), Steve Monfort and John Seidensticker, allowed me sabbatical leave during initial phases of the project and granted me *emeritus* status during its final years. Scott and Sue Derrickson offered generous accommodations at their home during our numerous visits to Front Royal while I was working on the manuscript. Alan Pine devoted a great deal of quality time and thought to the development of our ideas and models regarding factors governing migrant populations during different phases of the annual cycle, which appear in chapter 7 on population ecology and in two appendixes authored by him. Darren Fa, assistant director of the Gibraltar Museum, and the late Ralph Bingham, professor of mathematics and statistics at Texas A&M University, Kingsville, and R. Ciurylo, Atomic Physics Division, National Institute of Standards and Technology, provided thorough reviews of the reasoning and equations in chapter 7 and the appendixes.

The people who helped to bring out and shape my thought deserve a great deal of the credit for this work: They are a large part of the "we" in this book. It is no accident that much of the best work on migrants has been done by research groups—for example, the Lund University group, the Radolfzell group, the Grey Institute group, the Cornell group of the 1930s and 1940s, and so on. Obviously, good thinking is a group enterprise. The list of people who have helped me in this regard begins with my close friend and late colleague, Mario Ramos, a generous person who was surely one of the most brilliant minds and best on-your-feet thinkers in our field. Others who have directed and inspired my thinking include Gene Morton, Dwain Warner, Bonnie Rappole, Barbara Helm, Dave King, Bill McShea, and my former students Jorge Vega Rivera, Linda Laack, David Bergstrom, Kevin Winker, Dave (Swanny) Swanson, John Klicka, and Daniel Navarro Lopez.

Columbia University Press played an important role in this work. Jim Jordan, the director, worked with me on a former publication when he was at Johns Hopkins University Press. That pleasant association is responsible, at least in part, for this current partnership. Patrick Fitzgerald, publisher for life sciences at Columbia University Press, and Assistant Editor Bridget Flannery-McCoy, were extraordinarily helpful in providing the guidance and counseling necessary to see this project through to completion. Authors can, on occasion, be willful, proud, obstinate, paranoid, hysterical, lonely, and needy souls at various points over the years required to complete a project such as this, and it can take considerable skill to keep them relatively sane and on task. Patrick and Bridget possess that skill at a very high level or I doubt this book would ever have seen publication.

The copyediting team was also outstanding. Often this Augean stable–like chore of the publishing process is poorly or sloppily done by both author and copy-editors, which can result in errors that are horribly embarrassing such as misidentified photos, incorrect attributions, and incomplete sentences (or even paragraphs!). Ultimately, all errors are the author's responsibility, and they have the unhappy characteristic of permanence. As I write this (September 2012), it is premature to assume that no such errors exist. However, I have every confidence that the copyediting team (Chris Curioli and Ben Kolstad of Cenveo Publisher Services and Irene Pavitt of Columbia University Press) have done their very best to make as clean, clear, and correct a copy as possible.

Some years ago, I attended a meeting between potential grant recipients and a funding organization, which turned quite contentious when a colleague accused me of overstating the potential threat to migrants from wintering ground habitat destruction with the purpose of obtaining grant funds. I took the attack as personal, but after the meeting, he came over and asked me about some specific logistical aspects of conducting field work on migrants in a country where I had experience and he did not. We departed to discuss the project amicably over a beer. In thinking of this incident later, I realized that despite our differences in viewpoint, my colleague and I basically held similar values, lived similar lives, and were among the tiny number of people in the world who actually cared how or where migratory birds lived or evolved. I share this anecdote with the purpose in mind of reassuring my colleagues that although I may examine their work with a critical eye in this book, I know what is required to devote one's life to investigation of the abstruse minutia of biology, hold their work in the highest esteem, and acknowledge my deep debt to all of them.

THE
AVIAN MIGRANT

CHAPTER 1

INTRODUCTION

M IGRATION IS a form of dispersal involving regular movement and return between one place and another (Odum 1971:200; Fretwell 1972:130; Rappole 2005a), where "dispersal" is defined as a movement of an individual away from its place of birth or center of population density (Ricklefs 1973). For birds, the most typical form of migration involves an annually repeated, seasonal movement between the breeding range and those regions where breeding does not occur. The difference between migration for organisms in general and the phenomenon as it occurs in birds is principally a matter of scale in space and time. The purpose of this movement, regardless of the distances involved, is to exploit two or more environments whose relative suitability in terms of survival or reproduction changes over time, usually on a seasonal basis (Mayr and Meise 1930; Williams 1958; Rappole et al. 1983; Terrill 1991). A core concept for us is that the preponderance of field data that we will summarize in the course of this volume supports the view that initial movement by the first migrants (dispersers) was from a home environment of greater stability to one of lesser stability (i.e., greater seasonality); in other words, that most migrants derive from populations that evolved as breeding residents in their current wintering areas, not from populations originally resident on their current breeding areas.

Development of Human Understanding of Migration

Early human understanding of migration was based on extensive, but technically limited, observations subjected to limitless imagination. References to seasonal disappearance and reappearance of migrants, as well as to flocks of birds apparently in transit, is extensive in the classical literature—for example, the Bible, Pliny the Elder, and Aristotle—but most explanations were fanciful at best (Wetmore 1926; Hughes 2009). An exceptionally insightful work from the Middle Ages is the monograph *De Arte Venandi cum Avibus*, written by the Holy Roman Emperor Frederick II of Hohenstaufen (1194–1250), in which the author discusses feeding habits, morphology, and flight patterns of migrant versus resident birds. Sir Francis Bacon's publication of *The New Organon* in 1620 helped to establish a method for development and testing of hypotheses based on systematic accumulation of data, a concept that greatly enhanced investigation of natural phenomena, including migration. Ensuing developments are discussed in Birkhead's (2008) history of ornithology, *The Wisdom of Birds*. Among the landmarks was Catesby's (1746) presentation to the Royal Philosophical Society, which reflected some of this progress. Nevertheless, serious flaws in elementary comprehension remained (Catesby 1748). Even Linnaeus (1757) had some confused notions concerning migration, and as late as 1768, Samuel Johnson, as well educated and well read a person as existed at the time, could state with perfect confidence, "That woodcocks, (said he,) fly over the northern countries, is proved, because they have been observed at sea. Swallows certainly sleep all the winter. A number of them conglobulate together, by flying round and round, and then all in a heap throw themselves under water, and lye in the bed of a river" (Boswell 1791). In central Europe, bird fanciers noted behavioral changes in captive migrants that gave rise to ideas of "inborn migratory urge," with the first published description of migratory restlessness (*Zugunruhe*) appearing in 1707, and more thorough analyses of migratory activity in captive birds published by Johann Andreas Naumann in the late eighteenth century (Birkhead 2008).

The number and sophistication of observers increased rapidly during the nineteenth century, accompanied by wide dissemination of the results of their investigations through presentations at scientific meetings and publication. By the early twentieth century, the broad outlines of the timing, species components, and routes for avian migration systems in North America, Europe, and parts of Asia were fairly well understood (Middendorf 1855; Palmén 1876; Menzbier 1886; Gätke 1891; Clarke 1912; Cooke 1915; Wetmore 1926). Nevertheless, even the outlines remain somewhat dim for a few of the world's major migration systems—for example, the austral and intratropical migration systems of South America (Chesser 1994, 2005; Rappole and Schuchmann 2003) and the Himalayan–Southeast Asian systems (Rappole et al. 2011a). In addition, although the breadth and depth of scientific investigation of migration expands at a rapid pace in the early twenty-first century, significant questions remain, and new questions continue to appear (Bowlin et al. 2010).

ORIGINS OF MIGRATION

In the context of an evolutionary perspective of migration, consideration of the origins of migration are necessary. This topic has seen much debate (Zink 2002; Rappole et al. 2003a; Louchart 2008). For constructive discussion in the context of our analysis, we suggest that it is helpful to distinguish four meanings for the concept "origin of migration":

- The deep history of migration (i.e., its initial advent in geologic time in any organism)
- Its first occurrence in class Aves
- Its first appearance in a given avian taxon
- Its most recent occurrence in a given population or fraction of a population

In the first sense, migration likely appeared very early in the history of the development of life on Earth, which is why it has been found in such a wide range of organisms, including plankton, cnidarians, copepods, crustaceans, insects, fish, reptiles, birds, and mammals (Baker 1978; Ohman et al. 1983; Neill 1990; Dingle 1996; Bowlin et al. 2010). With regard to its first occurrence in class Aves (second sense of the meaning "origin of migration"), the earliest origins likely date back to the first evolutionary appearance of the group, with regular, large-scale movements probably being as old as flight and seasonality (Moreau 1972:xi; Alerstam 1990:6). The importance of seasonal change in habitat caused by variation in temperature or precipitation as the principal engine driving the development of long-distance migration is that such change provides powerful selection forces favoring those individuals capable of exploiting seasonal environments (Rappole and Tipton 1992; Rappole et al. 2003a; Rappole 2005a). Semiannual changes of habitat quality through effects on temperature or precipitation, the presumed environment favoring development of long-distance avian migration, is probably almost as old as terrestrial vegetation for a significant proportion of the planet's land surface. Seasonal habitats (e.g., deciduous forest) vary a great deal over the course of geologic time in their percentage of land cover and their location (Louchart 2008) according to whether or not an Ice Age is under way, but they have been a part of Earth's ecology since at least the Eocene Epoch of the early Tertiary, which predates the appearance of most modern bird families in the fossil record (Brodkorb 1971; Ericson et al. 2006; Chiappe 2007).

Holarctic environments during the Tertiary were, in general, much warmer than they are today, with subtropical climate at times extending as far as 50°N latitude (Louchart 2008; Finlayson 2011:15). Fossils dating from the Eocene (54–38 mya) and Oligocene (38–24 mya) of species representing avian families or their ancestors that are now considered tropical or subtropical (e.g., potoos [Nyctibiidae], trogons [Trogonidae], colies [Coliidae], and parrots [Psittacidae]) have been discovered in northern Europe (Mayr and Daniels 1998; Mayr 1999, 2001; Dyke and Waterhouse 2001; Kristoffersen 2002). Unfortunately, it is not possible based on their remains alone to tell whether or not these birds represented seasonal migrants on their summer breeding grounds or subtropical residents. There are modern examples of species of parrots and trogons known to migrate between temperate or subtropical portions of their breeding quarters and tropical or subtropical wintering quarters (e.g., Red-headed Trogon [*Harpactes erythrocephalus*], Elegant Trogon [*Trogon elegans*], and Burrowing Parrot [*Cyanoliseus patagonus*]) (Chesser 1994; Kunzman et al. 1998; Rasmussen and Anderton 2005a), which at least supports the possibility that these species could have been summer residents in the northern Holarctic during the Eocene and Oligocene, but there is no way of knowing for certain. Even if ancient DNA could be recovered, it is unlikely that a migrant individual could be clearly distinguished from a resident (Rappole et al. 2003a; Piersma et al. 2005a; Finlayson 2011:16), although some aspects of refinement for a migratory habit might be discernible from fossils (e.g., wing shape).

The third meaning for "origin" (development of migration in a particular avian taxon) is a much more recent event than either of the first two, at least for most species. Even so, the answer to the question of precisely how recently migration has appeared in any given taxon can be difficult to address. One reason for this difficulty is that resident species possess many if not most of the adaptations required of migrants, so that there are no known markers signaling a migratory habit whose appearance in a given phylogeny can be timed (Piersma et al. 2005a). Despite this qualification, it seems clear that for some groups or species, the origin of migration is much older than for others. For instance, a number of sandpiper species (Scolopacidae;—for example, the Eskimo Curlew *Numenius borealis*)—have breeding and wintering portions of the range separated by several thousand kilometers and lack conspecific populations that occur as year-round breeding residents in any portion of their current range (figure 1.1) (McNeil et al. 1994). Species of this sort evidently represent an evolutionary commitment to long-distance migration that is reflected not only in their lack of close relatives among resident populations in their home hemisphere but also in their wing structure (Pennycuick 1975) and transoceanic nocturnal navigation capabilities (Williams and Williams 1990). Nevertheless, it is important to remember that even for these species, many aspects of migration are fairly recent, as their current breeding areas were covered by several kilometers of ice up until a few thousand years ago (Finlayson and Carrión 2007).

An additional confounding factor for determination of time of origin of migration in a given species is that migration can appear or disappear in a population over very short time periods (Rappole et al. 1983; Able and Belthoff 1998; Helbig 2003; Rappole et al. 2003a; Pulido and Berthold 2003, 2010; Bearhop et al. 2005; Helm et al. 2005; Helm 2006; Helm and Gwinner 2006a; van Noordwijk et al. 2006). Hence, currently

FIGURE 1.1   Range of the Eskimo Curlew (*Numenius borealis*) in the mid-nineteenth century (Gill et al. 1998): *light gray* = breeding; *dark gray* = winter; *arrows* = migration route by season.

observed migratory behavior could have arisen from a fourth sense of the term "origin"—that is, as reinitiation of an activity that is well inside the evolved behavioral range (i.e., "reaction norm") of a species or population. The continental changes in range and migration pattern over very short time periods demonstrate the extraordinary flexibility in this dispersal/migration system, even in one that is newly developed (for a discussion on the evolution of migration, see chapter 8).

TYPES OF MIGRATION

There are no species of birds, even flightless ones, in which some part of the population does not undertake movement away from the breeding territory during some time of the year, whether through dispersal, extended foraging flights, or some sort of migration. In table 1.1, we list examples of major movement strategies of avian

TABLE 1.1   Major Movement Strategies of Bird Populations

MOVEMENT TYPE	DESCRIPTION
Local seasonal movements	This movement type includes several different seasonal movements found in supposedly sedentary species (e.g., postbreeding dispersal, distance foraging, and single- and mixed-species flocking).
Facultative migration	Movement during the nonbreeding season depends on environmental effects (e.g., weather, social interactions, and variation in food supply). As a consequence, individuals of the population move variable distances from the breeding territory; includes irruptive migrants.
Partial migration	Population consists of a migrant and resident fraction; some individuals undertake regular migrations while others remain on the breeding ground during the nonbreeding period.
Altitudinal migration	Individuals of the population migrate after breeding to a specific range at a different elevation.
Stepwise migration	Individuals of a population migrate from the breeding range to two or more specific nonbreeding ranges consecutively; a common example is molt migration.
Short-distance migration	All individuals of the population migrate to a specific nonbreeding range, but usually less than 2,000 km.
Long-distance migration	All individuals of the population migrate to a specific nonbreeding range, but usually greater than 2,000 km, and often across major barriers or to a different continent.
Differential migration	Different parts of a population migrate different distances, usually by age or sex.
Wanderers	Individuals breed at specific sites to which they return, and all individuals of the population depart the breeding site after reproduction is complete, but not to a specific nonbreeding range.
Nomadism	All individuals of the population move between breeding and nonbreeding areas, but neither is a fixed location.

populations, placing the different types of migration into this continuum, which we discuss in the sections that follow. We emphasize that several of these strategies are not mutually exclusive; that the grouping is somewhat arbitrary; and that in many species, local populations may differ among each other in terms of their migratory behavior (Terrill and Able 1988; Nathan et al. 2008; Newton 2008). Nevertheless, definition and depiction of these migration types serves the purpose of placing long-established categorization usage into the context of our continuum view of migrant evolution.

In table 1.1, we develop the idea of a continuum of migratory movements from sedentary resident to long-distance migrant and try to include all of the major migration types that have been described. Another way to approach this concept would be along the dichotomous divide between populations that undertake migratory movements only when environmental conditions force them to do so (i.e., "facultative" migrants) as opposed to those that undertake regular seasonal movements regardless of the environmental conditions (i.e., "obligate" or "calendar" migrants) (Newton 2008:334). However, we suggest that this dichotomy is less clear than is implied by these definitions (Berthold 1999, Newton 2012). For instance, while it is well known that many obligate migrants prepare for and undertake migrations based on an internal clock set by environmental light levels (Gwinner 1972), it is not known to what degree the ability of a facultative migrant to respond to weather may be under similar control; that is, can a facultative migrant in England respond by migrating to the same type of weather system that occurs in October as to one that occurs in January? Nor are similar situations understood for obligate migrants; that is, to what extent can the evolutionary program of an obligate migrant be modified by immediate individual circumstances having to do with age, food availability, weather, and so forth? For the present, we provide in the following sections a consideration of the major migration types as currently defined in the literature, with the understanding that we will address problematic issues concerning various aspects of these different types (e.g., what we consider to be a continuum between immediate versus evolutionarily programmed migratory response to individual circumstances or environmental change) as they arise in the context of our analysis later in this volume.

LOCAL SEASONAL MOVEMENTS

This category includes several different types of movement undertaken by a number of supposedly resident species on a regular, seasonal basis.

• *Sedentary resident dispersal.* This nonbreeding season movement strategy, principally among young birds, occurs in species that rely on year-round local food sources and is particularly common among a number of tropical birds, especially those living in some of the more stable environments (e.g., equatorial rainforest) (Skutch 1976). These habitats are not aseasonal, but variation is less than that found in many other environments at least in terms of some parameters (e.g., temperature), although others (e.g., rainfall) still may vary considerably by season (Moreau 1950). Typically, once a pair has established a territory, the adults remain on that territory until one or the other dies or the territory quality is altered (e.g., by burns, tree falls, and floods) (Willis 1967, 1972). The young raised each year, however, are forced to leave the territory some time after fledging but before the next breeding season begins (Skutch 1954, 1960, 1967, 1969, 1972; Fogden 1972; Willis 1972, 1973). Skutch (1969:419–421), for instance, describes the timing of this process for Northern Flickers (*Colaptes auratus cafer*) in the highlands of Guatemala. Adults dig the nest hole in February or March and the young leave the nest by June; fledged young remain with the family group until December or January when their parents drive them away as they begin preparations for the coming breeding season. For the Masked Tityra (*Tityra semifasciata*) (Tityridae) in Costa Rica where second broods may be raised, young from the first brood disappear from the parent's territory 3 weeks or so after fledging (Skutch 1969:35). Dispersal distances of the young of resident tropical birds are not well known, but Amazonian gene flow studies demonstrate that although low relative to Temperate Zone birds (Capparrella 1991), genetic exchange occurs between populations as much as 200 km apart (Bates 2000), even among some of the most sedentary of tropical understory species (e.g., antbirds [Thamnophilidae] [Terborgh et al. 1990]), demonstrating a clear capacity for long-distance dispersal in these sedentary residents.

• *Distance forager.* Colonially breeding, resident species of several different groups (e.g., some ardeids, ciconiids, and caprimulgids) move long distances between nest sites and feeding areas. For instance, Holland et al. (2009) found that Venezuelan Oilbirds (*Steatornis caripensis*) flew round trips of greater than 140 km between breeding caves and feeding areas.

• *Single-species flocking resident.* This behavioral pattern is similar to that described for sedentary residents except that once the chicks have fledged, a family with older young does not restrict its movements to the breeding territory and may range many kilometers distant from the breeding area in search of food in the form of fruiting trees or other temporary, clumped resources, a behavior prevalent among tropical frugivores (Levey and Stiles 1992). These flocks may be composed of family groups, groups of family groups, or combinations of both related and unrelated individuals as in the case of Passerini's Tanager (*Ramphocelus passerinii*) (Skutch 1954:163–164). The flocks can range many kilometers over a wide area during the nonbreeding season, but adults return to the home territory, lek, or colony site for breeding.

In other single-species flocking resident species, the nonbreeding season flocks may be composed of collections of individuals by different age or sex categories. For instance, the Great-tailed Grackle (*Quiscalus mexicanus*) is a resident of thorn forest and agricultural and urban areas in subtropical South Texas. Breeding colonies are concentrated in citrus orchards, thorn forest, and residential groves, but during the nonbreeding season, the birds form flocks of hundreds of birds, mostly by sex, that range over a large area between roosts in sugar cane fields or urban parks and feeding sites at agricultural fields and grain storage facilities, which may be 10–20 km distant from their breeding sites (Rappole et al. 1989a).

• *Mixed-species flocking resident.* Many resident birds that occupy distinct territories as pairs during the breeding season join mixed-species flocks during the nonbreeding season as individuals, mated pairs, or family groups in several different parts of the world (Moynihan 1962; Powell 1985; King and Rappole 2000, 2001a). For many participants, the flock's foraging area overlaps one or more of the breeding territories of the members, although for some the foraging area is outside of or well beyond the breeding territory boundaries (Powell 1985).

FACULTATIVE MIGRATION

There is a wide range of variation in terms of what facultative migration means with regard to timing, annual regularity, and distances traveled. For the purposes of our analysis, facultative migrants can be classified along an axis between two poles, likely with a complete gradation between them, perhaps even within the same species in different parts of their range:

- Species in which the majority of individuals remain on or near their breeding grounds throughout the nonbreeding period unless conditions become severe or food supplies fail; the most extreme case are irruptive migrants that emigrate from the breeding grounds only under extremely harsh conditions
- Species in which most of the population migrates away from the breeding ground every year, but settlement within the nonbreeding range varies according to variations in food supply and weather

In both cases, migrants occur on the nonbreeding grounds at irregular intervals and, if they do so in great numbers, are considered irruptive species (Newton 2008).

The Snowy Owl (*Bubo scandiacus*) is an oft-cited example of a resident species that undertakes periodic (irruptive) facultative migrations from its High Arctic breeding areas on a roughly 4-year cycle in synchrony with collapse of populations of its chief microtine food supply, the varying lemming (*Dicrostonyx groenlandicus*) and the brown lemming (*Lemmus trimucronatus*). However, Kerlinger et al. (1985) have challenged this assertion, noting that Arctic lemming populations are patchily distributed and not likely to vary in synchrony across the continent. Christmas bird count data from the species' North American wintering range show that although there is some indication of a 4-year cycle in eastern populations, the pattern in central and western populations is less clear, and all wintering populations show evidence of significant annual regularity of movement essentially beginning at the southern edge of the breeding range (figure 1.2).

In songbirds, classic examples of facultative movements are found among inhabitants of boreal forest, especially in the group of cardueline finches, but also

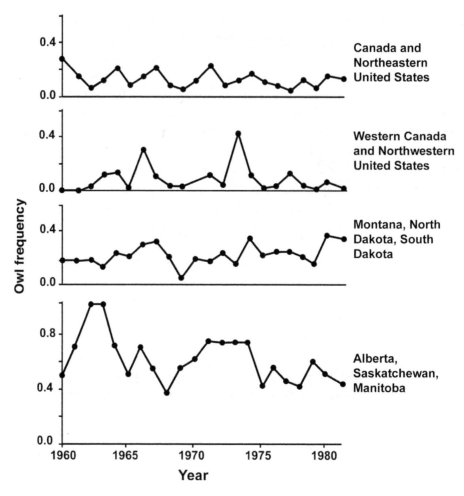

FIGURE 1.2   Snowy Owl (*Bubo scandiacus*) occurrence on Christmas bird counts for different parts of the winter range, 1960–1981 (Kerlinger et al. 1985).

in other groups such as tits and waxwings (Newton 2008:468, 561). The winter ecology of many of these species depends on food sources that are periodic (e.g., cycles of fruit, conifer cone, or mast production); these local or regional patterns affect the choice of wintering grounds in a given year, including years of local residency.

The Mallard (*Anas platyrhynchos*) is an example of a species that is a long- or short-distance migrant in northern parts of its breeding range and a facultative migrant in southern portions, where many individuals remain resident on breeding grounds until forced to move by actual freezing up of their foraging sites (hard weather movements) (figure 1.3) (Drilling et al. 2002). These movements can be, and often are, rapidly reversed when weather improves. In other species, part of the migratory movement may be obligate and a fixed part of the annual journey but may be followed by a facultative migration period that determines the precise location of the wintering grounds (Helms and Drury 1963; Terrill 1990).

FIGURE 1.3   Range of the Mallard (*Anas platyrhynchos*) in North America (Drilling et al. 2002): *medium gray* = summer resident; *dark gray* = summer and winter resident; *light gray* = winter resident.

PARTIAL MIGRATION

This type of migration involves residency on the breeding area throughout the year by one portion of the population, usually adult, territorial males, whereas the other portion migrates, usually females and young of the year. This migration strategy is found in a number of species that breed in Temperate Zone habitats that remain at least marginally usable during the winter period for the species but also occurs for many species from other migration systems. Nice (1937) reported partial migration in her population of Song Sparrows (*Melospiza melodia*) in Columbus, Ohio, as did Lack (1943) for his European Robins (*Erithacus rubecula*) in Devon, England. Marked birds are required to identify partial migration because in species that show different migration strategies in different populations, wintering birds from northern populations may replace southern breeding populations that migrate (figure 1.4). Only with marked birds can it be determined that the birds present during the nonbreeding period are the same as those that were present during the breeding period.

Factors controlling which portions of a given population migrate are not obvious and probably vary among species. Some authors have attributed differences in movement among different members of a given population to behavioral interactions in which dominant individuals force subdominant birds to migrate (Schwabl and Silverin 1990); others provide evidence of genetic dimorphism (Pulido and Berthold 2003). A third possibility is that the decision to migrate may be facultative and thus largely determined by environmental factors in combination with social interactions (Kalela 1954; Gillis et al. 2008), as has been suggested for Australian Silvereyes (*Zosterops lateralis*) and European Robins (Adriansen and Dondt 1990; Rappole et al. 2003a; Chan 2005; Helm 2006; van Noordwijk et al. 2006).

ALTITUDINAL MIGRATION

Altitudinal migration usually involves movement from a higher-elevation, seasonally occupied breeding habitat to a lower-elevation, nonbreeding or wintering habitat. Such movements have sometimes been documented by actual banding and recapture (Morton 1992; Gillis et al. 2008). Usually, however, their occurrence is surmised based on disappearance of the species from one site or region during a given season and its appearance at another site or region (Rappole and Schuchmann 2003). Nevertheless, accepting this caveat, altitudinal migration has been reported for many different avian groups from many parts of the world—for example, South America (Wetmore 1926; Roe and Ress 1979), Mexico and Central America (Monroe 1968; Ramos 1983, 1988; Winker et al. 1997), Africa (Moreau 1972; Curry-Lindahl 1981; Brown et al. 1982), and Southeast Asia (McClure 1974). An often-cited example involves hummingbirds that breed at a high elevation in the Andes but travel relatively short distances to lower elevations during the nonbreeding season (Schuchmann 1996, 1999). More than one-quarter of the more than 300 species of hummingbirds (Trochilidae) are thought to be altitudinal migrants (Rappole and Schuchmann 2003). Brown et al. (1982:15) report similar movements for several species in the Transvaal of South Africa. In addition, facultative altitudinal migration related to weather also has been reported in some areas. Ramos (1983, 1988), for instance, documented the occurrence of seven tropical highland (>800 m) species (e.g., White-throated Robin [*Turdus albicollis*] and Slate-colored Solitaire [*Myadestes unicolor*]) at lower elevations (<100 m) during the nonbreeding period in the Tuxtla Mountains of southern Veracruz, Mexico. He found a strong correlation between advent of cold fronts (*nortes*) and lowland captures of these birds. Presumed tropical

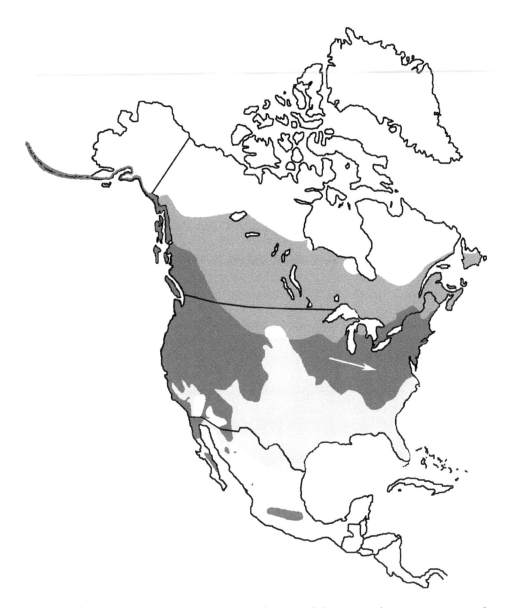

FIGURE 1.4   Range of the Song Sparrow (*Melospiza melodia*), a partial migrant (Arcese et al. 2002): *medium gray* = breeding; *dark gray* = breeding and winter; *light gray* = winter. Populations shown as "breeding and winter" have mixed migration strategies: The more northern of these populations are partial migrants in which adult males remain on territory throughout the year while females and young of the year migrate; more southern populations are sedentary residents. The arrow points to Nice's (1937) study area where she documented partial migration in this species.

hard weather movements across altitudes have been noted in Honduras (Monroe 1968), Bolivia (Wetmore 1926), and other regions (see review in Ramos 1983).

Well-documented altitudinal migration outside the tropics includes Appalachian populations of the Dark-eyed Junco (*Junco hyemalis carolinensis*), which move to neighboring lowlands in winter (Nolan et al. 2002), and American Dippers (*Cinclus mexicanus*), which move downstream into lower-elevation valleys (Gillis et al. 2008). In fact, some altitudinal migration likely occurs to a degree wherever major differences in elevation are present in habitable terrestrial regions.

There are also forms of altitudinal movement that are actually short- or even long-distance migration, in which members of a population breed in a highland habitat but travel long distances (>1,000 km) to winter at a lower-elevation habitat. Several species that breed in the Himalayas are thought to undertake these kinds of altitudinal migration, but banding data are lacking (Rasmussen and Anderton 2005a) (figure 1.5). The reverse of this type of movement also occurs, as in the Golden-cheeked Warbler (*Setophaga chrysoparia*), which breeds at elevations less

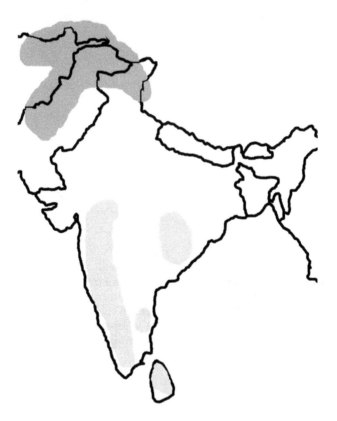

FIGURE 1.5   Long-distance altitudinal migration in the Hume's Whitethroat (*Sylvia althaea*) between high-elevation Himalayan breeding range (*dark gray*) and winter range (*light gray*) at lower elevations 1,500–3,000 km to the south in central and southern India and Sri Lanka (Rasmussen and Anderton 2005a).

than 600 m in Texas and winters in Central American highlands at elevations aver-aging greater than 1,600 m (Rappole et al. 2000a).

SHORT-DISTANCE MIGRATION

Obligate, short-distance migration in which all individuals of a population leave the breeding ground for a nonbreeding range in a different geographic area is a common strategy for species that breed in southern temperate and subtropical regions, but it also is found among a number of species in nearly all other migra-tion systems as well. However, different populations, and even individuals within populations, can follow different strategies, with some members acting as short-distance migrants and some as residents. Delineation of migration types by dis-tance is not straightforward. On the lower end of the distance range, short-distance migrations exceed those that we have called "local seasonal movements." Short-distance migration, as it is normally defined, leads to clear separation between a population's breeding and nonbreeding grounds and commonly follows latitu-dinal or climatic gradients. On the upper end, short-distance migrations are less extensive than long-distance migrations. The distinguishing characteristics of each type are not precise, but common cutoff criteria between short-distance and long-distance migration include the crossing of major barriers, distances that exceed 2,000 km, or changes of continent. These kinds of movement do not always occur on a north–south axis and are not necessarily related to temperate seasonal change. As in the case of altitudinal migration, documentation is often based on observa-tions of appearance of a particular species in one place at the same time that it is disappearing from another, rather than on mark–recapture data. Short-distance migration involving movement between different habitats at more-or-less the same latitude ("habitat migration") is best known in the tropics (South America [Zimmer 1938; Chesser 1994], Africa [Brown et al. 1982; Jones 1985, 1995, 1999; Urban et al. 1986; Fry et al. 1988], and Asia [McClure 1974; Rasmussen and Anderton 2005a]). In West Africa, where sharp differences in seasonal rainfall patterns can occur over relatively short distances, Brown et al. (1982) report that at least 70 species make annual seasonal movements between different habitats. For instance, the White-throated Bee-eater (*Merops albicollis*) breeds in subdesert steppe during the "north-ern tropical rainfall regime" (rainy season [April to October]; dry season [November to March]) and during the nonbreeding season moves into savanna, rainforest, and second growth habitats of the "equatorial bimodal rainfall regime" (rainy seasons [March to June and October to December]; dry seasons [the rest of the year]) (figure 1.6) (Fry et al. 1988:325).

LONG-DISTANCE MIGRATION

Long-distance migrants breed in areas that are separated by thousands of kilometers—often even on different continents—from the nonbreeding parts of their ranges. They

FIGURE 1.6   Range of the White-throated Bee-eater (*Merops albicollis*), a short-distance migrant between subdesert steppe breeding habitat (*light gray*) and savanna and evergreen forest winter habitat (*dark gray*) (Fry et al. 1988).

compose significant portions of higher-latitude avian communities during the breeding period (Powell and Rappole 1986; Newton 2003). In general, species that breed at the highest latitudes show the longest migration distances between breeding and nonbreeding portions of the range with the most extreme example being the Arctic Tern (*Sterna paradisaea*), which breeds in the high latitudes of the Arctic and spends the nonbreeding season in the high latitudes of the Antarctic (figure 1.7).

New microelectronic tracking devices now provide direct records of such long-distance movements of individual birds (Egevang et al. 2010). A recent, spectacular example was a transoceanic flight of a Bar-tailed Godwit (*Limosa lapponica*) that was recorded via satellite telemetry. Within 8 days, the bird flew more than 11,500 km over water from its Alaskan breeding grounds to the winter quarters in New Zealand, possibly nonstop (Gill et al. 2009).

Many long-distance migrants show similar, although perhaps not quite so extreme, separation between breeding and nonbreeding portions of the range.

As in other types of migration, whereas some populations may be long-distance migrants, other populations of the same species follow completely different migration strategies. For instance, different populations of the Snowy Egret (*Egretta thula*) show a wide variety of complex movement patterns including short-distance migration, long-distance migration, residency, and postbreeding wandering. Band

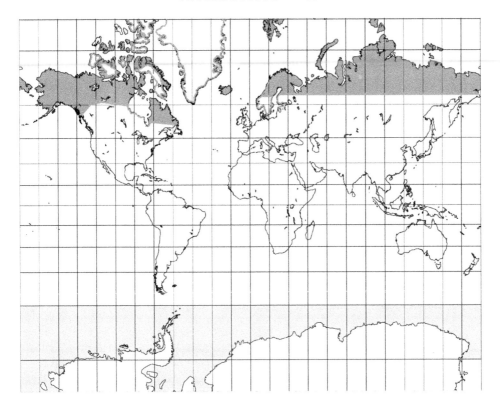

FIGURE 1.7   Range of the Arctic Tern (*Sterna paradisaea*), a long-distance migrant: *dark gray* = breeding; *light gray* = wintering.

return data show birds from the northeastern U.S. breeding range wintering in the Greater Antilles, whereas birds that are from the southeastern U.S. breeding range winter in eastern Mexico and Central America (Coffey 1948; Davis 1968). In both cases, long-distance migrants were found wintering in sites that have sedentary tropical resident populations of the same species.

The Yellow Warbler (*Setophaga petechia*) also shows mixed migration strategies according to the breeding latitude of the population, with more northern populations migrating the farthest south in a pattern referred to as "leap-frog migration" (Phillips et al. 1964:150). Conversely, in other species distinct populations from vastly different breeding areas meet on common wintering grounds—for example, the Palearctic migrant Yellow Wagtail (*Motacilla flava*), for which joint wintering in Zaire has been demonstrated (Curry-Lindahl 1981).

Differential Migration

This movement strategy is actually a subtype of nearly all categories of migration in which the population consists of individuals with different migratory behavior (Cristol et al. 1999). As will be discussed in detail in other chapters, each sex and age group, or even different individuals within groups, has its own physiologic attributes and needs that often result in differential expression of movement behavior, which, in turn, can result in entirely different mean movement timing and distance and in differential habitat use at stopover or wintering sites (Ketterson and Nolan 1976, 1983; Gauthreaux 1982; Mathot et al. 2007; Nebel 2007). The Dark-eyed Junco (*Junco hyemalis*) is exemplary. Nolan et al. (2002) report that as in the Song Sparrow, nearly all northern temperate and boreal breeding populations are short-distance migrants, although southern-breeding and montane populations may be partial migrants. For the short-distance migrant populations, adult males winter the farthest north on average, and young of the year tend to winter north of adult females.

An extreme case of age differences in migration is found in some long-lived species in which regular migration is delayed for years until sexual maturity is reached—for example, in raptors, shorebirds, and seabirds (Newton 2008). Sometimes differential migration strategies may vary by individual from one year to the next or even within a given season, apparently depending on environmental effects or population density in addition to age and sex (Hegemann et al. 2010).

STEPWISE MIGRATION

The extraordinary differences in seasonality that occur within tropical Africa have already been mentioned in the earlier discussion of short-distance migration. This pattern of seasonal variation in habitat quality and in the overall timing of the rains has resulted in a movement pattern in which some long-distance migrants from the Palearctic travel to at least two different nonbreeding sites sequentially, a pattern we define as stepwise migration ("step migration" of Curry-Lindahl 1981). Moreau (1972) defined this type of migration as "itinerancy," but this term carries an implication of wandering (see the next section, "Wanderers"). In contrast, it has been demonstrated by banding–recapture studies that some individuals of some long-distance migrant species return to the same nonbreeding sites, presumably sequentially, year after year (Curry-Lindahl 1981; Jones 1985). One of these sequential nonbreeding areas may be used for an annual molt (Yohannes et al. 2005, 2007). Indeed, "molt migration," in which birds migrate from breeding areas to specific molting sites during the postbreeding period prior to continuing on migration to wintering sites, is a well-known form of stepwise migration in North American waterfowl and some passerines (Bellrose 1976; Jehl 1990; Butler et al. 2002).

Recently, stepwise migration has been documented directly by use of satellite transmitters (Berthold et al. 2004) and indirectly by use of stable-isotope signatures that are incorporated into growing feathers in molting areas (Yohannes et al. 2005, 2007).

WANDERERS

For some species, especially pelagic birds (e.g., albatrosses and petrels), the breeding site, usually a colony attended by hundreds or thousands of breeding pairs, is

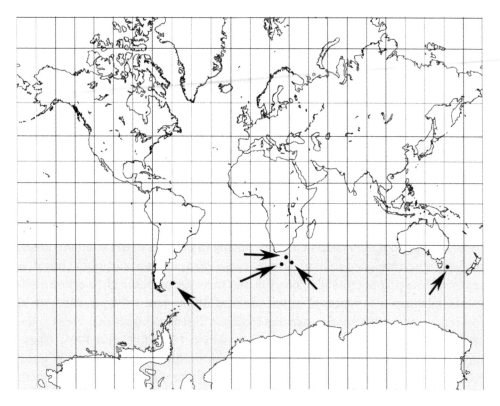

FIGURE 1.8   Breeding sites (*black circles*) and nonbreeding range (*light gray*) of the Wandering Albatross (*Diomedea exulans*).

a fixed location, often an oceanic island, to which the birds return year after year to raise their young. However, during the nonbreeding season, the birds range across a vast area of the world's oceans (figure 1.8).

NOMADISM

Nomads have no fixed breeding or nonbreeding areas (Dean 2004). They move throughout their range as weather patterns, usually rainfall, dictate, and breed when and where conditions are favorable in terms of food availability. This movement strategy is most common among species inhabiting tropical and subtropical deserts, especially in Australia and Africa. Frith (1969:35) notes,

> The best examples of large-scale nomadic movements are given by several of the waterfowl; the Grey Teal [*Anas giberifrons*], the Coot [*Fulica atra*] and the Pink-eared Duck [*Malachorhynchus membranaceus*] particularly. These birds are adapted to breed in shallow, temporary water, and their main breeding areas are on the inland plains of the Murray-Darling river system [southeastern Australia]. The extent of the habitat varies enormously in accordance with the degree of

flooding of the rivers, from month to month and year to year, so the birds must wander widely. They congregate briefly and then disperse widely in all directions.

Other examples are provided by the Budgerigar (*Melopsittacus undulatus*) (Frith 1969: 237) and by the African Red-billed Quelea (*Quelea quelea*) (Cheke et al. 2007), which gravitate quickly to sites receiving rainfall and commence breeding activities within a few days after precipitation events. Queleas, however, also show features of regular migrants, such as premigratory fattening and a tendency for directional preference (Ward and Jones 1977; Dallimer and Jones 2002). In some of their range, queleas are itinerant breeders; that is, they initiate clutches along the general route of their movements (Jaeger et al. 1986).

Not all nomads are desert species. The Red Crossbill (*Loxia curvirostra*) moves widely throughout the Holarctic region, mainly in boreal and mountainous areas, tracking their chief food source, conifer seeds (*Picea, Pinus, Pseudotsuga, Tsuga*). A remarkable aspect of crossbill biology is that there are at least eight North American populations that appear to be well defined in terms of call types and may in fact represent distinct species, as birds having different call types that breed at the same sites do not appear to interbreed (Benkman 1993; Adkisson 1996). These birds behave as though their range is wherever the flock is located at any given moment in time. Eurasian Crossbills show spectacular nomadic movements, with recorded distances of more than 3,000 km between breeding locations of individual birds (Newton 2006a).

PATTERNS OF MIGRATION WITHIN SPECIES

Populations of many migratory bird species are quite distinct, often being recognized as separate subspecies based on plumage differences. In addition to their distinctive appearance, these populations often differ in terms of their distribution. The principal categories of population distribution among migrant species are as follows:

- Common breeding and wintering area
- Common breeding area, separate wintering area
- Separate breeding areas, common wintering area (telescopic migration)
- Separate breeding and wintering area

Theoretically (i.e., not all types are known to occur), among those migrant species that have distinctive populations that have separate breeding and wintering areas, there are four major types:

- Longitudinal migration (with four subtypes, including chain migration)
- Parallel migration (with five subtypes)
- Leap-frog migration (with four subtypes)

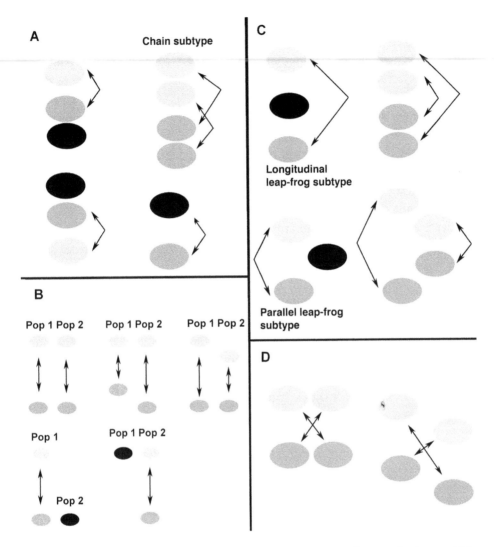

FIGURE 1.9   Potential distribution patterns for distinctive populations of migrants (Salomonsen 1955; Boulet and Norris 2006a:4): (A) four subtypes of longitudinal migration; (B) five subtypes of parallel migration; (C) four subtypes of leap-frog migration; (D) two subtypes of crosswise migration. *Light gray ovals* = breeding areas; *dark gray ovals* = wintering areas; *black ovals* = resident areas.

- Crosswise migration (with two subtypes) (Salomonsen 1955; Boulet and Norris 2006a; Newton 2008:675–696) (figure 1.9)

ADAPTATIONS FOR MIGRATION

Migratory birds present an extraordinary laboratory for observation of the process of evolution through adaptation. The term "adaptation" had a very specific meaning for Darwin (1859), who wished to restrict its use to refer to those characteristics formed through natural selection for the function they now serve. Feathers, for instance, are not "adaptations for migration" in the Darwinian sense because their evolution probably predates not only the development of migration but also the evolution of flight itself and in fact may have served a principally thermoregulatory function originally (Ostrom 1974; Prum and Brush 2002; Xu et al. 2009). Of course some kinds—for example, down and contour feathers—still do perform a mainly thermoregulatory function; but feathers of the wing and tail are critical for flight and certainly have undergone modification by natural selection for long-distance movement in those species that have been migrants for millennia (e.g., Eskimo Curlew [*Numenius borealis*]) (Pennycuick 1975). In these kinds of birds, the current structure of the feathers of the wing and tail can be considered to be "adaptations for migration."

Adaptiveness in the sense of enhanced probability of survival or reproduction for a given migratory habit is inferred from costs and benefits of particular feather and wing characteristics and from correlated evolution of migration with that of feather and wing (e.g., Bowlin et al. 2010). To focus on this specific meaning of "adaptation," Gould and Vrba (1982) developed an alternative term, "exaptation," to set aside characters evolved for other usages and later "co-opted" for their current role. Piersma et al. (2005a) attempted to assign the features of avian migrants to one of these two separate categories, adaptation versus exaptation, an exercise that we believe has value in terms of assessing how the characteristics of migrants differ from those of other birds ("preadaptation"; Helbig 2003). In the following sections, we take up the distinction between adaptation and exaptation to develop it into a conceptual theme of the book.

Many traits of avian migrants are plausible candidates for having been co-opted from other aspects of their evolutionary history to serve as the basis for a migrant lifestyle. These presumed exaptations can be assigned to major categories associated with theories of movement (Nathan et al. 2008). In table 1.2, we list five categories of exaptations that we believe make resident birds possessing them preadapted for migration. We discuss our arguments justifying these assignments in the text that follows.

LIFE HISTORY

In terms of life history, for the type of migration that is most prevalent in birds to develop requires a long-lived, K-selected organism the majority of whose breeding activities are restricted to a single season during the annual cycle. Evolutionary modifications to this basic form resulting from the selection pressures confronting migrants have been extensive, as will be discussed in later chapters.

MOVEMENT

Movement can cover a wide range of activities (e.g., flying, walking, crawling, and swimming); in addition, it can often be aided by wind or water currents. All of these

TABLE 1.2   Exaptations of Resident Bird Species for Migration

CATEGORY	EXAMPLES
Life history	K-selected (long-lived, few offspring); seasonal breeder
Movement	Flight (specialized circulatory system, skeleto-muscular system, pulmonary system, visual system)
Navigation	Directional orientation for long-distance movement; *ortstreue*
Energy storage and use	Hyperphagia; subcutaneous fat storage; rapid mobilization of fat reserves when needed
Timekeeping	Ability for keeping track of seasons at remote places that may differ greatly from those of the current environment

kinds of movement are used by different groups of organisms for the purposes of migration, although for birds, the principal movement mode is flight (Baker 1978; Dingle 1996). The adaptations for flight in birds affect most aspects of morphology and physiology. Although flight is not necessary for a migratory habit to develop, it is an extraordinary exaptation for migration; indeed, an exaptation that makes birds the animal taxon with the largest percentage of migratory species. Evolutionary adjustment of movement traits by migrants includes refinement and integration of various morphologic, physiologic, behavioral, and metabolic characteristics (Weber 2009).

NAVIGATION

In order for movement to be migration, it must involve at least two trips: a trip out to some site separate from the point of origin, and a trip back. Therefore, development of migration requires exaptations for navigation that must include a way for the organism to take in and store information about its environment and to use that information to guide itself from point A to point B and then back to point A, a process that requires a particular kind of navigation called "homing" (*ortstreue*). Nearly any piece of distinctive information that an organism can detect and remember, which can help to locate and distinguish one site from another, can be used in homing: What is used depends on the sensory capabilities of the organism. Sight, smell, taste, and hearing have been demonstrated to provide cues for homing navigation in various migratory organisms, as well as the ability to detect differences in Earth's magnetic field and polarized light (Storm 1966; Baker 1978; Wiltschko and Wiltschko 2003). Although the mechanism is not always understood, the ability to home has been discovered in a wide range of animal life, including many kinds of organisms in which migration is unknown (Rosengren and Fortelius 1986; Wiltschko 1992; Ramos and Rappole 1994).

ENERGY STORAGE AND USE

An additional category of exaptations for migrants includes energy storage (i.e., the laying down of subcutaneous fat reserves). Although considered by many authors to be actual adaptations for migration (e.g., Newton 2008:6), the capacity to store energy at times when patchy foods are abundant or when the organism is potentially confronted with an imminent need (e.g., seasonal change in resources) is found in many, if not most, resident birds as well (Rappole and Warner 1980). For instance, it is a common strategy in sedentary, resident, altricial birds to feed nestlings intensively prior to fledging so that when they leave the nest, the fledglings have significant fat reserves (e.g., Robinson et al. 1996; Phillips and Hamer 1999). It is also well known that young of at least some resident species lay down large amounts of subcutaneous fat prior to dispersal (Fogden 1972). Females of some resident species lay down fat reserves preparatory to egg-laying, and many also lay down fat reserves preparatory to undergoing molt or other "lean season" demands (McNeil 1971; Jehl 1997). Also, members of both wintering and resident tropical species that exploit temporary resource concentrations (e.g., fruiting trees) have been captured with moderate to heavy subcutaneous fat (Rappole and Warner 1980:381; Katti and Price 1999).

TIMEKEEPING ABILITIES AND RESPONSE TO ENVIRONMENTAL CUES

The selective value of movements over considerable distances improves significantly to the degree that such movements are properly timed to seasonal events (Gwinner 1972). Thus close timekeeping, including internal mechanisms that help migrants keep track of time in different, sometimes remotely distant areas, is critical. Furthermore, fitting of migratory periods into the annual cycle requires integration of other annual cycle stages. In principle, such long-term timing mechanisms are probably available to most higher, long-lived organisms, including early vertebrates. It seems reasonable to assume that in view of their general mobility, birds as a taxon have refined environmental tracking abilities, which have helped early migrants to time their movements. We argue that evolution of a progressively more migratory habit affects the entire annual cycle, as well as its organization, integration, and regulation.

CHAPTER 2

BREEDING PERIOD

F OR MIGRANTS, the onset of breeding is limited by time of arrival and the state in which they arrive. The factors that affect the timing of spring migration, from departure from the wintering grounds to arrival on the breeding grounds, are the topic of chapter 6, where we consider both the proximate factors that affect arrival (e.g., circannual rhythms) and the factors that affect progress along the route, in addition to ultimate factors that shape evolutionary refinement of a given population's migration biology. Here in chapter 2, we focus on that portion of the migrant life cycle commencing with arrival on the breeding grounds and terminating with completion of reproductive activities.

BEGINNING THE CYCLE: ARRIVAL AT THE BREEDING GROUNDS

Arrival times naturally differ for different kinds of migrants. For example, in temperate eastern Europe, long-term averages of first spring arrival dates for different species are spread out over almost 3 months, with short-distance migrants generally arriving back earlier and showing a much broader range of annual variation in arrival time than the long-distance migrants (Hubálek 2003; Tryjanowski et al. 2005). Many of these short-distance migrants are aquatic species that initiate migratory movements in apparent response to weather conditions. These birds are often referred to as facultative or weather migrants. In contrast to these species are many long-distance migrants whose spring arrival times are generally later and much less variable. These birds are often referred to as "calendar migrants" because, like swallows returning to Capistrano, the first arrival date can be quite similar from one year to the next, indicating possession of a programmed internal clock (Gwinner 1972). In reality, timing differences are much more complex than the dichotomous terminology would suggest (see chapter 6).

Species-specific arrival dates at a given locality indicate that birds time their return to meet a particular phase of the progressing spring phenology—for example, the "green wave" described for migratory geese (van der Graaf et al. 2006). Accordingly, arrival times for a given species can vary considerably between localities that have different seasonal phenologies, with generally later arrival at higher latitudes. This pattern has been demonstrated in a study with Bar-tailed Godwits (*Limosa lapponica baueri*) that were individually marked and equipped with miniature light loggers (geolocators) on the wintering grounds in New Zealand (Conklin et al. 2010). The geolocators collected locality information for the birds throughout the year and provided clear evidence that birds arrived at the highest-latitude breeding sites latest.

As they approach the breeding grounds, individuals of many migrant species increase their rate of movement during the last part of migration (Cooke 1915:43–47; Lincoln 1952; Dorst 1962; King 1972:211; Piersma et al. 2005b; Yohannes et al. 2009), although they may be limited by the progress of spring at staging and breeding sites (van der Graaf et al. 2006; Raess 2008; Tøttrup et al. 2010). This time period around first arrival is often considered a decisive point in, and even a potential bottleneck for, the annual cycle (Brown and Bomberger-Brown 2000; Buehler and Piersma 2008). Although there is convincing evidence that early arrival and early reproduction are rewarded by fitness benefits, there is also risk. For example, severe weather and storms in early spring can create mass mortality of a scale that is sufficient to affect population-wide arrival patterns in subsequent years (Brown and Bomberger-Brown 2000). These aspects of spring migration will be addressed more fully in chapter 6.

Arrival can differ considerably between sexes, age groups, and individual members of a given population (Helm 2006). We therefore begin consideration of the modifications to life history resulting from a migratory habit by comparing the major breeding ground arrival strategies found in migrants (table 2.1).

For many typically terrestrial migrants, the most common return strategy is that in which unpaired adult males are the first to arrive back on the breeding grounds, preceding adult females by several days or even weeks (table 2.1, no. 2) (Chapman 1894). This behavior is known as *protandry*, or *pioneering* in older literature (Cristol et al. 1999; Harari et al. 2000; Morbey and Ydenberg 2001; Newton 2008:427–432). Protandry is a frequent occurrence in taxa other than birds—for

TABLE 2.1   Arrival Patterns and Examples for Different Kinds of Migrant Birds

Arrival Pattern	Examples
1. Adult male remains on breeding territory throughout the year or migrates a shorter distance than the female and returns earlier (protandry).	Partial or differential migrants (see chapter 1)
2. Male precedes female in return to breeding territory (protandry).	Many long-distance migrants (see species accounts in Poole [2010] and del Hoyo et al. [1992–2011])
3. Female precedes male in return to breeding territory (protogyny).	Polyandrous species, for example, Spotted Sandpiper (*Actitis macularia*) (Oring and Lank 1982)
4. Sexes arrive on breeding ground simultaneously but unpaired.	Many species that winter and migrate as members of flocks of conspecifics of mixed age and sex groups, for example, many icterids (Brewer's Blackbird [*Euphagus cyanocephalus*] [Orians 1980]) and aredeids
5. Paired birds arrive together.	Anatids, for example, Northern Pintail (*Anas acuta*) (Austin and Miller 1995) and many others (Bellrose 1976; Batt et al. 1992)
6. Pairs arrive together with young of the previous year.	Many geese (Bellrose 1976; Batt et al. 1992) (e.g., White-fronted Goose [*Anser albifrons*]) (Barry 1966) and cranes (e.g., Sandhill Crane [*Grus canadensis*]) (Tacha et al. 1992)

example, arthropods, mammals, fish, nematodes, and amphibians (Morbey and Ydenberg 2001; Lodé et al. 2005). In birds, there are many variations on the basic protandry pattern (e.g., partial or differential migration, in which many of the adult males remain on or near the breeding territory throughout the year) but also diverse other patterns. In rare cases, the sex difference in arrival time is reversed. In migrant species in which the females are polyandrous, females precede males in arrival at breeding areas (protogyny) (e.g., Spotted Sandpiper [*Actitis macularia*], Wilson's Phalarope [*Phalaropus tricolor*]) (Oring and Lank 1982; Colwell and Jehl 1994). In species that winter or migrate as members of conspecific flocks including both sexes, males and females may arrive separately but at similar times (table 2.1). The degree and direction of sex differences in arrival falls on a continuous gradient and has been analyzed in detail for correlates with life-history patterns (Spottiswoode et al. 2006; Tøttrup and Thorup 2008).

Several hypotheses have been proposed to explain early arrival by one sex versus the other at breeding sites (Myers 1981; Morbey and Ydenberg 2001), at least

TABLE 2.2   Hypothetical Explanations for First Arrival at Breeding Sites

HYPOTHESIS	EXPLANATION
1. Different sex roles	Males and females differ in a variety of ways including morphology, physiology, and life history that affect endogenous timing of migration and arrival on the breeding ground (Myers 1981; Coppack and Pulido 2009). For instance, because of high nutritional investment in egg production and incubation, female condition is decisive for timing and success of reproduction (Caro et al. 2009). Reproduction of migrants often depends at least in part on reserves that females bring to the breeding grounds (Raveling 1979). Therefore, females may experience selection against maximization of time on the breeding ground at the cost of low body reserves at the time of arrival.
2. Intrasexual competition	Competition for quality breeding territories is intense, and early arrival is critical to obtaining the highest-quality territories. Therefore, territory defenders either remain on or near the breeding territory throughout the winter or arrive back at the earliest feasible moment to obtain an advantage over same-sex competitors (Myers 1981; Oring and Lank 1982).
3 Intersexual (epigamic) selection	Territory quality (habitat, potential nest sites, feeding areas) is an important factor in mate choice by females and breeding success, and early male arrival is critical to obtaining the highest-quality territories (Fretwell 1972).
4. Dominance	Defenders of the breeding territory, usually older males, are the dominant individuals in competition for all key resources. These individuals secure optimal wintering sites and are in better condition than other individuals in the population (e.g., females and younger males). Therefore, they migrate to breeding areas at the optimal time, arriving before less competitive individuals (Gauthreaux 1982; Marra and Holberton 1998; Marra et al. 1998).

four of which appear applicable to migrant birds (table 2.2). Field observations and modeling tend to support the sexual selection and general sexual difference hypotheses (table 2.2, nos. 1–3) (Myers 1981; Oring and Lank 1982; Francis and Cooke 1986; Kokko et al. 2006). In practice, it seems likely that both intrasexual and epigamic selection (table 2.2, nos. 2 and 3, respectively) affect early arrival time, along with general sexual differences in life history and reproductive responsibilities.

Kokko et al. (2006) point out that whereas sexual selection may explain early male arrival in many migrants, it does not explain why females do not arrive early as well, as females must compete among each other for the highest-quality males and territories. Their models, however, do not take into account the fact that early arrival has fitness costs as well as benefits, and this balance is likely to differ between the sexes, especially directly before the onset of egg production and incubation (table 2.2, no. 1). For instance, not only would early arriving females be taking the same risks of weather-related food shortages as males, but also they would be forced to compete with males for limited resources. At the same time, females are challenged by preparing for the subsequent production of eggs (Nager 2006). Eggs are often rapidly produced and, in many species, total clutch mass is high in relation to body mass. In addition, incubation, which is usually done mostly or entirely by females, requires considerable energy resources (Visser and Lessells 2001). Females therefore typically control both the timing of laying and clutch size (Ball and Ketterson 2008; Caro et al. 2009). The resources necessary for these activities may be difficult to accrue at the time of spring arrival when conditions are often still harsh. Female migrants may therefore partly or entirely rely on stores that they build up at wintering or staging sites (capital breeders, as opposed to income breeders) (Raveling 1979; Yohannes et al. 2010). Benefits of arriving in a good nutritional state thus may outweigh benefits of early arrival or rapid completion of migration ("sprint migration") (Alerstam 2006). Obviously, the optimal balance between these factors could differ significantly by sex and species.

The dominance hypothesis for protandry may have some relevance for explanation of age differences in terms of arrival on breeding territory (i.e., older males and females tend to arrive earlier than younger birds) (table 2.2, no. 4), but does not appear to provide a good explanation for differences in arrival time between males and females. These differences have been found to be genetically programmed in several species (Coppack and Pulido 2009), indicating a relationship between timing of arrival on the breeding ground and the different roles played by the different sexes during this period.

As noted in table 2.1, not all migrants follow separate return to breeding quarters strategies. In several migratory ducks and swans, pairing occurs on the wintering grounds, and paired males accompany females back to the female's birth site or region for breeding (Sowls 1955; Bellrose 1976; Rees 1987; Batt et al. 1992; species accounts in del Hoyo et al. 1992–2011; Poole 2010). Another pattern for timing of return to breeding sites is followed by some especially long-lived species (e.g., cranes and geese) in which parents invest significant care for young raised each year. In these species, adults allow the young to migrate with them to wintering sites, and the birds even return to breeding areas as a family group (Bellrose 1976; Batt et al. 1992; Tacha et al. 1992). Other species may perform spring migration as members of large flocks of conspecifics that consist of birds of different age and sex groups so that arrival of these groups at the breeding area is more or less simultaneous (Newton 2008).

Where individual birds are closely monitored, year-to-year arrival schedules are often highly repeatable between years (Rees 1989; Potti 1998; Gunnarsson et al. 2006). Furthermore, individual schedules of breeding partners may be closely correlated, regardless of whether or not the sexes travel together. An example is provided by findings obtained for the Black-tailed Godwits (*Limosa limosa islandica*) (Gunnarsson et al. 2004). Mates differ in precise over-wintering range, but their return schedules are nearly perfectly correlated. In this and other species, coordinated return schedules are thought to benefit the pair bond by facilitating a speedy onset of breeding activities (Davis 1988).

SITE FIDELITY AND DISPERSAL

As birds approach the breeding grounds, migration is accelerated but movements are increasingly scattered, possibly indicating a search for a suitable breeding habitat (Karlsson et al. 2010). However, whereas some birds may be searching for new breeding territories (see later), most migrants will return to breeding sites that they have experienced earlier as breeding or juvenile birds, and thereby display site fidelity (*ortstreue*) rather than dispersal (Greenwood and Harvey 1982). For many migrants, breeding habitat selection often involves return to the territory on which the bird bred the previous year. Return rates vary naturally from species to species and year to year but average around 50 percent for adult small passerines to as high as greater than 90 percent for some long-lived species like gulls (Roberts 1971; Pierotti and Good 1994; Sillet and Holmes 2002). Banding studies demonstrate that adult migrants of many taxa that have bred successfully in a previous year return to their breeding territories at a rate assumed to be comparable to that of survivorship (Lack 1966:111; Nolan 1978:37; Coulson and Butterfield 1986). In many songbirds, adult females tend to return at a rate somewhat considerably lower than that of males, for which there are at least two explanations:

1. Adult female survivorship is lower than that of adult males (Marra and Holberton 1998).
2. Females move to a different territory from that occupied the previous year at rates higher than those of males (Sillet and Holmes 2002:303).

At present, we know of no method for assessing the relative importance of these two possibilities for any single species of migrant, let alone for songbird migrants as a group. However, there are data demonstrating that adult females change mates (and territories) both within and between years, at fairly high rates in many studies, lending strong support to the second explanation as at least comprising a major portion of the apparent difference in adult male and female breeding site fidelity (Lack 1966:111; Nolan 1978:35; Vega-Rivera et al. 1999).

Dispersal of adult birds (breeding dispersal) differs from that of juveniles that return for their first breeding effort (natal dispersal) (Greenwood and Harvey 1982; Paradis et al. 1998; Arsenhault et al. 2005). The general pattern is a much larger dispersal distance in young compared with older birds although in some cases the opposite pattern has been reported (Dale et al. 2005). Natal philopatry to hatching sites for migrants shows a large range in variation among taxa. For those species that migrate as family groups, the return rate can be close to 100 percent of surviving offspring (Bellrose 1976; Tacha et al. 1992), although adults may force dispersal shortly after arrival at the breeding area. For colonial waterbirds—for example, the Herring Gull (*Larus argentatus*)—return rates to colony of origin vary by sex (male return rates are higher) and colony density (lower density colonies have higher return rates), and return often is delayed for several years until maturity is reached (Pierotti and Good 1994). In waterfowl, yearling females return to their natal areas at rates comparable to those of survivorship in several species whereas males do not, as they often pair in winter, and paired males accompany the female to her site of origin (Bellrose 1976; Rees 1987; Rohwer and Anderson 1988; Batt et al. 1992). At the other extreme are species that follow individual migratory schedules. For these species, which includes most migrant passerines, observed natal return rates are often less than 2 percent (Weatherhead and Forbes 1994).

Explanations that have been given for low observed return rates to natal sites by yearling passerine migrants include low survival, low fidelity, or a combination of the two (von Haartman 1971; Weatherhead and Forbes 1994). There is, however, an additional possibility: sampling bias. In other words, observed return rates could result from problems with the techniques used to measure them. Weatherhead and Forbes (1994) report on 17 studies of natal return rates in passerines based on recaptures or resightings of yearlings banded as juveniles. They found a median level of 2.1 percent, which they viewed as low if young were attempting to return to natal site. Certainly a 2.1 percent return rate is low compared with that of adults of the same species, which can be 30 to 70 percent (Roberts 1971; Sillet and Holmes 2002), but there is a potential difficulty in that to make this comparison, one must assume that yearlings will be found if they do return (Doligez and Pärt 2008). Here, it is important to note that there is a large difference between returning to the natal area and settling in the natal area. If adults are present, then returning young may be driven off from the site and its vicinity by parents or their neighbors as soon as they are found, as has been observed in Prairie Warblers (Nolan 1978:34). Studies of floaters (nonterritorial interlopers) on breeding territories demonstrate that large numbers of yearlings, especially males, occur in populations of many migrants (Hensley and Cope 1951; Stewart and Aldrich 1951; Morton 1977; Rappole et al. 1977; Vega-Rivera et al. 2003). Given the likelihood of being immediately displaced if adults are present, a 2.1 percent observed rate of return may be high relative to what it might be if natal return rates were random within even a small portion of the breeding range.

For instance, consider the probability of finding a yearling American Redstart (*Setophaga ruticilla*) returning to within 500 m of its natal territory, assuming random return within a tiny portion of its North American range—a single county in western New York:

DATA

Total forested area of Chautauqua County = 1,375 km² (Chautauqua County Soil and Water Conservation District 2010)

Total number of 1-km² blocks containing forest = 1,375

ASSUMPTIONS

At least one American Redstart pair per 1-km² block

Yearling bird survival rate = 50 percent (probably high)

One hundred percent nonrandom return by yearling birds to their natal region (Chautauqua County)

Complete random settlement by yearling birds within the natal region

CALCULATIONS

Probability that a given yearling redstart will settle within 500 m of its natal site = Yearling bird annual survivorship (0.5) / Total number of natal sites (1,375) = 0.0004

This figure of 0.0004 for the probability of return of a given surviving yearling redstart to within 500 m of its birth site is very conservative because although we have assumed random settlement within Chautauqua County, we have assumed 100 percent nonrandom return to this small part of the North American breeding range. Nevertheless, a comparison of this figure with actual median levels of return rates of small, migrant yearling passerines to their natal areas of 0.02 makes clear that actual return rates cannot be explained given random return to even a small portion of the breeding range. Therefore, we conclude that observed low return rates for relatively short-lived migrants may result from search bias; that is, an inability to find young birds on their return before they are forced to disperse results from lack of a suitable technique. It may be that young of the previous year of these species attempt to return to the site (or the vicinity) at which they were hatched just as young of most longer-lived migrants are known to do (excluding some groups; e.g., yearling male anatids).

The very few intensive studies of yearling long-distance migrant settling patterns support this view. For instance, Lack reported high return rates for female yearling Pied Flycatchers based on long-term, large-area studies of this species in Finland, England, and Germany (Lack 1966:114–115).

A fair question to ask in response to this analysis is, "So what?" Does it matter whether or not yearlings cannot return because they are dead, do not return because they do not have the capability or choose not to, or do return but are quickly forced to leave? If we are trying to understand migrant evolution and adaptations, then the answer to this question is most certainly, "Yes!" If yearling migrants return to natal sites at rates comparable to those of survival, it demonstrates that they have an adaptation (exaptation?) for the capability of moving from point to point across considerable distances. Also, it shows that they do not "choose" (in an innate, evolutionary sense) to settle on a breeding territory distant from the natal territory to avoid negative selective factors (e.g., inbreeding depression). The benefits of natal philopatry have been summarized by Bensch et al. (1998):

- Knowledge of where to find food
- Knowledge of where to find nesting sites
- Knowledge of predators and where and how to avoid them
- Possibility that kin may be less competitive neighbors (Greenwood 1980; Shields 1982; Waser and Jones 1983)

To this list could be added the certainty that appropriate habitat for reproduction exists at the natal site along with an existing population of conspecifics (access to mates). There are, of course, potential costs as well. Apparent displacement from the vicinity of the natal territory of returning offspring by a parent or parents could have as simple an explanation as expulsion of potential competitors for limited resources (e.g., food or mates) (Johnson and Grimes 1990). However, it is also possible that such displacement is an adaptation to reduce inbreeding (Moore and Ali 1984; Van de Casteele and Matthysen 2006; Doligez and Pärt 2008). We would argue that for yearlings of many long-distance migrants, the benefits of return to the natal site outweigh the costs. Allowing yearling offspring to settle nearby, however, has costs for adults that outweigh the benefits, which is why they normally chase returning young off as soon as they find them or displace them shortly after arrival, as often happens in species that migrate as family groups (Bellrose 1976).

Natal dispersal distance, whether chosen or forced, can vary even among individuals of the same sex of a given migratory species and, surprisingly, appears to be in part inherited. Evidence from several species suggests that relatives demonstrate similar dispersal distances and possibly even directions (Dale et al. 2005). Interspecific and intraspecific differences suggest that natal dispersal distance may have costs and benefits that differ both within and between species, possibly depending on such factors as sex, ecology, and environmental conditions (Rappole et al. 1977; Doligez and Pärt 2008).

HABITAT SELECTION

For most adult male migrant passerines, breeding habitat selection often amounts to return to the same territory as was occupied the previous year. Nolan (1978:35), for instance, found that 73 percent of returning male Prairie Warblers followed this pattern. For females, a key aspect of quality breeding habitat appears to be quality of the mate, which apparently is strongly associated with quality of the territory in terms of nest sites and perhaps other aspects (e.g., food or shelter from predators). Thus, female habitat selection appears to be influenced by prospective breeding partners, particularly in protandrous species, where female settlement decisions are likely to be affected by male advertisement (Lampe and Espmark 2003; Roth et al. 2009). If this surmise is true, then one might make two predictions regarding female breeding habitat selection/mating behavior:

1. Females would select different mates on returning to the breeding area if males of apparent higher quality (i.e., in terms of the male's individual attributes and the attributes of his territory) were available.
2. Females would desert mates given an indication of lower quality (e.g., nest failure).

Such behavior by females has been observed in a number of studies. For instance, in a 20-year study of Prairie Warblers in Indiana, 23 of 37 females settled on different territories with different mates on returning to their breeding areas; only 14 reoccupied their former territories, and two of these paired with different males although their mates from the previous year were present (Nolan 1978:35). In addition, 36 females moved onto the study area in mid-season (after most pairs had raised first broods), paired, and nested, while 62 females moved off the study area during the same period after initial nesting attempts (Nolan 1978:346).

For yearlings, the habitat-selection process is somewhat different, and in this regard comparison with the process in residents is instructive. Resident-bird habitat selection is seemingly a straightforward process in which a young dispersing individual, pushed away from its natal site by its parents and their neighbors, moves until it finds a place where it can compete successfully to satisfy its needs (e.g., Willis 1972:83). Initially, these needs include only food and shelter. However, eventually, the young must locate, identify, and settle in a habitat that satisfies not only survivorship but reproductive needs as well. In some cases, the habitat that provides the best opportunity for survival may not be the same habitat as the one that provides for highest reproductive capacity because of requirements specific to the reproductive period (e.g., availability of safe nesting sites). Thus, individuals of many resident species, confronted with the fitness benefits of control over a productive breeding site, may be forced to balance costs in the form of lower nonbreeding survivorship by remaining in the specific habitat where highest breeding success occurs. Migrants, however, are free to choose the optimal habitat for survival during the nonbreeding season and for reproduction during the breeding season. Thus, habitat selection during these different periods can be quite different for migrants as opposed to residents, even among closely related species.

Consider, for instance, the habitat use patterns of the resident Tropical Parula (*Setophaga pitiayumi*) and its closely related congener (conspecific? [Mayr and Short 1970]), the migratory Northern Parula (*Setophaga americana*). The Tropical Parula is a canopy-foraging, year-round resident of epiphyte-laden oak forest from south Texas to northern Argentina (Moldenhauer and Regelski 1997). In contrast, its migratory relative, the Northern Parula, released from the necessity of occupying a winter habitat that contains potential mates or nesting sites, occupies a much broader array of nonbreeding habitats from low scrub to mature forest (Mold-enhauer and Regelski 1996). During the breeding season, the Northern Parula inhabits a wide range of habitats across eastern North America, from broadleaf deciduous to coniferous forest, whose chief defining characteristic appears to be presence of epiphytes (e.g., Spanish moss [*Tilandsia*] or beard moss lichen [*Usnea*]) or other clumps of suspended plant materials required for nest construction or concealment (Moldenhauer and Regelski 1997).

Similarly, the Old World migratory Spotted Flycatcher (*Muscicapa striata*) breeds in an extraordinary array of woodlands, gardens, and parks across much of west-ern Eurasia, key aspects of which seem to be availability of perches, flying insects, and nooks (or open nest boxes) for nest placement (Dolphign 2000–2009, Taylor 2006a), whereas its tropical African resident relative (conspecific? [Mackworth-Praed and Grant 1973]), the Gambaga Flycatcher (*Muscicapa gambagae*), inhabits a restricted range in the open semi-arid woodlands and savannas of equatorial Africa year-round where it nests in stump hollows or similar crannies (Mackworth-Praed and Grant 1973, Taylor 2006b).

The point of this discussion is that migration may provide a form of ecologi-cal release in which optimal habitat can vary by season depending on whether or not the fitness emphasis is on survival or reproduction. This argument begs the question, of course, of how newly fledged migrants identify appropriate breeding and wintering habitat. For young migrants traveling as individuals (i.e., not as members of family groups) to wintering areas for the first time, critical aspects of habitat selection seemingly must include a programmed ability to travel to and identify key aspects of appropriate nonbreeding habitat (e.g., structure, food availability, and resident heterospecifics), guided in part perhaps by presence of conspecifics, at least as traveling companions (Morton 1990). Similar factors, assisted by site fidelity and with the additional requirement of mates and nest sites, likely guide yearlings to appropriate breeding habitat. Returning migrants that do not settle at their natal or former breeding sites may use prior experi-ence that they have gathered actively or passively during the previous reproductive season in order to select appropriate habitat (Reed et al. 1999; Pärt and Doligez 2003; Mukhin et al. 2005). In addition, they may use conspecific and heterospe-cific cues for landfall and for identification of breeding sites (Thomson et al. 2003; Hahn and Silverman 2006; Chernetsov 2008). For example, migrants could use cues regarding the presence of a particular community of resident bird species on the breeding grounds to assess habitat suitability. In a series of correlational and experimental studies, Scandinavian researchers made the surprising discovery that settlement decisions of migrants near particular resident species benefited both parties (Thomson et al. 2003).

A number of aquatic migrants appear to occupy very similar habitats in both the breeding and nonbreeding seasons. Several anatids and ardeids, for instance, choose breeding and nonbreeding habitats that are to all appearances at least struc-turally similar, although separated by many thousands of kilometers. For many

FIGURE 2.1   Sanderling (*Calidris alba*) on its nest in the Arctic tundra breeding habitat (*left*) and on a Texas coastal beach during the nonbreeding period (*right*).

terrestrial migrants, however, breeding habitats differ significantly from non-breeding habitats, at least for some parts of the population (e.g., adult females and juvenile birds in partial migrants) indicating the differential force of natural selection on breeding and nonbreeding habitat identification for migrants in contrast to their sedentary counterparts. In some species, the contrast between wintering and breeding habitat could hardly be more striking. The Sanderling (*Calidris alba*) spends the nonbreeding period on temperate and tropical coastal beaches but breeds in Arctic tundra (Macwhirter et al. 2002) (figure 2.1).

Similarly, some populations of the European Roller (*Coracius garrulus*) breed in mature pine forests of eastern Europe and winter in central and southern African savanna. In fact, 44 species that breed in European forests spend the nonbreeding period in open-country habitats in Africa (Mönkkönen et al. 1992; Rappole 1995). Even those species of migrants that use forest in both the breeding and wintering periods can use habitats that are markedly different. As an example, the Cape May Warbler (*Setophaga tigrina*) breeds in the crowns of mature spruce (*Picea*) in North American boreal forest and winters in a variety of habitats including scrub, gardens, pine, and broadleaf forest in the Caribbean (Baltz and Latta 1998).

That habitat structure can vary so markedly between breeding and nonbreeding periods for migrants and that the breeding habitat of the migrant can differ so strikingly from that of closely related congeners raises the question of just what aspects of habitat are most critical for a migrant. Obviously food, shelter, and mates are important during the breeding period, whereas food and shelter predominate during the nonbreeding period. But there is an aspect of shelter that is quite different during the breeding period than the nonbreeding period, which may help to explain how habitats that are seemingly disparate can offer comparable aspects. By "shelter" we mean protection from predators and the elements. During the nonbreeding period, fitness aspects of "shelter" apply only to the individual. However, during the breeding period, fitness aspects of shelter must include nest site as well because the individual's fitness is tied to nest success. Thus, even in those species that breed in temperate habitats that are completely different in most aspects from

the habitats in which their tropical counterparts breed, the nest site and structure often are remarkably similar (e.g., Spotted and Gambaga flycatchers, as cited
earlier).

INTERVAL BETWEEN ARRIVAL AND REPRODUCTION

Migrants that breed in highly seasonal habitats (e.g., at high latitudes) or specialize
on highly seasonal food sources (e.g., aerial insects) rush to breed and raise young
to independence before conditions deteriorate. Accordingly, a general pattern is
that migrants with short breeding seasons lose very little time between first arrival
and the onset of actual breeding. Rapid transition between migration and breeding
puts high physiologic demands on birds, including a need for an advanced state of
development of reproductive organs at time of arrival (Raess and Gwinner 2005;
Bauchinger et al. 2007), which requires rapid hormonal transitions to breeding
condition (Ramenofsky and Wingfield 2006; Ramenofsky 2010), and for maintenance of sufficient body reserves to power breeding (Raveling 1987; Drent et al.
2003; Alerstam 2006; Nager 2006; Yohannes et al. 2010). Thus, migrants demonstrate marked differences from residents in both physiologic aspects of preparation for breeding (more rapid growth rate and larger reproductive organ size) and
behavioral aspects (e.g., song rates).

This difference between migrants and residents in terms of the need among
migrants to transform rapidly from nonbreeding to reproductive mode has been
suggested as an explanation for the disproportionate number of dimorphic (or polymorphic [Rappole 1983]) species among migrants as opposed to residents (Hamilton 1961). If this hypothesis is correct, then the degree of dimorphism observed in
a long-distance migrant population might be expected to differ from that of tropical
resident populations of closely related congeners or conspecifics. Such differences,
if found, might be useful for calculating the length of time in which a migratory
strategy has been pursued in such groups based on genetic distance between the
populations. Two examples of superspecies groups among which such comparisons might be made would be (1) migrant Northern Parula (*Setophaga americana*)
compared with resident Tropical Parula (*Setophaga pitiayumi*) (figure 2.2) and (2)
migrant compared with resident populations of the Grace's Warbler (*Setophaga
graciae*)—assuming that the migrant and resident populations do not interbreed.
Many other possible comparisons exist (e.g., migrant compared with resident
Myioborus and *Polioptila*).

Overall in New World warblers (Parulidae), migratory populations are often
dimorphic or polymorphic, whereas the tropical residents are mostly monomor-
phic. However, this pattern has a strong phylogenetic bias. For instance, many
migrant icterids, emberizids, and cardinalids show strong sexual dimorphism or
polymorphism, whereas their tropical resident counterparts do not; yet migrant
and tropical resident members of many other groups show little or no dimorphism

FIGURE 2.2   Sexual dimorphism in temperate (*top*) versus tropical (*bottom*) resident populations of a parulid superspecies: (*top left*) Northern Parula (*Setophaga americana*) male; (*top right*) Northern Parula female; (*bottom left*) Tropical Parula (*Setophaga pitiayumi*) male; (*bottom right*) Tropical Parula female.

(e.g., turdids, muscicapids, and vireonids). In other groups (e.g., some trochilids), both migrants and residents are strongly dimorphic. Clearly, life history factors other than the need for rapid pair formation associated with a migratory habit must influence evolution of dimorphism or lack thereof (Rappole 1988).

REPRODUCTION

The contrast between breeding season adaptations for migrants as opposed to residents are obvious as applies to the first sections of this chapter, as residents do not experience most aspects of situations such as arrival, site fidelity, breeding habitat selection, and so forth, or at least not in ways comparable with those of migrants. The same cannot be said for most of the key aspects of reproduction, the key segments of which—territory, mating systems, egg-laying, and so forth—are shared by both migrants and residents. These elements require a closer comparison to identify aspects of them that may represent specific adaptations by migrants to a migratory lifestyle.

TERRITORY

In table 2.3, we define eight major types of avian breeding territories. There are many variations on these basic themes, and not all avian species fit well into a single category, but these eight classes seem sufficient for our purpose of considering how migrants might differ from residents in terms of breeding territory establishment, structure, and maintenance. Note that in the following discussion, the term "territory" is used in the sense of "a defended area" and therefore does not include the kinds of extraterritorial movements associated with visits to special feeding or roosting sites or extra-pair copulations (Whitaker and Warkentin 2010).

As table 2.3 illustrates, there are some striking differences between migrants and residents in terms of types of breeding territories. Among terrestrial residents, types C, E, F, and G are prevalent (Skutch 1954, 1960, 1967, 1969, 1972), whereas these types are rare in migrants (see species accounts in Poole [2010] and del Hoyo et al. [1992–2011]). In contrast, the most common territory type among terrestrial migrants is the type A territory where pairing, mating, nesting, and feeding of young all occur on the same plot of ground, which the family leaves once breeding has been completed (Nice 1941; Hinde 1956). For migrant species with type A territories, the usual pattern is that when males arrive, they establish a territory significantly larger than what will be defended after pairing takes place, which gets smaller as nest building and incubation progress, shrinks to its smallest size when the pair is feeding the nestlings, and may dissolve completely during periods of fledgling care (Stenger and Falls 1959; Catchpole 1972; Welsh 1975; Nolan 1978).

The purpose of territory has been stated as an attempt to sequester critical, defendable resources that are in short supply (Brown 1964). However, the fact that

TABLE 2.3   Mating Territory Types in Birds

TERRITORY TYPE	EXAMPLES
A. Mating, nesting, and feeding ground for young, breeding season only; in partial migrants, male may defend territory year-round	Many terrestrial migrants and some temperate and tropical residents
B. Mating and nesting, but not feeding ground, breeding season only	Many colonial species of both migrants and residents
C. Mating station only, breeding season only	Lek-forming species that are mostly resident (e.g., many otidids, ptilonorhynchids, trochilids, piprids, and phasianids)
D. Restricted to narrow surroundings of nest, breeding season only	Many swallows and pelagic species, mostly migratory
E. Year-round territory defended by pair	Many tropical and some temperate terrestrial residents
F. Nesting territory intensely defended by pair during the breeding season only, separate from or surrounded by feeding areas (e.g., ant swarms, fruiting trees) defended throughout the year	Many tropical and temperate terrestrial residents
G. Nest site defended year-round	Some tropical residents
H. No territory; mating takes place at ad hoc display areas	Some icterids, some of which are migratory (e,g, Brown-headed Cowbird [*Molothrus ater*]) and some of which are resident (e.g., Bronzed Cowbird [*Molothrus aeneus*])

breeding territories for many terrestrial migrants change significantly in size during different portions of the reproductive period raises the question of precisely what resources are being tracked by these variations in territory size? Indeed, there seems to be little relationship between size of migrant breeding territory and amount of food resources required, at least for those species in which these parameters have been measured (Catchpole 1972; Verner 1977). Several other explanations have been offered, including Verner's (1977) "super-territory hypothesis" in which a purpose of breeding territory includes the acquisition of interference phenomena whereby individuals impair the reproductive rate of competitors. Catchpole (1972) suggests that the adaptive significance of migrant breeding territory involves maintenance of a suitable area and nest site for successful breeding, relatively free from predation and intraspecific interference. A third possibility is that large, early season territory size in migrants is a result of epigamic selection in which female choice favors those males that offer territories with the largest amount of resources important for successful rearing of young (e.g., number of high-quality nest or feeding sites). Once the female has arrived and made her choice of partner and nest site, the focus of male attention then shifts from territory defense to nest-site defense and mate-guarding until the young are hatched, when male focus shifts to nestling provision. Regardless of what the various functions of type A migrant breeding territories may be, it is certain that the character, size, and defense periods usually differ in important ways from those of their resident counterparts (Stutchbury and Morton 2001), which raises the issue of what, how, and when modifications in territorial behavior occur as resident populations assume a migratory habit.

Intrasexual competition is a key aspect for returning migrants, although there are differences between the sexes and also between the different age groups (Rappole et al. 1977). For males, competition for quality territories can be intense, and young males often lose in these competitions, ending up in marginal habitats or as nonbreeding floaters (Darwin 1871:738–741; Moffat 1903; Groebbels 1937; Brown 1969; von Haartman 1971). We argued in the earlier section "Site Fidelity and Dispersal" that young males of short-lived migrants likely return in their second year to the site where they were hatched. However, adult males do not allow them to settle on their natal territory or neighboring sites, at least in those species in which the phenomenon has been studied in detail (e.g., Ficken and Ficken 1967). Failure by yearling males to obtain high-quality territories (i.e., those that have a high probability of attracting a mate and successfully producing offspring) has at least four results that vary according to species and year:

1. Yearling males occupy poor or marginal territories at the periphery of suitable habitats or range (Ficken and Ficken 1967; Graves et al. 2002).
2. Yearling males join a nonterritorial population called "floaters" whose members move continually through high-quality nesting habitat in search of territories whose male defenders have disappeared (predation, disease) or of females amenable to extra-pair copulations (Rappole et al. 1977).
3. Alternatively, yearling males do not return to the breeding area but remain at nonbreeding sites until their third year or later (e.g., Herring Gull [*Larus argentatus*] [Pierotti and Good 1994]).
4. Yearling birds of both sexes are allowed to stay in territories as helpers, most commonly with their own parents (Skutch 1976).

Yearling helpers are not uncommon among resident species but appear to be rare among migrants (although they may occur in the Chimney Swift [*Chaetura pelagica*]) (Skutch 1961). The female-like plumages seen in yearling males of a number of migrants may serve as status-signaling to reduce adult male aggression at the cost, of course, of reduced attractiveness to females (Rohwer et al. 1980; Rappole 1983).

Adult females of many migrant species also defend the breeding territory, particularly against intrusion by other females (Hinde 1956:355; Orians 1961:294; Nolan 1978:368–369). Adult males, however, often permit settlement of unmated (yearling?) females on or near their territory, resulting in a highly significant difference in the proportion of males and females in floating populations with males predominating by an order of magnitude or more (Hensley and Cope 1951; Lack 1966; Rappole et al. 1977).

MATING SYSTEMS

Breeding and mating systems can be classified in several ways, one of which derives from social interactions between the sexes that can be observed in the field. For instance, these may include monogamy, polygyny, polyandry, polygynandry, promiscuity, and forced copulation (Oring 1982; McKinney et al. 1983; Gill 2007). With the advent of molecular methods for determination of parentage, it has become clear that genetic parentage may differ strikingly from social partnership (Gill 2007). For example, whereas the majority of avian species appear to be monogamous (Lack 1968:4), extra-pair paternity and, to a lesser degree, intraspecific brood parasitism have been documented in many supposedly monogamous species investigated (Arnold and Owens 2002; Spottiswoode and Møller 2004). Proportions of broods that differed from social parentage varied across species and ranged between 0 and 95 percent (Arnold and Owens 2002). Extra-pair copulations are generally considered to offer males an opportunity to boost reproductive success above the capacity of a single breeding pair. Females in turn could increase fitness by mating with the highest-quality male available, whether or not that male is paired with the female (Wagner et al. 1996; Spottiswoode and Møller 2004). Otter et al. (1994), for instance, found that 17 percent of offspring in a Quebec population of Black-capped Chickadees (*Poecile atricapilla*) was sired by males to which the mothers were not mated and that 100 percent of their actual mates were subordinate to those males in competitive situations during the nonbreeding period. Similarly, Wagner et al. (1996) observed that the majority of males participating in extra-pair copulations in Purple Martin (*Progne subis*) colonies were adult males (>2 years old), while the majority of males whose mates participated in extra-pair copulations were yearling birds.

Several explanations have been put forward for why members of a given species have a particular mating system or mixture of different systems (e.g., monogamy and polygyny), usually having to do with the fitness trade-offs required for successful production of young, which, in turn, depend upon such factors as whether or not the young are altricial or precocial, the type of habitat occupied, and the type of foods required by the young (Crook 1964; Oring 1982). A point of interest in consideration of mating systems from our perspective is the apparent contrast seen between migrants and residents that are taxonomically or ecologically similar in many respects but that have significantly different mating systems. As mentioned earlier, comparisons between migrants and residents suffer the weakness of being confounded with differences in breeding latitude, food availability, and many other habitat characteristics. Therefore, the apparent differences could be an epiphenomenon of the temporarily high productivity of the habitats that many migrants exploit. For example, a suite of comparative studies of *Acrocephalus* warblers gave strong evidence that with increasing habitat productivity, paternal care decreased and polygyny increased, both along the tropical–temperate gradient and across Temperate Zone breeding habitats of Afro-tropical migrants (Leisler et al. 2002; Leisler and Schulze-Hagen 2012). It is nonetheless conceivable that in addition to ecological correlations, a migratory lifestyle could promote prevalence of mixed mating strategies (*sensu* Trivers 1972; Spottiswoode and Møller 2004). A high frequency of extra-pair copulations in populations of many supposedly monogamous species demonstrates that promiscuity is particularly common among principally monogamous migrants, in which offspring sired by a neighboring male can be greater than 50 percent (Westneat et al. 1990; Birkhead and Møller 1996; Westneat and Sherman 1997; Stutchbury and Morton 2001; Spottiswoode and Møller 2004). Notably, in a comparison between two sympatrically breeding, tropical flycatchers, extra-pair paternity was far less common among the resident species (Yellow-bellied Elaenia [*Elaenia flavigaster*]) compared with the migrant species (Lesser Elaenia [*Elaenia chiriquensis*]) (Stutchbury et al. 2007).

Several hypotheses have been put forward as explanations for the difference between migrants and tropical residents in terms of mating systems and extra-pair copulations, the principal one being latitudinal differences in relative length of the breeding season between the two groups (Emlen and Oring 1977; Birkhead and Møller 1992; Stutchbury and Morton 2001). Tropical resident birds in general have long breeding seasons, with pairs in a given population being relatively asynchronous in terms of timing of reproductive activities. In contrast, migrants have shorter breeding seasons, and pairs are relatively synchronous in terms of the timing of their reproductive activities. Stutchbury and Morton (2001:43) state that "low breeding synchrony decreases the benefits to individual males and females seeking EPFs [extra-pair fertilizations]." They argue that it is much easier for females to judge the quality of potential mates when there are several males to compare, as is presumably the case for migrants in which breeding is relatively synchronous, at least when compared with residents. However, Spottiswoode and Møller (2004) point out that relative patterns of synchrony compose only one of several factors that differ between migrants and residents. Other factors associated with a migratory habit that could be associated with increased amounts of extra-pair fertilization include "(1) hasty or (2) inaccurate mate choice, (3) facilitated assessment of male quality through the condition-dependence of arrival time, or (4) increased genetic variance in male quality" (Spottiswoode and Møller 2004:41). Some of these factors, for example potential hastiness of decision making among hurried female migrants, have been supported in field studies in related contexts (e.g., hybridization) (Randler 2002).

The synchrony hypothesis provides an explanation for a remarkable finding: the discovery of very high rates of male territory-holders raising the offspring of other males, rates that appear to co-vary with latitude. The reasoning behind the hypothesis is that increased synchrony of breeding among members of any given metapopulation, forced by a decrease in the amount of time available for successful rearing of offspring, results in the female member of the pair being unable to make an informed judgment as to the quality of her mate when she chooses to settle on his territory. This lack of informed judgment, in turn, results in mistakes (i.e., choosing a male of inferior quality). Discovery of this failure further results in prospecting for males of higher quality to sire her offspring, often among neighboring territory holders (Stutchbury and Morton 2001:43). We do not question the data on which the synchrony or related hypotheses are based or the reasoning. Rather, it seems to us that extraordinary rates of extra-pair paternity beg the question of the existence of a mating system (monogamy) whose fitness value to the pair members appears to be based on a high rate of intrapair paternity. If males have a 30 to 50 percent chance of raising the offspring of another male, why should they expend the effort (in a fitness sense) in obtaining and holding a territory and feeding the offspring born on that territory? A lek system would seem to be more appropriate for both males and females in that males would compete only to fertilize females wasting no time defending a large space and raising offspring that are not their own, and females could choose the highest-quality males without the danger of being confused by such factors as territory quality and lack of time in which to make an evaluation. Fewer offspring might be raised per individual female, but individual fitness might actually be enhanced by such a system. We suggest that the reason monogamous behaviors persist in what is obviously a promiscuous system is that something has changed in recent years (decades? centuries?). These changes have made what was previously a selectively valuable system into one of questionable value (at least for the territorial male).

One possible reason for differences in extra-pair paternity between migrants and residents may be related to the very different life history challenges confronting adult tropical resident birds as opposed to their migratory counterparts and how those life history differences might have been affected by recent environmental changes. For instance, destruction of breeding habitat presumably confronts adult and juvenile migrants in ways similar to that for adult and juvenile residents. However, for migrants, destruction of winter habitat may have a disproportionate effect on adults. Adults of many species of migrants move between known and established wintering and breeding sites; indeed, adults of several migrant species travel to specific winter territories, which they defend against conspecifics (see chapter 5). In contrast, young must search for and find adequate wintering sites or territories during their first winter, a difference that presumably results in much higher losses among juveniles as opposed to adults. However, if good winter habitat is destroyed or becomes altered through climate change, adults may be placed into a situation more similar to that of young individuals in that they must search for new wintering sites or function as nomadic wanderers due to decline in winter habitat quality. Early arrival at a wintering site has been demonstrated to be an important factor in competitive outcomes, apparently irregardless of age (Snell-Rood and Cristol 2005), and forced movement by adults displaced by winter habitat destruction or degradation could have an effect on this dynamic. If so, then, winter habitat loss or decline in quality could result in an increased percentage of yearling males entering the breeding population and obtaining higher-quality breeding territories as a result of lowered adult male survivorship. These young birds might be much more vulnerable to extra-pair copulation strategies by their mates, because they are of lower quality in terms of experience and competitive ability (Wagner et al. 1996).

We acknowledge that the hypothesis that we have put forward as an explanation for differences in terms of extra-pair copulations in migrants versus tropical residents is not supported by any data of which we are aware. There are, however, ways in which a key aspect of the hypothesis might be tested, namely that the change is recent. One way would involve the comparison of evidence of extra-pair paternity rates at present versus those existing in the past. Many egg collections exist (dating back a century or more) that could provide information on this question. In any event, the main point of our analysis is that high rates of extra-pair paternity are not "normal" in the sense that such promiscuity is unlikely to have favored the evolution of a monogamous system. Therefore, promiscuity is likely to be the result of some recent change.

A second example of the mating system differences between migrants and residents is the successful settlement and reproduction by unpaired females on or near the territories of paired males. This mating system has been referred to as "successive polygyny," or "serial monogamy" (Lack 1968a:5; Rappole et al. 1977). Although common among terrestrial migrant species, often occurring at rates of about 10 percent (Hann 1937; Stewart 1952; Lack 1968:31; von Haartman 1971; Welsh 1975; Ezaki 1990), this system is rare to our knowledge among resident species. We suggest that it may result from a situation similar to that described earlier for increased levels of extra-pair copulation in migrants compared with tropical residents: the disproportionate entry of yearling females into the breeding population. Normally, unmated females are forced to leave the territories of mated females (Hinde 1956:355). However, if these mated females include a higher number of yearling females than is normal, it may be that these yearlings will have lower success in forcing female interlopers out.

A third contrast in mating system types between migrants and residents is the rarity of primarily promiscuous systems (e.g., leks) among migrants (although the Ruff [*Philomachus pugnax*] is an exception). In most of these systems, males gather at display areas for the purpose of attracting and mating with females; all other reproductive activities, nesting, brooding, and care for the young, are undertaken by the female. The main theories proposed for promiscuous or polygynous systems include

- Types of foods required by the young—those species in which offspring require insect provisioning rather than those requiring fruits or seeds appear to need both mates for successful rearing of offspring (Crook 1964)
- Whether or not the young are precocial or altricial, with precocial species normally requiring less parenting

Promiscuous systems occur among several taxa of residents (e.g., phasianids, cotingids, paradisaeids, ptilonorhynchids, and piprids) but are rare among migrants, one example being the Brown-headed Cowbird (*Molothrus ater*; a social parasite that lays its eggs in the nests of other species), as well as some migratory hummingbirds. The presumed reason for the near absence of promiscuous systems among altricial migrants is that participation by both male and female increases the number of offspring that can be raised to fledging or independence (Lack 1968). This factor may be relaxed in altricial tropical resident birds that produce, on average, a smaller number of offspring per nesting attempt, perhaps to reduce risk of predation resulting from increased provisioning rates for larger clutches (Fogden 1972; Ricklefs 1972; Skutch 1976; Stutchbury and Morton 2001).

NEST BUILDING

Among most groups of non-passerines, residency status apparently is of little or no importance with regard to which sexes are involved in nest building and to what extent; males assist the female for the most part (Skutch 1976:99). The situation is more complicated in passerines in which it appears that male assistance, while rare among migrant species, is somewhat more prevalent among both tropical and temperate residents (Skutch 1969; Poole 2010). Nevertheless, a true comparison would involve tropical residents with migrant congeners or conspecifics, the data for which are lacking at present.

COPULATION

Forced copulation, in which males copulate with apparently unwilling females that are not their mates, has been reported in many avian species—for example, 39 species of waterfowl as well as the Red-winged Blackbird (*Agelaius phoeniceus*) and Purple Martin (*Progne subis*) (Heinroth 1911; Raitasuo 1964; Derrickson 1977; McKinney et al. 1983; Wagner et al. 1996). McKinney et al. (1983) suggested that this behavior resulted from a "mixed male reproductive strategy" engaged in opportunistically by males of socially monogamous pairs. Wagner et al. (1996) proposed that apparently forced copulations were not, in fact, forced, but could be controlled by females (i.e., that such copulations were a female reproductive strategy), in which case forced copulations are effactually no different from other types of extra-pair copulations as discussed earlier. It is possible that both explanations are correct, depending on the circumstances and species. In any event, we find it interesting that the vast majority of species in which forced copulation has been reported are migrants. We suggest that, like serial polygyny and extra-pair fertilization, this behavior could result from a disproportionate number of yearling birds of migratory species entering the breeding population, specifically young males that are not able to mate-guard properly or whose quality is doubted by their mates who then seek, or allow, copulation with males who are not their mates.

EGG LAYING

Egg laying patterns generally are similar between migrants and residents of the same taxonomic group. However, there is a notable exception in that some tropical resident tyrannids lay an egg every other day until the clutch is complete, whereas temperate, migrant tyrannids lay an egg every day until clutch completion (Stutchbury and Morton 2001:34). Whether or not this pattern is genetically fixed or results from differences in prey availability affecting embryo growth rates is not known. Nevertheless, the patterns demonstrate a difference between some migrants and their resident congeners.

CLUTCH SIZE

Clutch size in tropical resident species is smaller on average than that for northern Temperate Zone migrants and residents (Hesse 1922; Skutch 1985). In the majority of passerines, clutch size is one or two in the tropics (Skutch 1954, 1960; Lack and Moreau 1965; del Hoyo et al. 1992–2011), whereas songbird clutch sizes at higher latitudes are usually four to six (Poole 2010; del Hoyo et al. 1992–2011). Several explanations have been offered for this difference, but basically they fall into four categories:

1. Available resources are lower in the tropics, resulting in reduced capacity for provisioning of young (Hesse 1922; Lack 1968).
2. Stability is higher in the tropics compared with the northern Temperate Zone, favoring evolution of K-selected reproductive strategies in populations that hover close to carrying capacity (e.g., longer-lived adults having a larger number of nesting attempts over the individual's life span with fewer offspring per attempt) (Cody 1966; Skutch 1985; Young 1996).
3. Nest predation rates are higher in the tropics while at least in some habitats multiple breeding attempts are possible (Fogden 1972). Thus, the best strategy may be low investment in any single clutch and production of multiple clutches.
4. The (metabolic) "pace of living" is lower in the tropics, implying low instantaneous effort (Ricklefs and Wikelski 2002).

These explanations are fundamentally different. The first explanation maintains that clutch size is somewhat flexible and can change from season to season or even for different nesting attempts within a season for the same female (Lack 1947–1948, 1968; Haywood 1993a). The second, however, presents a situation in which mean clutch size for a given species is genetically set within narrow limits. Both explanations could be valid, of course, for different species, and there is evidence that this is the case (Gwinner et al. 1995). For instance, clutch size in columbids and trochilids appears to be set genetically at two, whether the species is a tropical resident or a migrant (see individual species accounts in Poole [2010]). However, provisioning rate studies and artificial manipulation of nestling number have shown that adult provisioning abilities can and do limit number of offspring that can be fledged by tropical residents for several species (Stutchbury and Morton 2001:30). Knowing whether or not clutch size is broadly versus narrowly flexible for a given species is important. A major claim by those who propose rapid evolution of migration to temperate regions by tropical species is that those individuals capable of making the journey would reap significant fitness benefits because of increased food availability resulting in more offspring than could be produced by their tropical relatives (Rappole et al. 1983; Rappole and Tipton 1991; Rappole 1995). If clutch size is narrowly fixed, for example, in doves and pigeons (at two), then this benefit presumably would be restricted, although lowered predation rates or increased provisioning rates in temperate regions could still result in higher offspring production per nesting attempt.

We are aware of two ways for testing the degree of clutch size flexibility:

1. Compare mean clutch size for tropical resident versus temperate migrant species pairs.
2. Hand-rear members of temperate migrant and tropical resident species pairs, and see what they produce under comparable conditions of prey availability and day length.

The second test has been carried out on an Old World songbird, the Stonechat (*Saxicola torquata*), which has both long-distance migrant and tropical resident populations. Results from this study gave strong support for genetic determination of clutch size in this species (Gwinner et al. 1995). The first idea (i.e., comparison of clutch size in temperate and tropical species pairs) can be tested using data from the literature (table 2.4). These data show that temperate migrant clutch size averages one or more eggs larger than that for their tropical relatives in most cases, even when those relatives are suspected of being conspecifics. Certainly, one would predict that flexibility in clutch size allowing the female to match number of eggs produced to number of offspring likely to be raised to independence based on resource availability would be a major asset for relatively short-lived migrants and a likely target on which natural selection could act to produce significant differences between migrants and residents in life history trade-offs between clutch size, number of annual nesting attempts, and survivorship, given sufficient evolutionary time. In any event, mean clutch size represents

TABLE 2.4   Comparison of Reproductive Activities for Taxonomically Related Tropical Resident and Northern Temperate Migrant Species Pairs

Species	Clutch Size[1]	Incubation	Nestling Parental Care
Tropical Pewee (*Contopus cinereus*)	2–3	Female only; male feeds female on nest, 15–16 d	?
Eastern Wood-Pewee (*Contopus virens*)	**3**	**Probably female only; male feeds female on nest; 12–13 d**	**16–18 d**
Yellowish Flycatcher[2] (*Empidonax flavescens*)	3	Female only; ?	17 d
Cordilleran/Pacific Slope Flycatcher (*Empidonax difficilis/occidentalis*)	**3–4**	**Female only, 13–16 d**	**15–17 d**
Black Phoebe (*Sayornis nigricans*), tropical resident (Costa Rica)	2–3	?	?
Black Phoebe (*Sayornis nigricans*), temperate resident (California)	4	Mostly female, but male assistance reported, 15–18 d	18–21 d
Eastern Phoebe (*Sayornis phoebe*)	**5**	**Female, 16 d**	**16 d**
Myiarchus flycatchers (*Myiarchus* sp.)	2–4	?	14 d?
Great Crested Flycatcher (*Myiarchus crinitus*)	**5**	**Female, 13–15 d**	**13–15 d**
Tropical Kingbird (*Tyrannus melancholichus*)	2, 3	Female, 15–16d	18–19 d
Western Kingbird (*Tyrannus verticalis*)	**4**	**Female, 14 d**	**16 d**
Gray-breasted Martin[2] (*Progne chalybea*)	2–4	?	?
Purple Martin (*Progne subis*)	**4–5**	**Female, 15–18 d**	**28–29 d**

(continued)

TABLE 2.4   (continued)

Species	Clutch Size[1]	Incubation	Nestling Parental Care
Southern Rough-winged Swallow[2] (*Stelgidopteryx ruficollis*)	4.7 in Costa Rica (De Jong 1996)	Female, 16–18 d	20–21 d
Northern Rough-winged Swallow (*Stelgidopteryx serripennis*)	**6.3 in southern Michigan (De Jong 1996)**	**Female, 15–17 d**	**19–20 d**
House Wren[2] (*Troglodytes aedon musculus*)	4, 3	Female, 17 d	18–19 d
House Wren (*Troglodytes aedon aedon*)	**6**	**Female, 12–13 d**	**15–18 d**
Tropical Gnatcatcher (*Polioptila plumbea*)	3–2	Both sexes, 13 d	?
Blue-gray Gnatcatcher (*Polioptila caerulea*)	**4.5**	**Both sexes, 13 d**	**10–15 d**
Orange-billed Nightingale-Thrush (*Catharus aurantiirostris*)	2	Female, 13–15 d	13–17 d
Veery (*Catharus fuscescens*)	**4**	**Female, 10–14 d**	**10–12 d**
Clay-colored Thrush (*Turdus assimilis*)	3	Female, 12–13 d	15–16 d
American Robin (*Turdus migratorius*)	**3.3**	**Female, 12–13 d**	**13 d**
Tropical Parula[2] (*Setophaga pitiayumi*)	2–3	?	?
Northern Parula (*Setophaga americana*)	**4**	**Female, 12–14 d**	**10–11 d**
Masked Yellowthroat (*Geothlypis [chiriquensis] aequinoctialis*)	2	Female only, 15 d	

Species	Clutch Size[1]	Incubation	Nestling Parental Care
Common Yellowthroat (*Geothlypis trichas*)	**4**	**Female, 12 d**	**12 d**
Rufous-collared Sparrow (*Zonotrichia capensis*)	2, 3	?	10–12 d
White-crowned Sparrow (*Zonotrichia leucophrys*)	**4**	**Female, 12 d**	**8–10 d**
Great-tailed Grackle (*Quiscalus mexicanus*)	3, 2	Female, 13–14 d	20–23 d
Common Grackle (*Quiscalus quiscula*)	**4.8**	**Female, 13.5 d**	**12–15 d**
Yellow Oriole (*Icterus nigrogularis*)	3	?	?
Baltimore Oriole (*Icterus galbula*)	**4–5**	**Female, 12 d**	**12–13 d**

Note: Migrants appear in boldface type. Data for tropical species are from Skutch (1954, 1960, 1967, 1985), whereas those for temperate migrants are from species accounts in Poole (2010). Where the migrant–resident species pairs have been considered to be conspecific, the source is footnoted. For discussion, see the section "Territory."

[1]Data on tropical resident clutch size are from Skutch (1985), unless otherwise cited. He states, "The most frequent clutch size is given first, separated by a comma from the next most frequent. When one size is not clearly more frequent than another, a dash separates the numbers" (Skutch 1985:581). Data on temperate migrant clutch size are from species accounts in Poole (2010); these usually are given as the average or modal amount.

[2]Species pairs considered to be conspecific by one or more authors (sources in parentheses): *Empidonax flavescens/E. difficilis* (Phillips et al. 1964); *Progne chalybea/P. subis* (Ridgway 1901–1950); *Stelgidopteryx serripennis/S. ruficollis* (American Ornithologists' Union 1957); *Troglodytes aedon/T. musculus* (American Ornithologists' Union 1998); *Setophaga pitiayumi/S. americana* (Mayr and Short 1970).

an important difference between migrants and their resident relatives. Whether or not these differences represent an adaptation (genetic) or an adjustment to different environmental circumstances in terms of food availability has yet to be determined. As in the case of other such differences, it may be that for species in which a migratory habit is a relatively recent development, clutch size differences from tropical resident relatives are a result of flexibility, whereas for those species with a longer history of migratory behavior, clutch size norms are fixed genetically at a higher level than that for tropical-resident congeners, as was found by Gwinner et al. (1995).

In attempting to understand differences between migrants and residents, we have discussed clutch size and compared this trait between tropical resident and temperate migrant congeners. A comparison between tropical residents and their intratropical or tropical–subtropical migrant congeners would be much more revealing, but data are mostly lacking (although see Stutchbury and Morton 2008).

INCUBATION

The female is responsible for incubation in most of the migrant/tropical resident species pairs considered; however, incubation period is shorter in general for migrants (table 2.4), which could result from greater prey availability in temperate regions (i.e., greater prey availability = reduced absence from nest = reduced egg cooling periods = shorter total incubation period).

PREFLEDGING PARENTAL CARE

Both sexes care for the young in most of the species pairs considered, although the nestling period is shorter for most migrant representatives (table 2.4). This disparity could result from greater prey availability or longer daily feeding periods in temperate as opposed to tropical breeding areas, resulting in more rapid nestling growth per unit of time.

POSTFLEDGING PARENTAL CARE

Broadly speaking, parental care lasts for only a few days or weeks after fledging in most migrants (geese, cranes, etc., are obvious exceptions), whereas long-term parent–offspring associations lasting weeks or months are more common among tropical and subtropical residents (Skutch 1976:349). One type of data confirming this difference derives from studies summarizing occurrence of cooperative breeding ("helping") in which young from a previous clutch assist their parents in raising young of a subsequent clutch. This behavior is found in a number of resident species but is much less common among migrants (Skutch 1961, 1976;

Koenig and Dickinson 2004). An even rarer occurrence in migrants is the situation in which young of the previous year assist adults in raising offspring. This behavior has been recorded in several resident species (Skutch 1976:351). A second type of data confirming differences in postbreeding treatment of offspring by adults comes from studies of mixed-species flocks, which document the fact that resident conspecifics occurring in the same flock often are family members, whereas migrants are either solitary or, when more than one individual is present, unrelated to conspecific flock members (Munn 1985; Powell 1985).

CHAPTER 3

POSTBREEDING PERIOD

THE POSTBREEDING period is that portion of the annual cycle beginning with the termination of reproductive activities (adults) or independence from adults (juveniles) and ending with departure on migration. For most Temperate Zone migrants, this period involves a significant segment of time during which adults and juveniles undergo the prebasic molt and prepare for their departure (Dwight 1900; Stresemann and Stresemann 1966; Ginn and Melville 1983; Pyle 1997; Leu and Thompson 2002). This period generally is treated either as part of the breeding or the transient phase of the annual cycle (see species accounts in del Hoyo et al. [1992–2010] and Poole [2010]). One reason for this treatment is that any given migrant population often will have some individuals in one of these phases whereas others are in the postbreeding period, and it can be difficult to tell the difference. A second complicating factor is that individuals in the postbreeding period often are not easy to study because molting birds are quiet, relatively inactive, and may move considerable distances and occupy habitats completely different from those used during the breeding period (Cherry 1985; Rappole and Ballard 1987; Rappole 1995; King et al. 2006). Nevertheless, birds in this phase have physiologic and behavioral adaptations unique to this period, which in our view justifies its recognition and consideration as an important and separate portion of the avian migrant annual cycle (Rappole 1996:396).

Purposes of the Postbreeding Period

We see five principal purposes for the postbreeding period:

- Conduct of the prebasic molt
- Exploratory movements by juveniles to familiarize themselves with location of feeding and potential breeding areas
- Prospecting movements by adults anticipatory of the next breeding season
- Preparation for departure on fall migration
- Optimization of feeding strategies to limit predation while engaged in other postbreeding activities

Conduct of the Prebasic Molt

All birds undergo a periodic renewal of their plumage to maintain its thermoregulatory and aerodynamic properties (Stresemann and Stresemann 1966). Molt plays a large role in the postbreeding period for most migrants. A brief overview of the key terminology and most commonly observed sequences is provided in table 3.1, following Palmer (1972).

Most species experience at least one complete molt each year as adults (Humphrey and Parkes 1959). Because feathers wear at different rates in different species and can have special additional functions at different times of the year (e.g., crypsis or communication), there often are additional molts into alternate or supplemental plumages in various species (Palmer 1972:66–72).

Molt has costs in terms of metabolic demands on the individual, including production of the replacement plumage; regulation of body temperature in the absence of complete body feathering; and loss of aerodynamic qualities, or even complete loss of flight, due to flight feather loss (Ginn and Melville 1983; Walsberg 1983; Murphy 1996). These energetic costs can increase metabolic demands by greater than 45 percent (Lustick 1970; Gavrilov 1974; Chilgren 1975; Lindström et al. 1993). In addition, there are costs resulting from increased predation due to decreased flight capabilities (Chandler 2007). Both residents and migrants must balance these costs against other demands that arise during the course of the annual cycle. For residents, this normally means delaying molt until after breeding is completed and then undergoing a prolonged prebasic molt, perhaps lasting several months (Ginn and Melville 1983; Jenni and Winkler 1994; Kjellén 1994; Berthold 2001) (figure 3.1).

Most migrants are limited in their ability to minimize short-term energy demands due to molt. They must balance the costs not only of breeding and molt

TABLE 3.1   Molt and Plumage Terminology

Molt or Plumage	Description
Natal plumage	The first plumage developed by a bird in the egg or nest, often composed of feathers lacking a central vane (down).
Prejuvenal molt	Molt from the natal plumage to a complete vaned or pennaceous plumage. In most altricial birds, this molt begins while the young are still nestlings and is completed some time after fledging.
Juvenal plumage	Usually the first complete covering of vaned plumage for a recently hatched bird. The flight feathers (remiges and rectrices) of the juvenal plumage usually are the same in structure and appearance as those of adult birds. The body feathers, however, often differ from the adult basic plumage. The structure of juvenal body feathers is somewhat looser and fluffier in many birds, providing different thermodynamic qualities from the plumages of adults. The appearance also often differs from that of adult birds, perhaps signaling their status as fledglings to adults.
First prebasic molt	Molt from juvenal plumage into a plumage similar in both structure and, often, appearance to that of an adult bird, usually the female, if there is a difference between adult male and female in appearance of the basic plumage. In many birds, the first prebasic molt does not include the flight feathers because these were just recently grown as part of the prejuvenal molt and have not experienced significant wear.
First basic plumage	The first plumage acquired by newly hatched birds that is similar in both structure and, usually, appearance to that of adults.
Basic plumage	The definitive feather generation in a molt cycle.
Prealternate molt	Molt from the basic plumage into a second plumage generation within a single molt cycle. Often, this molt is not complete, involving only some portion of feather tracts.
Alternate plumage	The second feather generation in a molt cycle, often the breeding-season plumage.
Presupplemental molt	Molt from the alternate plumage into a third plumage generation within a single molt cycle. Often, this molt is not complete, involving only some portion of feather tracts.
Supplemental plumage	The third feather generation in a molt cycle. Known for only a few species, for example, the Long-tailed Duck (*Clangula hyemalis*). A fourth feather generation (second supplemental plumage) may occur in the Willow Ptarmigan (*Lagopus lagopus*) (Johnsen 1929).

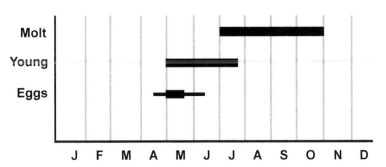

FIGURE 3.1   Timing of prebasic molt and breeding activities in a North American, southern Temperate Zone resident, the Ladder-backed Woodpecker (*Picoides scalaris*).

but also of migration within a relatively circumscribed period. These pressures result in a wide variety of molting strategies among migrants depending on the latitudes in which they breed and winter, the kinds of foods they use, and a wide variety of other life history factors (table 3.2). Thus, many of the differences in timing of molt between migrants and residents, especially for closely related species or subspecies, represent adaptations by migrants in response to selective pressure imposed by a migratory habit (Stresemann and Stresemann 1966; Helm and Gwinner 2006a) (figure 3.2).

The timing of molt activities for species that follow pattern 1 (table 3.2), the most common among temperate, passerine migrants, normally involves a complete prebasic molt for adults after completion of breeding activities and prior to migration (see species accounts in Poole [2010]). This molt normally lasts 6 to 8 weeks for the individual, but extends over 3 to 4 months for the species because of differences in timing of completion of breeding activities (Vega-Rivera et al. 1998a; Heise and Rimmer 2000). Newly hatched young of these species undergo a molt from their natal down into complete pennaceous feathering, called the juvenal plumage, while still on the nest or shortly after fledging (Palmer 1972). Most then undergo a prebasic molt into their first basic plumage (basic I), although this molt usually does not include flight feathers (i.e., remiges and rectrices), which normally are not molted until the first prealternate or second prebasic molt when the birds are a year old (Dwight 1900:107; Pyle 1997).

Perhaps the most extraordinary aspect of the variations that occur even within pattern 1 and all other major molt patterns is that important differences in timing, process, and extent can occur in members of closely related species (Johnson 1963), within a single species by subspecies, age, or sex groupings (Foster 1967; Helm and Gwinner 2006a), as well as from individual to individual within the same population (Young 1991; Helm and Gwinner 2006a). There are two ways to look at the existence of different molt strategies in the same population: They can either result from flexibility built into the genetic control of molt (i.e., "reaction norms" *sensu* Van Noordwijk [1989]) allowing individuals under different environmental

TABLE 3.2 Principal Timing Patterns for Major "Breeding Season" Activities in Various Migrant Species

PATTERN	CHARACTERISTICS BY AGE AND SEX	COMMON EXCEPTIONS	EXEMPLARY SPECIES	COMMENTS
1	Adult males and females: breeding/prebasic molt/migration Juveniles: fledging /prebasic molt of all but flight feathers (rectrices and remiges)/ migration	Nest failures or production of multiple broods results in overlap of breeding activities with the early stages of body feather molt in adults; in some species (e.g., Willow Warbler [*Phylloscopus trochilus*] and Common Whitethroat [*Sylvia communis*]) there is evidence to indicate that some adults may begin molt after rearing a first brood, suspend molt while rearing a second brood, and then complete it after raising of the second brood (Ginn and Melville 1983:19).	Many Temperate Zone passerines	This pattern is the most common among Temperate Zone migrant passerines.
2	Adult males and females: breeding/migration/ prebasic molt Juveniles: fledging/prebasic molt of all but flight feathers (rectrices and remiges)/migration	Juvenile birds interrupt molt to depart on fall migration, completing body molt after arriving at the winter quarters.	Yellow-bellied Flycatcher (*Empidonax flaviventris*), Reed Warbler (*Acrocephalus scirpaceus*)	
3	Adult males and females: breeding/prebasic molt/migration Juveniles: fledging/extensive dispersal followed by migration in flocks, often greater than 100 km/prebasic molt of all but flight feathers (rectrices and remiges) occurs over an extended period evidently during movement periods		Many ardeids, for example, Cattle Egret (*Bubulcus ibis*), Tri-colored Heron (*Egretta tricolor*)	Most pattern 3 species breed in colonies. Adults leave the colonies for feeding and roosting areas, often before the young fledge. Juveniles depart colonies in flocks with other juveniles.

	Sequence of events	Comments	Examples
4	Adults and juveniles: breeding/migration in family groups to molting sites ("staging areas")/prebasic molt/migration in family groups to wintering sites	Unpaired adults or adults with failed nesting attempts either remain on wintering grounds or at some point along the migration route or leave breeding areas early on "molt migrations."	Geese, several accipitrids
5	Adult males: breeding/migration to molting sites/prebasic molt/migration to wintering sites Adult females: breeding/ prebasic molt/ migration Juveniles: Independence from female/dispersal in flocks to molting sites (as much as several hundred kilometers)/migration to wintering sites		Several anatids
6	Adults and juveniles: breeding/migration to molting sites/prebasic molt/migration to wintering sites	Similar to pattern 3 except that members of all age and sex groups migrate as individuals rather than as members of family groups and may move to completely different molting sites.	Various anatids, Lazuli Bunting (*Passerina amoena*)? Great Reed Warbler (*Acrocephalus arundinaceus*), Northern Lapwing (*Vanellus vanellus*), Green Sandpiper (*Tringa ochropus*)
7	Adults: prebasic molt begins before or during breeding Juveniles: prebasic molt often partial and conducted during or after completion of fall migration		Many Arctic-breeding species

Note: Major breeding season activities are breeding, molt, and departure for migration.

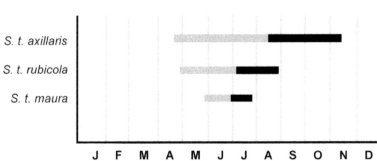

FIGURE 3.2   Timing of egg laying (*light gray bar*) and prebasic molt (*black bar*) under native photoperiod conditions for three populations of the Stonechat (*Saxicola torquata*): African resident (*S. t. axillaris*), European migrant (*S. t. rubicola*), and Siberian migrant (*S. t. maura*) (Helm and Gwinner 2006a).

circumstances to modify their molting programs or could represent different alleles for the genes controlling the process. It is likely that both explanations apply to a different extent in different populations or species (Helm and Gwinner 2006a).

Migrants as a group differ from residents in terms of molt in the following ways: (1) duration, (2) geographic location, and (3) number. The amount of difference between a given migrant population compared with a related resident population for each of these classes varies from nonexistent to considerable and, in addition to various ecological and life history factors, is probably related to two additional aspects that are pertinent specifically to migration: length of evolutionary time for which the migratory population has been migratory and migration distance.

• *Duration of molt.* Prebasic molt tends to be shortest in duration for populations migrating the longest distances, intermediate for shorter-distance migrants, and more or less prolonged for residents (Helm and Gwinner 2006a). We have proposed that the wintering ground is the region from which migratory populations of most species derive (see chapter 1). If this hypothesis is correct, then for those species of migrants portions of whose populations remain throughout the year and breed in these regions (23 to 48 percent of migrants), residents may, at least in some cases, represent an ancestral state with regard to timing of life history events. If so, then a prediction would be that wintering ground resident populations of migratory species would demonstrate characteristics of timing of prebasic molt that would be more similar to that of other local resident species than to that of migratory populations of their own species. Helm and Gwinner (2006a) found evidence for such timing differences in migrant and resident populations of juvenile Stonechats (*Saxicola torquata*) in which migrants show a shorter period of prebasic molt than that of tropical residents (figure 3.2).

• *Geographic location of prebasic molt.* Most migrants undergo the prebasic (i.e., postbreeding) molt on the breeding ground or at special sites separate from either the breeding or wintering areas at times when critical resources at these sites are seasonally superabundant (Pearson 1973, 1975, 1990; Bellrose 1976; Pearson and Backhouse 1976; Pearson et al. 1980; Pearson and Lack 1992; Jones 1995; Rohwer et al. 2008). Residents, of course, undergo prebasic molt at areas where they are resident. Thus for related species of migrants and residents, the physical location of the prebasic molt takes place at sites hundreds or even thousands of kilometers apart, an example being the Stonechats discussed by Helm and Gwinner (2006a) in which prebasic molt for the migratory populations occurs in Europe, whereas that of resident populations occurs in Africa.

• *Number of molts.* A third difference in molt patterns between migrants and residents is that many migrants of particular families (e.g., Anatidae, Parulidae, Icteridae, and Cardinalidae) have more than one molt per year (Rohwer and Butcher 1988). These alternate plumages are rare among resident species (Pyle 1977). The prealternate molt for migrants usually takes place on the wintering ground. Often, this molt is partial and results in an alternate (breeding) plumage that is brighter or fresher for those parts of the body associated with breeding displays (e.g., the head, throat, and breast), at least in adult males. Among the most extreme examples is that of the adult male Scarlet Tanager (*Piranga olivacea*), which molts from a cryptic green winter (i.e., basic) plumage to a brilliant scarlet and black breeding (i.e., alternate) plumage prior to spring migration.

EXPLORATORY MOVEMENTS BY JUVENILES

A second purpose of the postbreeding period in migrants involves juvenile exploration. Postbreeding movements are poorly known for most migrants because they require intensive field work—often supplemented by special technology—to be documented. Vega-Rivera et al. (1998b) were able to observe movements of Wood Thrushes (*Hylocichla mustelina*) during the postbreeding period using radio tracking. They found that, in addition to those movements associated with apparent normal daily activities (e.g., foraging and roosting), 15 of 28 juvenile birds performed movements of greater than 300 m away from their site of origin, lasting less than 3 days before return to site of origin. It has been noted that young birds of many migrant species probably return from their wintering areas to the vicinity of the territory on which they were fledged the previous year, at least initially (see chapter 2). We hypothesize that the movements described for young Wood Thrushes are exploratory, serving purposes suggested by a number of authors of familiarizing the young bird with feeding sites (Morton et al. 1991) as well as possible nesting sites for the next year's breeding season (Brewer and Harrison 1975; Seastedt and Maclean 1979; Gluck 1984; Pärt 1990; Morton et al. 1991) and enhancing their ability to find their way back to the area (Löhrl 1959; Wiltschko and Wiltschko 1978). Documentation of such movements in the Wood Thrush suggests that intense investigation of juvenile postbreeding movements in other migrant species may reveal similar kinds of exploratory efforts.

PROSPECTING FOR NEW MATES OR TERRITORIES

A third purpose of the postbreeding period may involve preparation for the next year's breeding season by adults. As in the case of other aspects of the postbreeding period, the behavior of adults is poorly known. However, several lines of evidence indicate that, at least in some species, some adult behavior (adult females prospecting for new mates; adult males prospecting for new territories) may be related to this purpose. Evidence of this behavior for adult males includes the following:

1. Adult males tend to remain on or near their breeding territory during the postbreeding period in greater proportion than that of other sex or age groups (Nice 1937; Lack 1943; Nolan 1978; Sykes et al. 1989; Vega-Rivera et al. 1999).
2. Song is often heard in North American Temperate Zone and Boreal Zone breeding habitat during the postbreeding period when most adults are physiologically in a postreproductive state (Chandler et al. 2011).
3. Adult males have been observed in actual advertisement (i.e., song) and defense (chases) of territory during the postbreeding period in several species (e.g., Song Sparrow [*Melospiza melodia*; Nice 1937], European Robin [*Erithacus rubecula*; Lack 1943], Yellow Warbler [*Setophaga petechia*; Rimmer 1988], Prairie Warbler [*Setophaga discolor*; Nolan 1978:436], House Wren [*Troglodytes aedon*; J. H. Rappole, personal observation]).
4. Males of several migrant species undergo fall gonadal recrudescence (Nice 1937), whose function may be establishment of territory for the next year's breeding season prior to departure on fall migration.
5. Male House Wrens construct dummy nests during the postbreeding period (J. H. Rappole, personal observation).
6. Males of some species respond to playback during the postbreeding period (e.g., Common Yellowthroat [*Geothlypis trichas*]) (J. H. Rappole, personal observation).

PREPARATION FOR DEPARTURE TO THE WINTERING GROUND

Fall migrants preparing to depart to the wintering ground is an obvious purpose of the postbreeding period and, although not well investigated in the field, has been studied extensively in the laboratory. Groebbels (1928) proposed the existence of two distinct physiologic states common to all migrants: *Zugdisposition*, during which the bird prepares for undertaking a migratory flight by feeding intensively (hyperphagia), increasing food intake by as much as 40 percent (Berthold 1975) and laying down energy reserves in the form of subcutaneous fat; and *Zugstimmung*, in which the bird actually begins and sustains a period of migratory activity or actual flight on the way to its ultimate destination. What portions of the timing of initial *Zugdisposition* are under endogenous versus exogenous control apparently depends upon the species and its environment (Gwinner 1975). Unfortunately, most of what is known about preparation for migration derives from studies of birds in the laboratory (e.g., King 1972) or in transit (i.e., already engaged in the migratory journey) (e.g., Rappole and Warner 1976; Moore et al. 2005). Little work has been done on birds undergoing the initial transition from the postbreeding activities discussed earlier to preparation for departure on migration. In fact, it has been suggested that many species may depart on migration prior to undergoing the kinds of preparation described by Groebbels (Odum et al. 1961; Helms and Drury 1963; McNeil and Carrera de Itriago 1968). However, some field studies have found that migrants remaining in the vicinity of the breeding area undergo significant increases in mass and/or subcutaneous fat storage at the completion of molt and just prior to departure on fall migration (Gray Catbird [*Dumetella caro-linensis*] [Heise and Rimmer 2000], Yellow Warbler [*Setophaga petechia*] [Rimmer 1988]), and several authors have now demonstrated that although many migrants may shift habitat use in passing from the breeding into the postbreeding period, they tend to remain in postbreeding habitats until departure on fall migration, as documented by detailed studies of some species and implied by habitat shift studies as discussed later.

OPTIMIZING FEEDING STRATEGIES TO MAXIMIZE FOOD INTAKE WHILE LIMITING PREDATION

Migrants undergoing molt, familiarizing themselves with possible future breeding sites, or initiating preparation for fall migration may have different needs in terms of predator avoidance, food sources, and thermoregulation from those of birds in the reproductive period (Rappole 1995:75–78; Vega-Rivera et al. 1998b; King et al. 2006; Chandler 2011). This observation is obvious, and the purpose of mentioning it is to emphasize the point that the process of balancing the various needs of the bird during the postbreeding period may require use of different habitats from those used during the breeding period.

MOVEMENTS AND HABITAT USE DURING THE POSTBREEDING PERIOD

Despite the different needs confronting migrants during the postbreeding period as contrasted with those during the breeding period and the fact that many migrants disappear from their breeding areas shortly after completion of reproductive activities, it has long been assumed that postbreeding habitat requirements for migrants were equivalent to those of the breeding period. This assumption is clearly reflected in the habitat descriptions provided in hundreds of species accounts for migrants in Bent (1919–1958), Poole (2010), del Hoyo et al. (1992–2010), and all other such summaries of which we are aware, which divide migrant habitat needs into three classes at most: breeding, transient, and wintering.

Understanding that the needs for migrants may be different during the post-breeding as opposed to the breeding period may help to explain why many migrants disappear from their breeding territories well before they appear on migration. It has long been assumed that such disappearance reflected early departure on migration (Hann 1937; Bent 1953; Holmes et al. 2005). Certainly, some individuals found south of the breeding range are transients in the midst of migration, as indicated by the fact that whereas migration peaks for many species of long-distance migrants occur in October based on Florida television-tower kill data, the same data show large numbers of individuals killed at night in presumed migratory flight much earlier in the season (Crawford 1981). These transients could represent members of southern populations, which often begin and complete breeding, complete prebasic molt, and commence migration earlier than northern populations of the same species (Nolan 1978; Rappole et al. 1979) or they could be young birds from first clutches or adults that failed to breed, both of which are likely to complete molt and preparation for migration well before the majority of a population. Nevertheless, for most migrants, early disappearance from breeding habitat probably is not reflective of migration, although the degree to which disappearance from the breeding area represents a form of gradual migration as opposed to dispersal to habitats optimal for molting is not known for any passerine migrant to our knowledge. Several studies document that many migrants move to habitats different from those in which they bred, often but not necessarily within the vicinity of their breeding areas. This fact has now been confirmed for at least 38 species of North American Temperate Zone, mostly forest-breeding passerine migrants based on data gathered using a variety of field techniques (table 3.3) and probably represents normal behavior for members of many other forest-breeding migrant species.

There are striking differences in terms of the kinds of resources that are available in different habitats during the breeding period as opposed to the postbreeding period. When forest-breeding species arrive on the breeding ground in temperate North America, insect larvae are the most abundant food source in forest environments. However, this situation changes in many habitats later in the summer season, when flying insects, fruits, and seeds increase in amounts, whereas caterpillars decrease (Tauber et al. 1986). Also, the distribution of these resources by habitat changes, so that greater amounts of resources are found in second-growth than in forest later in the season.

The complexity of postbreeding movements is indicated by the work of Graves et al. (2002) on the Black-throated Blue Warbler (*Setophaga caerulescens*). Using stable isotope technology, they were able to document that in Appalachian populations of Black-throated Blue Warblers, older males (>2 years) underwent prebasic molt in a relatively narrow altitudinal band, presumably comparable to the location of their highland breeding areas, whereas younger males showed no evidence of altitudinal patterns. The authors attributed this difference to natal dispersal into molting areas in different elevational zones from sites where they fledged, whereas older males tended to molt on or near their highland breeding areas.

TABLE 3.3   Breeding Habitat and Postbreeding (Molting) Habitat Use for Selected Species of North American Migrants

Species	Breeding Habitat	Postbreeding Habitat(s) Sampled	Postbreeding Habitat Use by Age (Adult/ Juvenile/ Unknown), n	Sampling Techniques, Region, and Citation
Whip-poor-will (*Caprimulgus vociferus*)	Deciduous and mixed forest	Clear-cuts[1]	Recorded; no numbers given	Mist nets; Ohio (Vitz and Rodewald 2006)
Whip-poor-will (*Caprimulgus vociferus*)	Deciduous and mixed forest	Clear-cuts[2]	Recorded; no numbers given	Mist nets; Virginia and West Virginia (Marshall et al. 2003)
Ruby-throated Hummingbird (*Archilochus colubris*)	Deciduous and mixed forest	Old field,[3] forest[4]	Old field 2/0/1; forest 0/0/0	Mist nets; Georgia (Rappole and Ballard 1987)
Ruby-throated Hummingbird (*Archilochus colubris*)	Deciduous and mixed forest	Clear-cuts[2]	8/0/0	Mist nets; Virginia and West Virginia (Marshall et al. 2003)
Ruby-throated Hummingbird (*Archilochus colubris*)	Deciduous and mixed forest	Clear-cuts[1]	3/77/0	Mist nets; Ohio (Vitz and Rodewald 2006)
Eastern Wood-Pewee (*Contopus virens*)	Deciduous and mixed forest	Clear-cuts[1]	4/11/0	Mist nets; Ohio (Vitz and Rodewald 2006)
Eastern Wood-Pewee (*Contopus virens*)	Deciduous and mixed forest	Clear-cuts[2]	Recorded; no numbers given	Mist nets; Virginia and West Virginia (Marshall et al. 2003)
Acadian Flycatcher (*Empidonax virescens*)	Deciduous and mixed forest	Old field,[3] forest[4]	3/2/0	Mist nets; Georgia (Rappole and Ballard 1987),
Acadian Flycatcher (*Empidonax virescens*)	Deciduous and mixed forest	Clear-cuts[1]	17/19/0	Mist nets; Ohio (Vitz and Rodewald 2006)

(continued)

TABLE 3.3 *(continued)*

Species	Breeding Habitat	Postbreeding Habitat(s) Sampled	Postbreeding Habitat Use by Age (Adult/ Juvenile/ Unknown), *n*	Sampling Techniques, Region, and Citation
Acadian Flycatcher (*Empidonax virescens*)	Deciduous and mixed forest	Forest,[5] clear-cuts[6]	Some adults in forest; most adults and all juveniles in clear-cuts	Point counts and mist nets; Missouri (Pagen et al. 2000)
Acadian Flycatcher (*Empidonax virescens*)	Deciduous and mixed forest	Clear-cuts[2]	Recorded; no numbers given	Mist nets; Virginia and West Virginia (Marshall et al. 2003)
Least Flycatcher (*Empidonax minimus*)	Deciduous and mixed forest	Clear-cuts[2]	Recorded; no numbers given	Mist nets; Virginia and West Virginia (Marshall et al. 2003)
Great Crested Flycatcher (*Myiarchus crinitus*)	Deciduous and mixed forest	Clear-cuts[2]	Recorded; no numbers given	Mist nets; Virginia and West Virginia (Marshall et al. 2003)
Yellow-throated Vireo (*Vireo flavifrons*)	Deciduous forest	Clear-cuts[1]	5/5/0	Mist nets; Ohio (Vitz and Rodewald 2006)
Blue-headed Vireo (*Vireo solitarius*)	Boreal coniferous and mixed forest	Wildlife openings,[7] clear-cuts,[8] forest[9]	Unknown; wildlife openings preferred over clear-cuts and forest	Point counts and mist nets; New Hampshire (Chandler 2011)
Blue-headed Vireo (*Vireo solitarius*)	Boreal coniferous and mixed forest	Clear-cuts[2]	Recorded; no numbers given	Mist nets; Virginia and West Virginia (Marshall et al. 2003)
Red-eyed Vireo (*Vireo olivaceus*)	Deciduous and mixed forest	Wildlife openings,[7] clear-cuts,[8] forest[9]	Unknown; clear-cuts preferred over wildlife openings and forest	Point counts and mist nets; New Hampshire (Chandler 2011)
Red-eyed Vireo (*Vireo olivaceus*)	Deciduous and mixed forest	Clear-cuts[1]	95/115/0	Mist nets; Ohio (Vitz and Rodewald 2006)

Species	Breeding Habitat	Postbreeding Habitat(s) Sampled	Postbreeding Habitat Use by Age (Adult/ Juvenile/ Unknown), n	Sampling Techniques, Region, and Citation
Red-eyed Vireo (*Vireo olivaceus*)	Deciduous and mixed forest	Forest,[5] clear-cuts[6]	Some adults in forest; both adults and juveniles in clear-cuts	Point counts and mist nets; Missouri (Pagen et al. 2000)
Red-eyed Vireo (*Vireo olivaceus*)	Deciduous and mixed forest	Clear-cuts[2]	63/12/0	Mist nets; Virginia and West Virginia (Marshall et al. 2003)
Winter Wren (*Troglodytes hiemalis*)	Deciduous and mixed forest	Clear-cuts,[8] wildlife openings[9]	Unknown/4 birds recorded singing in mature forest; 1 bird captured in clear-cuts; 2 birds captured in wildlife openings	New Hampshire (Chandler et al. 2011)
Blue-gray Gnatcatcher (*Polioptila caerulea*)	Deciduous and mixed woodlands	Clear-cuts[3]	Recorded; no numbers given	Mist nets; Virginia and West Virginia (Marshall et al. 2003)
Veery (*Catharus fuscescens*)	Deciduous and mixed woodlands	Clear-cuts[3]	13/4/0	Mist nets; Virginia and West Virginia (Marshall et al. 2003)
Veery (*Catharus fuscescens*)	Deciduous and mixed woodlands	Clear-cuts[1]	Recorded; no numbers given	Mist nets; Ohio (Vitz and Rodewald 2006)
Swainson's Thrush (*Catharus ustulatus*)	Coniferous and mixed forest	Wildlife openings,[7] clear-cuts,[8] forest[9]	Unknown; clear-cuts preferred over wildlife openings and forest	Point counts and mist nets; New Hampshire (Chandler 2011)
Swainson's Thrush (*Catharus ustulatus*)	Coniferous and mixed forest	Mature mixed forest and seral stages	Adults remained mostly in forest whereas juveniles moved to earlier seral stages	Radio tracking; New Hampshire (Chandler 2011).

(continued)

TABLE 3.3 *(continued)*

Species	Breeding Habitat	Postbreeding Habitat(s) Sampled	Postbreeding Habitat Use by Age (Adult/ Juvenile/ Unknown), N	Sampling Techniques, Region, and Citation
Hermit Thrush (*Catharus guttatus*)	Boreal coniferous and mixed forest	Wildlife openings,[7] clear-cuts,[8] forest[9]	Unknown; clear-cuts and forest preferred over wildlife openings	Point counts and mist nets; New Hampshire (Chandler 2011)
Wood Thrush (*Hylocichla mustelina*)	Deciduous and mixed forest	Old field,[3] forest[4]	Old field 1/2/1; forest 0/0/0; 33 birds captured in forest neighboring old field from April 17 to July 30; none from July 30 to September 17	Mist nets; Georgia (Rappole and Ballard 1987)
Wood Thrush (*Hylocichla mustelina*)	Deciduous and mixed forest	Forest,[10] early successional stages[11]	Some adults in forest; some adults and nearly all juveniles (92%) in early successional stage habitat	Radio tracking; northeastern Virginia (Vega et al. 1998a, 1998b, 1999)
Wood Thrush (*Hylocichla mustelina*)	Deciduous and mixed forest	Closed canopy forest,[12] open forest,[13] clear-cuts[14]	Of 10 juveniles followed, 5 used mostly clear-cuts, 4 used mostly open forest habitats, and 1 used closed canopy forest	Radio tracking; Missouri (Anders et al. 1998)
Wood Thrush (*Hylocichla mustelina*)	Deciduous and mixed forest	Clear-cuts[2]	26/3/0	Mist nets; Virginia and West Virginia (Marshall et al. 2003)
Wood Thrush (*Hylocichla mustelina*)	Deciduous and mixed forest	Clear-cuts[1]	16/66/0	Mist nets; Ohio (Vitz and Rodewald 2006)

Species	Breeding Habitat	Postbreeding Habitat(s) Sampled	Postbreeding Habitat Use by Age (Adult/ Juvenile/ Unknown), n	Sampling Techniques, Region, and Citation
Northern Parula (*Setophaga americana*)	Coniferous, mixed, and deciduous forest	Old field,[3] forest[4]	Old field 1/4/0; forest 0/0/0	Mist nets; Georgia (Rappole and Ballard 1987)
Northern Parula (*Setophaga americana*)	Coniferous, mixed, and deciduous forest	Forest,[5] clear-cuts[6]	Adults and juveniles in clear-cuts	Point counts and mist nets; Missouri (Pagen et al. 2000)
Black-throated Blue Warbler (*Setophaga caerulescens*)	Deciduous and mixed forest	Wildlife openings,[7] clear-cuts,[8] forest[9]	Clear-cuts preferred over forest or wildlife openings	Point counts and mist nets; New Hampshire (Chandler 2011)
Black-throated Blue Warbler (*Setophaga caerulescens*)	Deciduous and mixed forest	Appalachian forest and successional habitats	Adult males molt predominantly in highlands, presumably at similar elevations as breeding sites; young males molt in a mixture of sites across a broad range of elevations	Stable isotope ratios in feathers; southern Appalachian region, Graham County, North Carolina (Graves et al. 2002)
Black-throated Blue Warbler (*Setophaga caerulescens*)	Deciduous and mixed forest	Clear-cuts[2]	Recorded; no numbers given	Mist nets; Virginia and West Virginia (Marshall et al. 2003)
Yellow-rumped Warbler (*Setophaga coronata*)	Coniferous forest	Wildlife openings,[7] clear-cuts,[8] forest[9]	Wildlife openings preferred over forest and clear-cuts	Point counts and mist nets; New Hampshire (Chandler 2011)

(*continued*)

TABLE 3.3 (*continued*)

Species	Breeding Habitat	Postbreeding Habitat(s) Sampled	Postbreeding Habitat Use by Age (Adult/Juvenile/Unknown), n	Sampling Techniques, Region, and Citation
Golden-cheeked Warbler (*Setophaga chrysoparia*)	Open oak (*Quercus*)–Juniper (*Juniperus*) woodlands	Bald cypress (*Taxodium distichum*) groves along Camp Verde Creek, Kerr County	Adults and juveniles	Observations; Texas (Pulich 1976:110)
Black-throated Green Warbler (*Setophaga virens*)	Coniferous, mixed, and deciduous forest	Clear-cuts[1]	Recorded; no numbers or age ratios given	Mist nets; Ohio (Vitz and Rodewald 2006)
Black-throated Green Warbler (*Setophaga virens*)	Coniferous, mixed, and deciduous forest	Wildlife openings,[7] clear-cuts,[8] forest[9]	Clear-cuts and wildlife openings preferred over forest	Point counts and mist nets; New Hampshire (Chandler 2011)
Black-throated Green Warbler (*Setophaga virens*)	Coniferous, mixed, and deciduous forest	Clear-cuts[2]	Recorded; no numbers given	Mist nets; Virginia and West Virginia (Marshall et al. 2003)
Blackburnian Warbler (*Setophaga fusca*)	Coniferous and mixed forest	Wildlife openings,[7] clear-cuts,[8] forest[9]	Clear-cuts and wildlife openings preferred over forest	Point counts and mist nets; New Hampshire (Chandler 2011)
Blackburnian Warbler (*Setophaga fusca*)	Coniferous and mixed forest	Clear-cuts[2]	Recorded; no numbers given	Mist nets; Virginia and West Virginia (Marshall et al. 2003)
Yellow-throated Warbler (*Setophaga dominica*)	Mature bottomland woodlands and upland pine–oak forest	Clear-cuts[1]	Recorded; no numbers or age ratios given	Mist nets; Ohio (Vitz and Rodewald 2006)

Species	Breeding Habitat	Postbreeding Habitat(s) Sampled	Postbreeding Habitat Use by Age (Adult/ Juvenile/ Unknown), N	Sampling Techniques, Region, and Citation
Pine Warbler (*Setophaga pinus*)	Coniferous forest	Old field,[3] forest[4]	Old field 0/1/0; forest 0/0/0	Mist nets; Georgia (Rappole and Ballard 1987)
Pine Warbler (*Setophaga pinus*)	Coniferous forest	Clear-cuts[2]	Recorded; no numbers given	Mist nets; Virginia and West Virginia (Marshall et al. 2003)
Cerulean Warbler (*Setophaga cerulea*)	Deciduous forest	Clear-cuts[1]	Recorded; no numbers or age ratios given	Mist nets; Ohio (Vitz and Rodewald 2006)
Black-and-white Warbler (*Mniotilta varia*)	Deciduous and mixed forest	Clear-cuts[1]	23/34/0	Mist nets; Ohio (Vitz and Rodewald 2006)
Black-and-white Warbler (*Mniotilta varia*)	Deciduous and mixed forest	Old field,[3] forest[4]	Old field 0/1/1; forest 0/0/0	Mist nets; Georgia (Rappole and Ballard 1987)
Black-and-white Warbler (*Mniotilta varia*)	Deciduous and mixed forest	Clear-cuts[2]	26/16/0	Mist nets; Virginia and West Virginia (Marshall et al. 2003)
American Redstart (*Setophaga ruticilla*)	Deciduous and mixed forest	Clear-cuts[1]	10/13/0	Mist nets; Ohio (Vitz and Rodewald 2006)
American Redstart (*Setophaga ruticilla*)	Deciduous and mixed forest	Old field,[3] forest[4]	Old field 0/1/1; forest 0/0/0	Mist nets; Georgia (Rappole and Ballard 1987)
Worm-eating Warbler (*Helmitheros vermivorum*)	Deciduous and mixed forest	Forest,[5] clear-cuts[6]	Adults and juveniles mainly in clear-cuts	Point counts and mist nets; Missouri (Pagen et al. 2000)
Worm-eating Warbler (*Helmitheros vermivorum*)	Deciduous and mixed forest	Clear-cuts[2]	67/37/0	Mist nets; Virginia and West Virginia (Marshall et al. 2003)

(*continued*)

TABLE 3.3   *(continued)*

Species	Breeding Habitat	Postbreeding Habitat(s) Sampled	Postbreeding Habitat Use by Age (Adult/ Juvenile/ Unknown), N	Sampling Techniques, Region, and Citation
Worm-eating Warbler (*Helmitheros vermivorum*)	Deciduous and mixed forest	Clear-cuts[1]	66/156/0	Mist nets; Ohio (Vitz and Rodewald 2006)
Ovenbird (*Seiurus aurocapilla*)	Deciduous and mixed forest	Clear-cuts[1]	108/164/0	Mist nets; Ohio (Vitz and Rodewald 2006)
Ovenbird (*Seiurus aurocapilla*)	Deciduous and mixed forest	Clear-cuts[2]	35/14/0	Mist nets; Virginia and West Virginia (Marshall et al. 2003)
Ovenbird (*Seiurus aurocapilla*)	Deciduous and mixed forest	Forest,[5] clear-cuts[6]	Adults and juveniles mostly in clear-cuts	Point counts and mist nets; Missouri (Pagen et al. 2000)
Ovenbird (*Seiurus aurocapilla*)	Deciduous and mixed forest	Old field,[3] forest[4]	Old field 1/0/0; forest 0/0/0	Mist nets; Georgia (Rappole and Ballard 1987)
Ovenbird (*Seiurus aurocapilla*)	Deciduous and mixed forest	Mature forest and seral stages	Juveniles occupied earlier seral stage habitats with denser understory than breeding habitat	Radio tracking; New Hampshire (King et al. 2006)
Louisiana Waterthrush (*Parkesia motacilla*)	Deciduous and mixed forest	Clear-cuts[1]	2/11/0	Mist nets; Ohio (Vitz and Rodewald 2006)
Kentucky Warbler (*Geothlypis formosus*)	Deciduous forest	Forest,[5] clear-cuts[6]	Adults and juveniles in both forest and clear-cuts	Point counts and mist nets; Missouri (Pagen et al. 2000)
Hooded Warbler (*Setophaga citrina*)	Deciduous forest	Clear-cuts[1]	42/115/0	Mist nets; Ohio (Vitz and Rodewald 2006)
Hooded Warbler (*Setophaga citrina*)	Deciduous forest	Clear-cuts[2]	62/16/0	Mist nets; Virginia and West Virginia (Marshall et al. 2003)

Species	Breeding Habitat	Postbreeding Habitat(s) Sampled	Postbreeding Habitat Use by Age (Adult/Juvenile/Unknown), N	Sampling Techniques, Region, and Citation
Green-tailed Towhee (*Pipilo chlorurus*)	Montane shrub-steppe and sagebrush (*Artemesia*)	Subalpine meadow	Juveniles moved considerable distances (>3 km) from breeding habitat to montane meadows to undergo prebasic molt	Mist nets; California (Morton 1991)
Summer Tanager (*Piranga rubra*)	Deciduous forest	Clear-cuts[1]	Recorded; no numbers or age ratios given	Mist nets; Ohio (Vitz and Rodewald 2006)
Scarlet Tanager (*Piranga olivacea*)	Deciduous and mixed forest	Clear-cuts[1]	17/101/0	Mist nets; Ohio (Vitz and Rodewald 2006)
Scarlet Tanager (*Piranga olivacea*)	Deciduous and mixed forest	Clear-cuts[2]	6/2/0	Mist nets; Virginia and West Virginia (Marshall et al. 2003)
Western Tanager (*Piranga ludoviciana*)	Coniferous and mixed forest	Southwestern riparian and western montane habitats	Adults migrate to riparian habitats in the southwestern United States after breeding, undergo prebasic molt, then continue south to wintering areas in Mexico and Central America; juveniles move from natal areas into neighboring montane habitats for prebasic molt	Museum specimens; western United States (Butler et al. 2002)

(*continued*)

TABLE 3.3   *(continued)*

Species	Breeding Habitat	Postbreeding Habitat(s) Sampled	Postbreeding Habitat Use by Age (Adult/ Juvenile/ Unknown), N	Sampling Techniques, Region, and Citation
Rose-breasted Grosbeak (*Pheucticus ludovicianus*)	Deciduous and mixed forest	Clear-cuts[2]	7/2/0	Mist nets; Virginia and West Virginia (Marshall et al. 2003)
Rose-breasted Grosbeak (*Pheucticus ludovicianus*)	Deciduous and mixed forest	Clear-cuts[1]	Recorded; no numbers or age ratios given	Mist nets; Ohio (Vitz and Rodewald 2006)
Lazuli Bunting (*Passerina amoena*)	Brushy hillsides, riparian woodlands, thickets and scrub			Museum specimens (Young 1991)
Baltimore Oriole (*Icterus galbula*)	Open deciduous woodland	Clear-cuts[1]	21/55	Mist nets; Ohio (Vitz and Rodewald 2006)

[1]Low, shrubby second growth with emergent saplings, 4 to 7 years after clearing. Saplings and shrubby second growth, 3 to 10 years after clearing.

[2]Low, herbaceous growth, saplings, and shrubby second growth, 1 to 7 years after clearing, 8.2 to 13.4 ha in size.

[3]Low (<2 m) herbaceous growth with scattered emergent saplings (<3 m).

[4]A strip of riparian forest, 50 to 100 m in width bordering the Oconee River on one side and old field habitat on the other dominated by American elm (*Ulmus americana*) and red maple (*Acer rubrum*); canopy height 15 to 20 m.

[5]Mature upland or riparian forest.

[6]Nine to 10 years after cutting and 3 to 4 years after cutting.

[7]Low herbaceous growth, greater than 3.5 ha, surrounded by forest.

[8]Six to 8 years after clearing, dominated by saplings.

[9]Mixed forest dominated by American beech (*Fagus grandifolia*), sugar maple (*Acer saccharum*), red maple (*Acer rubrum*), paper birch (*Betula papyrifera*), and eastern hemlock (*Tsuga canadensis*); canopy greater than 20 m, greater than 50 years old.

[10]Mature deciduous and mixed forest dominated by American beech (*Fagus grandifolia*), oak (*Quercus* sp.), Virginia pine (*Pinus virginianus*), and tulip poplar (*Liriodendron tulipifera*), greater than 70 years old.

[11]Dense, low herbaceous growth and shrubs with emergent saplings.

[12]Mature oak–hickory forest.

[13]Burned-over pine plantations, high-graded oak forest, and riparian forest with open canopies and dense understory of herbaceous growth.

[14]Saplings greater than 5 m with dense understory of *Rubus* and other herbaceous growth.

The number of species employing the strategy of using different habitats during the postbreeding period from those used during the reproductive period actually is probably far larger than indicated by available information for three reasons:

1. Temperate seasonal phenology produces different resource superabundances in different habitats at different times over the course of the summer season (Tauber et al. 1986).
2. Habitat use is likely to be as different as needs, and the needs during the breeding period (e.g., advertisement and mate attraction, nest placement, care of young) (Martin 1995; Barg et al. 2006) are quite different from those during the postbreeding period (e.g., molt, preparation for migration) (King et al. 2006).
3. Intensive study is required to document habitat use change during the postbreeding period, which has not been done for most migrant species.

In addition to optimization of habitat use, changes in social behavior are also observed in some migrants during the postbreeding period. For example, depending on how food resources and predators are distributed, it may be advantageous for them to associate in single-species or mixed-species flocks as a behavior to reduce probability of predation (Powell 1985; King and Rappole 2000, 2001a). Vega-Rivera et al. (1998b) found that most juvenile Wood Thrushes (25 of 28) during the early postbreeding period tended to associate in temporary, small, loose flocks of conspecifics while feeding mainly on fruits (e.g. *Rubus*). Such flocking behavior has been observed in other migrants during the postbreeding period (Morton et al. 1991; Young 1991). Later in the season, some individuals became solitary and more aggressive, perhaps responding to changes in the kinds of foods available (invertebrates versus fruits?) or in defense of next year's breeding territory.

DURATION OF THE POSTBREEDING PERIOD

Most Temperate Zone migrants demonstrate a relatively precise date after which initiation of reproductive activities (e.g., copulation, nest building, and egg laying) is rare. Whether a species is a calendar or facultative migrant does not appear to make much difference in terms of the existence of this date, which often occurs during the middle of the Holarctic summer when conditions presumably are still suitable for production of offspring. Such timing is indicative of genetic programming (i.e., an adaptation for optimal survivorship through the periods of molt and migration), rather than maximum production in the short term.

Thus, a block of time exists between completion of breeding and commencement of migration, but information on duration of the postbreeding period is incomplete or lacking for most species. In fact, a major difficulty involved in calculating duration of this portion of the life cycle has been that many migrants simply disappear from their breeding area shortly after completion of reproductive activities (Rappole 1995:75). As noted earlier, where the subject of this disappearance is addressed at all, it is usually assumed that the birds have initiated a gradual southward movement, perhaps representing incipient migration. Three types of information argue against this conclusion for most migrants and support the concept of a distinct postbreeding period:

- Detailed studies of postbreeding movements and activities for a few species
- Recent research documenting postbreeding habitat shifts in several migrants
- Timing of actual migration peaks in migrants occurring weeks or months after their disappearance from breeding territories

We discuss each of these points in an attempt to clarify duration of the postbreeding period in migrants.

DETAILED STUDIES OF SELECTED SPECIES

• Prairie Warbler (*Setophaga discolor*). Few, detailed studies of postbreeding movements and activities have been done. One of the best of these for a Temperate Zone passerine migrant derives from Nolan's (1978) extraordinary 20-year study of the Prairie Warbler. Timing of the major events of the annual cycle, including the postbreeding period for the southern Indiana population of the Prairie Warbler, is shown in figure 3.3. This timing is for the entire species of course, the north–south extent of whose breeding range covers 1,600 km in eastern North America. Julian dates (day numbers) for non-leap years are given in parentheses after calendar dates in the following discussion.

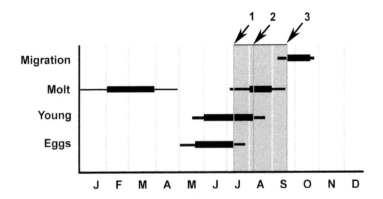

FIGURE 3.3   Postbreeding period (*gray shading*) for the Prairie Warbler (*Setophaga discolor*) in southern Indiana (Nolan 1978): (*1*) mean date for independence of first brood (juveniles begin postbreeding period); (*2*) mean date for independence of second brood (adults begin postbreeding period); (*3*) mean date for departure on fall migration by adults (end of postbreeding period).

At Nolan's study area in Indiana, the average date over a 14-year period for laying of the first egg was May 14 (134); mean clutch size was four with one egg laid per day, so mean date for commencement of incubation was May 17 (137). Incubation lasted an average of 12 days, so mean hatching date for first clutches was May 29 (149). Average age at fledging was 10 days, so mean fledging date for successful first nests was June 8 (159). After fledging, each member of the pair assumed responsibility for a portion of the fledglings whose care continued for an average of 30 days after fledging; so mean date for independence for the first brood would be July 8 (189), for an average total length of time required to raise offspring from first egg laid to independence of about 56 days. Thus, for these newly independent juvenile birds, the postbreeding period commences on average at the beginning of the second week in July. Not so for adults, the majority of which attempt to raise second broods or re-nest if the first nest fails. Mean date for laying the first egg for a second brood was June 21 (172), 2 weeks or so before the first brood has reached independence, at which point the male assumes care for all of the surviving first brood members (Nolan 1978:321). Calculating 56 days from laying of the first egg for the second brood until the young reach independence gives a mean date for beginning of the postbreeding period for most adult birds of August 15 (227). Attempts to re-nest after failures can make this date much later for some pairs. Nolan (1978:426) notes that these attempts end by early July to mid-July, which would push the beginning of the postbreeding period for at least some adults to late August or early September. Thus, the beginning date for the postbreeding period can extend from early July for some juveniles to early September for some adults and juveniles.

The ending date for the postbreeding period is even more difficult to pinpoint than the starting date for Nolan's Prairie Warblers because 21 percent of adult males, 57 percent of adult females, and 89 percent of juvenile birds disappear from the breeding area long before the commencement of fall migration. Nolan documents this fact by contrasting sighting data of marked individuals by age and sex at his Indiana study area with television-tower data for Prairie Warblers from north Florida killed during nocturnal flights, presumably representing southbound migration (table 3.4). Southern populations of this species evidently begin fall migration weeks or months before northern populations, so there is an early peak in late August of presumed southern-breeding birds followed by later peaks in mid-September to early October, thought to represent birds from more northern populations. The assumption that these later peaks are composed of northern birds is supported by the fact that most birds in these kills show little or no molt. Because molt begins about a week before young reach independence, requires roughly 40 days to complete, and does not overlap significantly with fall migration in this species (Nolan and Mumford 1965), few northern adults could complete breeding and molt activities before mid-September. Thus, duration of the postbreeding period for adult Prairie Warblers from southern Indiana likely averages from about the middle of August until the third week of September.

TABLE 3.4   Observations of Banded Prairie Warblers (*Setophaga discolor*) by Age and Sex During the Postbreeding Period in Southern Indiana

Age/Sex	Number Banded	Number Found at Least Once on the Study Site	Number Found Repeatedly on or Near (<200 m) Breeding Territory	Peak for Late Sightings on Study Area
Adult male	63	50	31	September 22–28; 6 into early October
Adult female	54	23	10 (3 of which left and then returned)	Late September; 8 into early October
Juvenile (independent, fledged young)	246	26	0	No data

Note: Post-breeding period is defined as July 10 to October 10.
Source: Nolan (1978).

The information on duration of the postbreeding period for juvenile birds is less solid, but Nolan (1978:442) suggests that the first prebasic molt in juvenile birds likely is complete at 60 to 70 days of age, after which presumably they initiate fall migration.

• Wood Thrush (*Hylocichla mustelina*). Although data are not nearly so complete, several aspects of the postbreeding movements and activities for the Wood Thrush can be compared with the Prairie Warbler data based on a radio-tracking study performed at the Quantico Marine Base in eastern Virginia, which is summarized here (Vega-Rivera 1997; Vega-Rivera et al. 1998a, 1998b, 1999, 2000). Nestling Wood Thrushes fledge at 12 to 15 days of age, in late May to early June for first clutches; late July to early August for second clutches. Both adults care for the fledglings, although female assistance ceased with initiation of the second clutch. Fledglings remained with and were cared for by a parent until about 32 days after fledging when they became independent and dispersed. As is clear from table 3.5, the postbreeding residence pattern for Wood Thrushes by age and sex was similar to that found by Nolan for his Prairie Warbler population, although there are some obvious differences. As was found for the Prairie Warbler, most male Wood Thrushes (85%) remained in the general area of the breeding territory (<7 km distant) for completion of the prebasic molt and presumptive preparation for fall migration (late September to early October), although a smaller percentage of these (30%) stayed on, or in the immediate vicinity of, the breeding territory.

TABLE 3.5   Persistence of Wood Thrush (*Hylocichla mustelina*) by Age and Sex During the Postbreeding Period in Northeastern Virginia

Age/Sex	Number Radio-Tagged	Number Departed from Study Site, Late June to Early August	Number Remaining on Study Site (<7 km from Breeding Territory) Until Mid-September (Juveniles) or Late September to Early October (Adults)	Number Remaining on or Near Breeding Territory Until Mid-September (Juveniles) or Late September to Early October (Adults)
Adult male	23	3	20	9
Adult female	25	15	10	6
Juvenile (independent, fledged young)	39	14	25	2

Note: Post-breeding period is defined as July 1 to October 10; data from one study site.

A larger percentage of Wood Thrush adult females remained on the study area (30% as opposed to 19% for Prairie Warblers), but the difference could be attributed to the fact that Vega-Rivera had a higher probability of finding birds (using aerial radio-tracking) than Nolan, who searched for color-banded individuals on foot. The most striking difference between the species is in the behavior recorded for juvenile birds. Nolan found that 89 percent of juveniles disappeared from the study area after reaching independence, whereas Vega-Rivera found that 64 percent (25 birds) remained on the study area through completion of molt (mid-September); again, however, a significant part of the difference may involve the search techniques. Of these 25 birds, 23 dispersed a mean distance of 1.5 km from the breeding territory when they reached independence (32 days after fledging) to one or more dispersal sites on the study area where they remained for most of the time until evident departure on migration in September. More than half of young (53%) performed short-term movements (>300 m) away from their dispersal sites, returning in <3 days. Mean distance traveled from the dispersal site for 11 (28%) of these "exploratory" movements was 1.7 km; however, the signal was lost until their return to the dispersal site, presumably due to movement well beyond the study area, for 29 (72%) of these movements. Another interesting observation of these young birds was that of 25 individuals tracked until departure on fall migration, 10 were recorded making movements a mean distance of 1.8 km from their dispersal site 3 to 10 days before departure, where they remained for 1 to 4 days before disappearance.

Recent Research Documenting Postbreeding Habitat Shifts in Several Migrants

We have already discussed the kinds of habitat shifts observed for many species of migrants, which may serve to accommodate shifts in terms of needs for food or cover (table 3.3). Typical of these is the study by Morton (1991), who reported shifts in habitat use for juvenile Green-tailed Towhees (*Pipilo cholorura*) from sagebrush (*Artemesia*) shrub-steppe habitat in which they were raised to alpine meadows during the postbreeding period in apparent search for optimal feeding habitat, a reason for postbreeding behavior suggested for several other migrant species as well (Morton et al. 1991). These kinds of studies demonstrate that many migrants spend a significant portion of time on or near the breeding grounds after completion of reproductive activities and before departing on migration.

Migration Peaks

The third type of information indicating the existence of a distinct postbreeding period for Temperate Zone migrants is the difference in timing of the end of reproductive activities compared with the actual appearance of migrants along the migration route. There are many good data sources documenting peak appearance of migrant passerines as transients on actual migration—for example, television-tower kills (Crawford 1981), banding station activities (Mackenzie and Friis 2006), long-term regional data sets (Rappole and Blacklock 1983), and radar studies (Richardson 1976). For most migrants, these data document that the peak of migration for birds en route to where they will spend the majority of the nonbreeding period (i.e., "winter quarters" for Temperate Zone species) occurs long after completion of breeding activities (Rappole 1995). We summarize data on peak of migration for selected migrant species in table 3.6. Assuming that the postbreeding period begins when breeding activities cease for most individuals and ends a few days before the peak of fall migration is observed, a rough calculation of mean postbreeding duration for many migrants can be made. We provide such a calculation in table 3.6, using breeding cessation data from species accounts in Poole (2010) and basing fall migration peak on television-tower kill data from north Florida (Crawford 1981). We include species from our "Detailed Studies of Selected Species" section (Prairie Warbler, Wood Thrush) in this table to show the basic similarity in timing of the postbreeding period among these species, for which the details are well known, and several others whose details are not well known. For the 21 migrant species shown in the table, representing three different orders and

TABLE 3.6   Comparison of Peak Date for Completion of Breeding and Date for Midpoint of 10-Day Period of Peak Fall Migration for Selected Species of North American Migrants

SPECIES	PEAK COMPLETION OF BREEDING	PEAK FALL MIGRATION: NORTH FLORIDA	DIFFERENCE BETWEEN COMPLETION OF BREEDING AND PEAK OF FALL MIGRATION IN DAYS
Yellow-billed Cuckoo (*Coccyzus americanus*)	August 28 (240)	October 16 (289)	49
Chimney Swift (*Chaetura pelagica*)	August 1 (213)	October 16 (289)	76
Yellow-throated Vireo (*Vireo flavifrons*)	August 31 (243)	October 6 (279)	36
Blue-headed Vireo (*Vireo solitarius*)	August 31 (243)	November 6 (310)	67
House Wren (*Troglodytes aedon*)	August 11 (223)	October 6 (279)	56
Marsh Wren (*Cistothorus palustris*)	August 15 (227)	October 6 (279)	52
Veery (*Catharus fuscescens*)	July 5 (186)	September 16 (259)	73
Gray-cheeked Thrush (*Catharus minimus*)	August 8 (220)	October 6 (279)	59
Swainson's Thrush (*Catharus ustulatus*)	August 4 (216)	October 6 (279)	63
Wood Thrush (*Hylocichla mustelina*)	July 26 (207)	October 6 (279)	75
Gray Catbird (*Dumetella carolinensis*)	August 3 (215)	October 16 (289)	74
Tennessee Warbler (*Oreothlypis peregrina*)	August 11 (223)	October 6 (279)	56
Magnolia Warbler (*Setophaga magnolia*)	August 3 (215)	October 16 (289)	74

(continued)

TABLE 3.6   (*continued*)

Species	Peak Completion of Breeding	Peak Fall Migration: North Florida	Difference Between Completion of Breeding and Peak of Fall Migration in Days
Blackburnian Warbler (*Setophaga fusca*)	July 16 (197)	September 16 (259)	62
Prairie Warbler (*Setophaga discolor*)	July 24 (205)	October 6 (279)	74
Bay-breasted Warbler (*Setophaga castanea*)	August 1 (213)	October 16 (289)	76
American Redstart (*Setophaga ruticilla*)	July 7 (188)	October 6 (279)	91
Eastern Towhee (*Pipilo erythrophthalmus*)	July 28 (209)	October 26 (299)	90
Scarlet Tanager (*Piranga olivacea*)	July 26 (207)	October 6 (279)	72
Indigo Bunting (*Passerina cyanea*)	August 21 (233)	October 6 (279)	46
Bobolink (*Dolichonyx oryzivorus*)	June 27 (178)	September 26 (269)	91

Note: Peak date for completion of breeding is based on species accounts in Poole (2010), and date for midpoint of 10-day period of peak fall migration is based on 25 years of television-tower kill data from northern Florida (Crawford 1981). Julian dates are given in parentheses.

10 families, the average length of time from completion of breeding until peak of southbound fall migration is 67 days.

TIMING AND SITING OF THE PREBASIC MOLT IN RELATION TO OTHER MAJOR LIFE CYCLE EVENTS

The data presented in this chapter demonstrate that many species of migrants experience the postbreeding period as a major part of the annual cycle, distinct from reproduction, migration, or wintering. A correlate of this finding is that close consideration of the events of this period, especially when compared with the same period in related resident species, may provide insights into how a migratory habit can shape adaptations through natural selection in different species of migrants. We have proposed that the chief benefit of migration for individuals in any given population is the ability to contribute more offspring to the next generation than sedentary individuals of the same population (see chapter 1). According to this hypothesis, migrants move to places with seasonally greater food availability than the sites from which they originated and are therefore able to produce more off-spring (Rappole and Tipton 1992). This movement has costs, but these costs can be reduced through the process of adaptation shaped by natural selection over time. A major set of costs resulting from an initial migratory movement presumably results from the need to balance energy demands of breeding, molt, and migra-tion. Examination of how costs and benefits are balanced in various migratory spe-cies may be instructive in terms of how long they have been exposed to the need for balancing these costs, which, in many cases will be the same as how long a population has been composed of migrants.

The majority of adult resident birds undergo a single molt each year, which normally occurs over a prolonged period after completion of breeding activities (Ginn and Melville 1983; Pyle 1997; Poole 2010). Migrants are quite a different story, in which the pattern of relationships between breeding, molt, and migra-tion are extremely complex and varied. We suggest that the variety of molting patterns seen in migrants represent various forms of adaptation for migration. It is the interaction between two major factors that produce the wide variety of breeding/molt/migration patterns seen in migrants: (1) different life history demands for different species (e.g., pelagic versus forest-related) and (2) time since a migratory habit first appeared in a species, presuming that selection will favor reduction of overlap between major events due to energy constraints (Farner 1958:18).

If, as we have suggested, migration can occur in any group of resident individu-als, then perhaps the molt pattern seen in a resident population can be considered as the likely pattern for the first migratory individuals derived from the resident group. We assume that the first migrant members of a resident population dis-persed or migrated to their new breeding area some time after fledging (Rappole and Tipton 1992). We also assume that they will migrate back to point of origin after completion of their first breeding season's activities. Because the prebasic molt occurs after completion of breeding, we can further assume that molt and return migration overlap in the first migrants and that it is likely that natural selec-tion will serve to modify this pattern over time, balancing the selective pressures of increased efficiency of flight resulting from molts providing new remiges and rectrices against increased energy demands resulting from a more rapid molt. We envision a scenario for the evolution of the relationship between breeding, molt, and migration something like that presented in table 3.7.

TABLE 3.7   Hypothetical Phases in the Evolution of an Adaptive Balance Between Molt and Migration for Adult Birds

PHASE	DESCRIPTION
1	Molt follows breeding and overlaps with migration
2	Reduction of overlap between molt and migration by (a) interrupting molt during migration (b) delaying molt until after migration
3	Reduction of time over which molt occurs
4	Molt prior to migration
5	Refinement of molt timing and patterns for different age and sex groups
6	Migration to special sites for the sole purpose of molt

Note: Evolutionary phases described assume increasing genetically based endogenous control over timing of breeding, molt, and migration specific for each age and sex group within a population.

The hypothesized sequence of events in the evolution of balance between demands of reproduction, molt, and migration shown in table 3.7 is for adults of a given migrant population assumed to be derived from an ancestral sedentary resident ancestor with a single, prolonged prebasic molt after completion of breeding activities. Juvenile birds confront a different balancing problem than that of adults for two reasons:

1. Selection is likely greater on juvenile birds than adults to undergo the prebasic molt of all body feathers prior to migration because the body feathers of the juvenal plumage lack the resistance to wear and thermodynamic properties of the first basic plumage.
2. The need to molt flight feathers before migration is much less in juvenile birds because the flight feathers of the juvenal plumage are essentially equivalent in structure to those of adults and have very little wear compared with those of adults, whose flight feathers will have been subjected to several months of wear, including at least one migration.

For these reasons, differences between adult and juvenile birds in timing and extent of the prebasic molt relative to fall migration are likely to appear very early in the process of development of a migratory habit in any given population. Mixtures of different timing patterns within the same age and sex group should be typical of populations in the early stages of the development of a migratory habit,

"difficillis"   AHY & HY

oberholseri   HY
                AHY

flaviventris   HY
                 AHY

hammondii   HY
              AHY

Jun Jul Aug Sep Oct Nov Dec

FIGURE 3.4   Timing of migration (*dark gray bars*) and prebasic molt (*light gray bars*) for hatching year (HY) and adult (AHY) members for four species of *Empidonax* flycatchers. Black bars indicate overlap of migration and molt. Note that *"difficilis"* refers to the superspecies complex containing *E. difficilis* and *E. occidentalis*, as explained in the text (data from Johnson 1963; species accounts in Poole 2010).

whereas near uniformity of timing within the reaction norms for a given pattern for a particular age or sex group should typify populations in which migration has been under way for longer periods.

With these considerations in mind, we examine what is known regarding the timing of molt and migration for selected members of the genus *Empidonax*, a group of New World flycatchers (Tyrannidae). This genus includes a combination of Nearctic–neotropical migrants and neotropical residents, including some species that have both migrant and neotropical resident representatives. Several of the possible variations in resolving the problem of overlap between migration and molt can be found in this genus (figure 3.4) (Johnson 1963; Poole 2010).

The "species" whose molt and migration patterns are shown in figure 3.5 (*"difficilis"* actually includes at least two species as part of a superspecies complex) are presented in order of proposed length of time for which each species has been a migrant, from most recent at the top (*"difficilis"*) to most ancient at the bottom (*hammondii*). Our reasoning for this arrangement is as follows:

• Western Flycatcher (*Empidonax "difficilis"*). This group, formerly referred to as the Western Flycatcher, is now split into two species: Pacific-slope Flycatcher (*Empidonax difficilis*) and Cordilleran Flycatcher (*Empidonax occidentalis*). Both members of this superspecies complex have breeding populations in the western Nearctic (figure 3.5) and winter ranges in the neotropics that overlap with resident populations of conspecifics (figure 3.6). We suggest that overlap of distribution between wintering migrant and tropical resident populations may be a sign that development of migration in such species is in an early stage. We further suggest that another indication of the recent nature of migration in the group is the

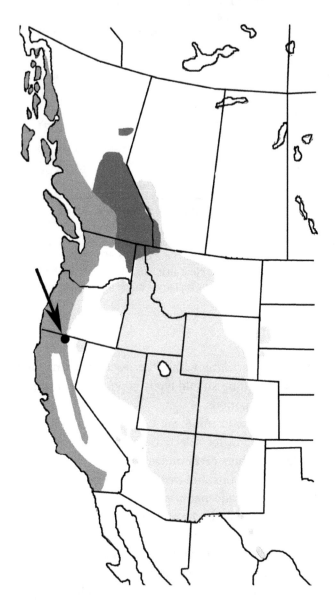

FIGURE 3.5   Breeding distribution of the Pacific-slope (*Empidonax difficilis*) and Cordilleran (*Empidonax occidentalis*) flycatchers in western North America (Lowther 2000): *medium gray* = Pacific-slope breeding; *light gray* = Cordilleran breeding; *dark gray* = area of suspected sympatry; *black circle (arrow)* = area of known sympatry.

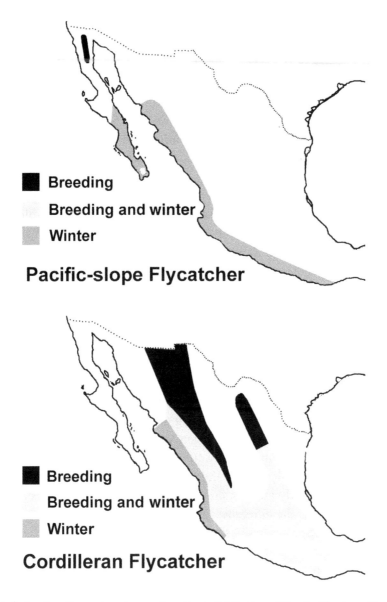

FIGURE 3.6   Breeding, wintering, and resident distribution of the Pacific-slope (*Empidonax difficilis*) and Cordilleran (*Empidonax occidentalis*) flycatchers in southwestern North America (Lowther 2000).

extensive overlap between two energetically costly events, molt and migration, in both adults and juveniles.

• Dusky Flycatcher (*Empidonax oberholseri*). Both adults and juveniles of this species delay molt until after migration. This timing means that juvenile birds migrate in juvenal body plumage that is structurally weak and relatively poor from a thermoregulatory perspective. Adults migrate with worn flight feathers that are months old and have already undergone a migration.

• Yellow-bellied Flycatcher (*Empidonax flaviventris*). Juvenile birds begin molt before migration but still overlap molt and migration to some extent, a situation perhaps marginally better from an energy-cost perspective than completing migration in juvenal plumage. Adults migrate before molting with worn flight feathers that are months old and have already undergone a migration.

• Hammond's Flycatcher (*Empidonax hammondii*). Both adults and juveniles complete prebasic molt prior to migration, which means that juveniles migrate in basic plumage, which presumably is structurally better for this purpose than the rather loose juvenal plumage, and adults migrate with fresh flight feathers. This sequence (table 3.2, pattern 1) is the one most commonly found in long-distance migrant passerines and, we suggest, represents the longest exposure to a migratory habit within the group considered.

Obviously, the relationship presented between length of time as a migrant for a given population and differences in timing of molt and migration is highly speculative. The reasoning behind it is largely circular, as we have presented only a single explanation for a pattern that could have a wide range of explanations. In fact, Johnson (1963), in discussing the different molt patterns among the species of this same group of flycatchers, provides a completely different explanation for the patterns based on the differences in food availability and habitats occupied at breeding, stopover, and wintering sites by the different species, which clearly could affect the relative importance of the relationship between molt and migration. He also points out that it should make little difference whether a bird migrates twice on flight feathers formed on the wintering grounds, as is the case for adult Yellow-bellied Flycatchers (i.e., spring and then fall migration) or the breeding grounds for adult Hammond's Flycatchers (i.e., fall and then spring migration). A third possible explanation for the differences in pattern has to do with differences in intensity of intraspecific competition on the wintering ground among the different species; that is, earliest possible departure from the breeding ground (pre-molt) is favored in those species in which competition for quality wintering sites is intense (Winker and Rappole 1992).

It is not our purpose to argue which of these explanations for timing patterns of molt and migration is more likely or to exclude consideration of other ideas on the subject. Rather, we wish to suggest that length of time for which a population has been migratory should be added to the various life history constraints when considering critical factors that could affect the balance of timing between molt and

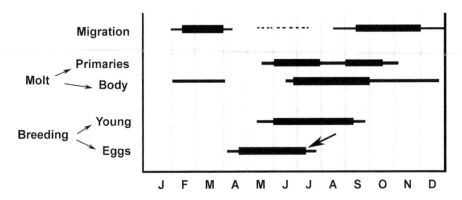

FIGURE 3.7   Annual cycle summary for the Mallard (*Anas platyrhynchos*), a facultative migrant. Arrow points to normal cessation date for egg laying. *Dashed line* = adult male molt migration.

migration for different age and sex groups within a given migrant population. For instance, consider the pattern found by Rohwer and his colleagues in some western Nearctic migrants in which various age groups molt at stopover sites or on the wintering grounds (Butler et al. 2002; Rohwer et al. 2009). The authors suggest that this pattern conforms to the highly evolved special "molt migrations" seen in phase 5 species (table 3.7) like the Mallard. We would argue that these patterns are more likely indicative of migrants in phase 2 (table 3.7), relatively new migrants, early in the process of resolving optimal balance between timing of migration and molt.

We further suggest that consideration of some facultative migrants from the perspective of when migration began in a group can be instructive. Facultative migration, in which the southward movement of populations appears as an ad hoc, discretionary, or optional movement directly related to deteriorating conditions in temperate or boreal environments (see chapter 1), has often been cited as an initial step in the evolution of all types of migration (e.g., Cohen 1967; Gauthreaux 1982). However, examination of the highly evolved timing and location of breeding, molt, and migration for facultative migrants raises questions concerning this interpretation of the phenomenon. Consider for instance this annual cycle summary for the facultative migrant, the Mallard (*Anas platyrhynchos*), based on life history information in Drilling et al. (2002) (figure 3.7).

Mallards pair during winter and migrate together to a breeding site where the female bred previously or was raised. They establish a territory and mate. The female constructs a nest and lays eggs. The male migrates to a molting site, usually by early June in central North America, often some distance from the breeding site (hundreds of kilometers) where he stays until molt is complete. He remains at the molting site until the weather deteriorates and then moves southward to a wintering area where he meets and pairs with a female, beginning the process for the new annual cycle. The female incubates and raises the offspring on her own. When the young are able to fly, they may or may not molt on or near the breeding area or leave the breeding site and migrate to a molting site where they remain until the weather deteriorates, when they move southward. The female remains at, or in the general vicinity of, the breeding site to molt and then moves southward when the weather deteriorates.

Timing of breeding and molt are evidently under precise endogenous control for the Mallard, which is different for adult males, females, and juvenile birds. At a given latitude, no nest is initiated after a certain date. Similarly, molt begins and ends within well-defined seasonal parameters. Clearly, migration is a highly evolved strategy for the Mallard, whose success is based on natural selection occurring over millennia and shaping the contours of breeding, molt, and migration for each age and sex group. The only ad hoc aspect of the strategy is initiation of southward migration in response to weather. But, as any duck hunter knows, the Mallard usually is well prepared for this eventuality with heavy fat reserves (Heitmayr 1987; Bluhm 1988). Thus, we propose that the example of the Mallard shows that facultative migration is not necessarily an intermediate step in the evolution of a migration strategy.

CHAPTER 4

FALL TRANSIENT PERIOD

THE FALL transient period begins with the end of the postbreeding period as the bird departs on fall migration. From an evolutionary perspective, this is the most critical portion of the annual cycle because, whereas it is relatively easy to understand how a dispersing individual might leave its natal territory to move a considerable distance to a new breeding site, it is difficult to understand how this first migrant gets back to its natal area after successful completion of breeding, and even more formidable to comprehend how its offspring complete such a journey. We will address these and related issues in this chapter and will continue to focus on distinguishing between exaptations and adaptations for the movement from the breeding to the wintering area.

PREPARATION FOR DEPARTURE

Groebbels (1928) pioneered investigation of the series of physiologic and associated behavioral changes that a migrant undergoes in preparation for, and during, its migratory journey. He observed that birds in preparation for migratory flight (*Zugdisposition*) ate intensively and laid down subcutaneous fat reserves at a remarkable rate, increasing their mass by as much as 50 percent in a matter of days.

At some point, this intensive eating (hyperphagia) response was "turned off," to be followed by a very different behavioral state in which the bird was no longer hyperphagic but was actually undertaking migratory flight (*Zugstimmung*). Captive birds, prevented from departure, nevertheless demonstrated this behavior in the form of migratory restlessness (*Zugunruhe*). Groebbels's observations have now been documented under both field and laboratory conditions for many different species (see reviews in King 1972; Gwinner 1990; Berthold 1993; Berthold et al. 2003).

The ultimate cause for these changes seems to be fairly clear; namely, to place the individual in the optimal environment for survival and/or reproduction in a system in which the geographic location of that optimal environment changes sharply over the course of an annual cycle (Mayr and Meise 1930; Williams 1958; Rappole et al. 2003). For those temperate-breeding species in which it has been thoroughly investigated (see reviews in Berthold 1988; Gwinner 1990; Pulido and Berthold 2003; Newton 2008:337–347), the factors triggering the physiologic and behavioral responses facilitating migration appear to be a combination of photoperiod and genetically programmed endogenous rhythms (Gwinner 1968; Gwinner and Helm 2003), although a relatively small number of species has been studied, most of which breed in the Palearctic (e.g., Helm 2003).

The internal, physiologic controls over a migratory bird's responses are even less well understood than the environmental triggering mechanism (Holberton and Dufty 2005:294). Recent studies have addressed the question of hormonal control over the hyperphagic response (*Zugdisposition*), focusing on the role of corticosterone in mediating foraging activity and fat deposition. Holberton et al. (1996) compared plasma levels of corticosterone in Gray Catbirds (*Dumetella carolinensis*) captured during molt (i.e., prior to onset of the behavioral and physiologic changes associated with migration) with corticosterone levels of fattening birds apparently preparing for migratory flight (i.e., in *Zugdisposition*). They found that birds captured in apparent *Zugdisposition* showed significantly higher baseline levels of corticosterone than birds captured during molt. On the basis of these findings, they proposed a "migration-modulation" hypothesis, which states that elevated corticosterone levels during the migratory period facilitate hyperphagia and lipogenesis independent of short-term changes in energetic condition, and that further elevation of corticosterone in response to acute stress is suppressed during this premigratory period to protect skeletal muscle needed for flight (Holberton et al. 1996:558). Despite these and similar studies, many questions remain regarding the relationships between environmental cues, hormonal mediators, and physiologic and behavioral responses involved in the phenomenon called migration.

FAT DEPOSITION

Migrant birds exhibit fattening on at least two different timescales: seasonal fattening (e.g., prior to migratory flights in the fall and spring, which are distinct life history phases) and daily fattening in response to short-term food shortages (Biebach 1996).

In recent decades, both kinds of fattening have been the subject of many experimental, laboratory studies (Biebach 1996), as well as a few in natural populations (e.g., Gosler 1996; Holberton et al. 1996; Katti and Price 1996). Although these studies have demonstrated some of the ecological and physiologic correlates of fattening, the underlying mechanisms are not yet fully understood (Holberton and Dufty 2005). For instance, migratory fattening is known to be a distinct physiologic state (the *Zugdisposition* of Groebbels [1928]) characterized by hyperphagia, often accompanied by extensive modification of body tissues (Lindström and Piersma 1993), and under endogenous control (Gwinner 1986, 1990). Yet, it is not clear how this process relates to daily fattening, which appears to be a short-term, adaptive response to environmental stress (Biebach 1996). It is likely that at least some of the physiologic mechanisms of lipid production and deposition are shared between these two responses with differences perhaps in regulatory mechanisms involving endocrine control.

INTERACTIONS BETWEEN FAT DEPOSITION AND MOLT

In general, the mechanisms involved in mediating fat storage and molt have been examined separately, whereas in many birds these processes often overlap in time and probably involve trade-offs among competing nutritional and energy needs. For example, in the Green Leaf Warbler (*Phylloscopus nitidus*), which molts on the wintering grounds without overlap with migratory fattening (Katti and Price 1999), it was found that reduction in food availability caused by drought resulted in increased daily fattening in the period before molt, as well as a delay in the onset of molt. Further, total body mass did not increase even when fat levels went up significantly, indicating a decrease in lean mass (Katti and Price 1996). These results may reflect a corticosterone-based response to the stress of starvation: Corticosterone can increase foraging activity leading to greater fat deposition, while at the same time, if present chronically, it can induce breakdown of muscle proteins (Wingfield and Silverin 1986). The loss of protein reserves may then cause a further delay in molt (Katti and Price 1999). In contrast, Lindström et al. (1994), in an experimental study of the effect of light regimes on postbreeding Bluethroats (*Luscinia svecica*), in which molt overlaps with fall migration, found that onset and rate of molt appeared to be endogenously set (prior to the start of the experiments), but the onset and rate of premigratory fattening was flexible and responded to the manipulation of light regimes. These studies suggest that whereas both fat storage and molt may be plastic processes that respond to environmental change, the degree of plasticity and the nature of the response probably depend on the species-specific life history context, as well as the particular environmental regimes.

ENDOCRINE RELATIONSHIPS

Thyroid function has been implicated in a number of physiologic control mechanisms in passerines including metabolic rate, molt, gonadal function, and migratory behavior, but species differences preclude broad generalizations (for a review, see Wingfield and Farner 1993). Although the direct causal linkage between thyroxine secretion and prebasic molt is controversial, there does appear to be either a reciprocal or parallel relationship between thyroid function and gonadal activity in passerines (Jallageas and Assenmacher 1979; Smith 1982).

With respect to adrenal cortical activity, it appears that adrenal activation may peak at the end of the period of maximal gonadal activation, just before the onset of the prebasic molt (Fromme-Bouman 1962; Lorenzen and Farner 1964). Additionally, stress-induced corticosterone production may actually delay the onset of molt (Katti and Price 1996), whereas increased interrenal cell activity (and corticosterone secretion) has been associated with premigratory hyperphagia and increased subcutaneous fat deposition (Gorman and Milne 1971; Chester-Jones et al. 1972; Wingfield et al. 1982; Gwinner 1986; Wingfield and Silverin 1986; Astheimer et al. 1992).

Prolactin secretion is believed to be mediated, in part, by photoperiod and the expression of an endogenous seasonal rhythm that can be further modulated by cues from nests, eggs, and young during the period of parental care (for a review, see Wingfield and Farner 1993). Although not well characterized, prolactin probably mediates many physiologic events, and its potential impact on the complex behavioral changes associated with migration warrant assessment. For example, there is strong evidence that both prolactin and corticosterone may act synergistically to affect premigratory fat deposition and migratory activity (Meier et al. 1965, 1980; Meier and Ferrel 1978).

Despite these and similar studies, resolution of the roles played by specific hormones in the physiologic processes involved in *Zugdisposition* have not yet been clarified completely for any migrant species, let alone for the entire cohort of avian migrants.

TIMING, DURATION, AND AMOUNT OF PREMIGRATORY FATTENING

The characteristic behaviors and physiologic changes associated with premigratory hyperphagia begin suddenly, and, at least in those small passerines that have been studied, last for 6 to 9 days prior to reaching a plateau in terms of percentage of body mass when the bird is seemingly ready to depart (King 1972). The value of this percentage varies by species, as well as by season for the same species (table 4.1). These kinds of interspecific seasonal differences in timing, duration, and amount of premigratory fattening demonstrate clearly that the process is under a highly evolved mechanism of endogenous control triggered presumably by environmental cues.

In addition to fat, energy can be stored as protein (i.e., muscle). Protein provides less energy per unit of mass than fat, but it has two advantages that may be pertinent in certain situations:

1. It contains 60 percent water, which may be necessary for birds crossing large expanses of water-less regions or exposed to inordinately high temperatures (Biebach 1990; Lindström et al. 2000; Klaassen and Biebach 2000).

TABLE 4.1   Seasonal Premigratory Fattening Patterns by Species

SPECIES	PATTERN	CITATIONS
White-crowned Sparrow (*Zonotrichia leucophrys*)	Fattens more rapidly preparatory to spring migration and stores greater reserves	King et al. 1963; King and Farner 1965
Several North American finches and thrushes	Fattens more rapidly preparatory to spring migration and stores greater reserves	Weise 1963
Common Whitethroat (*Sylvia communis*) and others	Fatter in fall than in spring	Merkel 1966
Several shorebirds	Fatter in fall than in spring	McNeil 1969
Brambling (*Fringilla montifringilla*)	Roughly the same amount of fattening in fall and spring	Dolnik and Blyumental 1964:289
European Greenfinch (*Carduelis chloris*)	Fattens more in spring	Dolnik and Blyumental 1964:289
Scarlet Grosbeak (*Carpodacus erythrinus*)	Fattens more in fall	Dolnik and Blyumental 1964:289

2. It may be useful for female spring migrants to arrive with sufficient protein reserves to initiate egg formation (Raveling 1979; Ankney 1984; Davidson and Evans 1989; Lindström and Piersma 1993; Bairlein and Gwinner 1994).

CUES FOR DEPARTURE

CUES FOR THE FIRST MIGRANTS

A central theme of our treatment is that migration originates as dispersal by young individuals from sedentary populations. We hypothesize that these individuals have the necessary exaptations to be able to move from their natal site to a new breeding area where they then proceed to breed and raise offspring. In addition, we envision these birds as having exaptations in terms of genetic programming to enable both them and their offspring to return to the original natal region where the population evolved if necessary. This scenario raises the question of what it is that pushes these first migrants to depart from the newly colonized breeding site once reproduction has been accomplished. We propose that there are at least two possible, related answers to this question: (1) decline in available food resources presumably related to seasonal weather changes or (2) density-dependent, intra-specific competition for declining resources. Which explanation is likely to best fit a given first-migrant's situation may be related to the kinds of resources on which they depend and the ways in which those resources are distributed and exploited. Thus, for a shorebird migrant that breeds at high latitudes and feeds mainly on littoral or soil invertebrates, the period between change in weather and disappearance of food supplies may be quite short, leaving little time for competition to play a role, at least on the breeding site. However, for seed-eaters breeding in more temperate climates, the time between onset of weather that negatively affects food resources and the actual disappearance of those resources may be prolonged, allowing competitive interactions to determine which individuals will depart first.

Whether the first migrants leave the breeding area as a result of resource deple-tion or competition, it seems logical that the place toward which they should go is the place from which they originally came; namely, the region in which the adults were born, as that is an area they can be relatively certain has appropriate habitat and which they know how to find, as the ability to return to sites previously occu-pied is an exaptation common not only to many avian residents but also to many kinds of organisms (Storm 1966; Rosengren and Fortelius 1986; Wiltschko 1992; Ramos and Rappole 1994). However, we expand on this idea of return capability in our treatment of avian migration to propose that knowledge of, and ability to return to, the region in which a population originally evolved is part of the first migrants' genetic makeup, as well as that of their offspring and the resident mem-bers of the population from which the first migrants were derived. We will develop this hypothesis elsewhere in the life history chapters as appropriate and will pres-ent a more formal model in chapter 8.

For the first migrants in any population, the environmental stimulus favoring movement (e.g., decreased food availability) and the bird's response (departure from the breeding area) likely are not separated by very much time (days?). How-ever, it is clear that selection should favor rapid changes in the genetics of the popu-lation so that internally programmed mechanisms will allow for proper behavioral and physiologic preparation for departure to occur before the need arises. Delaying departure from the breeding grounds until forced to leave by resource declines is likely to have potentially high costs in terms of survivorship probabilities. There-fore, selection should act quickly on two aspects of a population's adaptation to a migratory lifestyle:

- Anticipation of departure in terms of detection of environmental cues adjust-ing timing of major life history events (e.g., reproduction and molt) so that the individual can be ready to depart at an optimal time for survival
- Physiologic preparation for departure in terms of storage of energy reserves (Gwinner 1990)

Evolved Cues Governing Migration Departure

At least two classes of environmental cues are evident in control over the evolved physiologic and behavioral responses of migrants:

- *Distal cues*. Environmental changes that may occur weeks or months before the physiologic or behavioral response is evident
- *Proximal cues*. Those environmental changes that occur days, hours, or even minutes before the physiologic or behavioral response is evident

Ramenofsky and Wingfield (2007:140–141) also recognized two classes of environmental cues governing migratory movement:

- *Initial predictive*. Those cues that trigger the first transition from one life history stage to another
- *Local predictive*. Those cues that provide for adjustment to local conditions

These definitions conform fairly well to our "distal" and "proximal" classes, respectively, but we will use our definitions of these two classes in the discussion that follows because they focus attention on what we believe is the key difference between the classes: separation in timing between receipt of the cue and the subsequent physiologic or behavioral response of the individual.

Evolved Distal Cues Governing Movement

Evidence of the existence of distal cues for the timing of migrant departure is extensive for most species whose life histories have been investigated and includes the following:

1. Adults cease nesting activities months before departure (see chapters 2 and 3).
2. Adults and young of most migrant species either complete or interrupt molt prior to departure or delay molt until after wintering ground arrival (see chapter 3).
3. *Zugdisposition* is initiated after completion, interruption, or delay of molt and prior to any obvious changes in the environment in terms of availability of critical resources (King 1972).
4. Actual departure on migration also occurs prior to obvious changes in the environment for many species, especially long-distance migrants (see species accounts in Poole 2010).
5. Sharp differences in timing of migration departure for different sex and age groups within a given population of most migratory species (e.g., Nolan and Mumford 1965).

Sources documenting most of this indirect evidence for existence of some distal cue controlling the timing of major life history events for a large number of migratory species can be found in species accounts in Poole (2010) and del Hoyo et al. (1992–2011), the exception being no. 3, initiation of *Zugdisposition*, which has been documented and studied in free-living birds for only a few species of migrants (e.g., Rimmer 1988; Heise and Rimmer 2000).

Direct evidence testing the existence and function of distal cues is also extensive, although the number of migrant species investigated in studies is relatively small (<30?). Most of this experimental work has been conducted in the laboratory and involves testing the effects of photoperiod manipulation on the timing of physiologic changes associated with major life history events (e.g., gonadal development, molt, and premigratory fat deposition) (see reviews in King 1972; Gwinner 1990; Berthold et al. 2003). Many studies have also examined the effects of photoperiod manipulation on onset of migratory restlessness (*Zugunruhe*), based on the assumption that such restlessness can be equated with actual departure on migratory flight (*Zugstimmung*) (see reviews in Emlen 1975; Berthold 1975, 2001). In addition, studies of a few species have been done by testing the genetic nature of the onset and duration of *Zugunruhe* through captive breeding and artificial selection (Pulido and Berthold 2003; Pulido 2007).

Photoperiod is not always the distal cue governing the timing of movement and other major life history events for migrants. For instance, some factor or factors related to the onset of rain appears to serve as perhaps both distal and proximal cue in arid environments where rainfall is unpredictable (Dean 2004). Other such cues probably exist, especially for the large number of intratropical migrants, in which the factors triggering movement are almost completely uninvestigated.

In summary, both direct and indirect evidence indicate that the timing of major life history events is under genetic control and that the distal cue controlling the timing for most species of migrants is photoperiod, at least for those few North American and European Temperate Zone migrants that have been tested (Gwinner 1968, 1990).

EVOLVED PROXIMAL CUES

Conclusions concerning proximal cues triggering migration have been based mostly on correlations between environmental variables and actual departure on migratory flight as observed or recorded by radar (see reviews in Lack 1960; Berthold 1975; Richardson 1978, 1990; Alerstam 1981). Results of this work document that one or more aspects of weather constitute the most critical proximal cues for initiation of migratory flight, at least in temperate North America and Europe where most of the research has been done (e.g., Cochran and Wikelski 2005:278). The determination of precisely which aspect has been hindered by tight relationships between weather variables, and different authors have come to different conclusions as to which was most important. Bagg et al. (1950) concluded that barometric pressure was the key proximal cue triggering migratory flight; Lack (1960) suggested temperature whereas others have mentioned humidity (Nisbet and Drury 1968), atmospheric stability (Kerlinger 1982), direction of winds aloft (Richardson 1978), and cloud cover (Beason 1978; Alerstam 1978). Each of these variables could serve as the key trigger, and, indeed, one may serve as the proximal cue for one particular group of birds, while another could serve that role for another group. For instance, many hawks and other migrant, soaring species depend upon thermals—rising columns of warm air generated by uneven solar heating of Earth's surface—for migratory flight, so atmospheric stability (or rather lack thereof in the form of thermals) constitutes a key aspect of weather for them. Because thermals are generated by short-term temperature change (warming air after sunrise) (Pennycuick 1975:57), rising temperature could serve as the proximal cue for migration departure for soaring species (Pennycuick 1998).

Wind direction and other weather variables including temperature, relative humidity, precipitation, and barometric pressure occur in obvious correlative relationships in association with low- and high-pressure systems. Passing generally from west to east, these systems constitute the predominant weather patterns affecting migrants during the fall period (July to December) in the northern temperate regions of the Western Hemisphere. *Fronts* are literally the leading edge of these systems. Taken together, these aspects of weather are referred to as "synoptic weather features" (figure 4.1). Extensive studies, performed mostly in northern latitudes, have demonstrated that favorable (i.e., following) winds aloft serve as the most common weather factor associated with migratory flight for the majority of species that depend upon powered horizontal flight (as opposed to gliding or soaring) (Richardson 1978, 1990). Winds blow clockwise around high-pressure systems and counterclockwise around lows. Thus, in fall in the Northern Hemisphere, northerly winds serve as those most favorable for southbound migrants, and these winds occur with the advent of cold fronts. However, even if wind direction is the key aspect of weather determining departure for most migrants, that does not mean that synoptic weather features associated with wind direction could not serve as proximal cues indicative of winds aloft for migrants.

DEPARTURE CUES FOR FACULTATIVE, IRRUPTIVE, AND CALENDAR MIGRANTS

A distinction often is made between birds that appear to migrate in response to sharp changes in weather (weather migrants) or food supply (irruptive migrants), both of which are often referred to as facultative migrants, versus those that migrate in apparent response to cues that are genetically programmed to produce migration according to specific annual timetables ("calendar" or "obligate" migrants) (Newton 2008:334, 468, 2011). This distinction is useful insofar as it reflects marked differences between group members in terms of timing and proximal cues for departure. The principal differences between these groups are

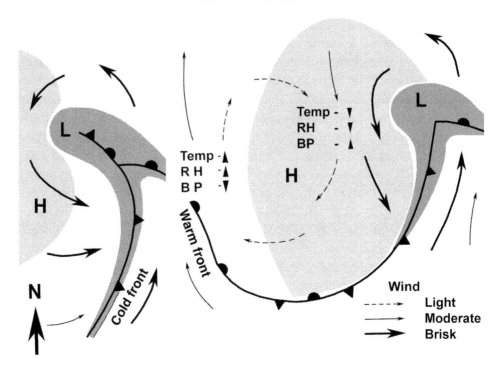

FIGURE 4.1   Synoptic weather features of high (H) and low (L) pressure systems and associated avian migration (based on Richardson 1978, 1990). *Dark gray* = precipitation; *light gray* = high density of flying migrants; Temp = temperature; RH = relative humidity; BP = barometric pressure; *arrows* = wind direction.

(1) when during the migratory period they begin their journeys, and (2) which aspects of weather they use as the proximal cue for departure. Calendar migrants tend to depart earlier in fall and return later in spring than facultative migrants, and the standard deviation around their mean departure time is much narrower; indeed, some members of facultative species may not migrate at all during mild winters (Tryjanowski et al. 2005).

Nevertheless, the distinction misrepresents differences in terms of the importance of distal cues for both groups. It is clear from extensive information derived from life history studies that both calendar and facultative migrants prepare for migration according to responses to distal cues (e.g., photoperiod). For instance, species in both groups undergo behavioral and physiologic preparations for the possibility of migration well in advance of when they must actually depart, including early termination of the reproductive period (i.e., long before critical resources are diminished), completion of molt, and deposition of subcutaneous fat. Thus, members of both groups are genetically programmed through the timing of key life history parameters for the possibility of undertaking migration. Proximal cues signaling the need to initiate migratory flight differ between the two, as does the timing and duration of the genetically programmed window of dates during the season when the birds will respond to such cues.

Timing of Departure Within the Fall Transient Period

The changes in behavior associated with going from a feeding state (*Zugdisposition*) to a flying state (*Zugstimmung*) as the fall transient period begins are obvious to the most casual observer and have been commented upon for millennia (see chapter 1):

1. The activity pattern of the individual shifts from daily short-distance foraging or roosting movements of tens or hundreds of meters to long-distance movements of tens or hundreds of kilometers, usually in the direction of the wintering area.
2. Diurnal migrants (e.g., waterfowl, herons, and swallows) associate in loose flocks that move throughout the course of the daylight hours, often forming concentrations called "leading lines" along the edges of certain obstacles like large bodies of water; deserts, or mountain ranges.
3. Nocturnal migrants change from a diurnal activity period to nocturnal, flying through the night and feeding or resting during the day (Mukhin et al. 2009).
4. Members of normally solitary species associate with conspecifics in loose flocks, apparently held together by visual and auditory cues during migratory flight (Hamilton 1962, 1966; Larkin and Szafoni 2008), and often continuing association while resting or foraging during the day (Rappole and Warner 1976).
5. Sudden appearance of species, often in large numbers, at sites or regions where they are not known to breed (Rappole and Warner 1976).

Each of these kinds of observations is indicative of the presence of birds in a migratory state, and they are made ad hoc or systematically by thousands of observers from many parts of the world every year, a number of which have been published as long-term summaries by site or region (e.g., Stone 1937; Dolnik and Blyumental 1964; McClure 1974; Rappole and Blacklock 1983; Rappole et al. 2011a, 2001b). Therefore, we have excellent data for most North American and European species and a few Asian species documenting when fall migration begins, reaches its peak, and ends for many different geographic regions, at least on a population basis. These data reveal patterns that are quite consistent from year to year by latitude for a given population by sex and age.

To place migration timing into some meaningful context, a standard scale is needed. Study of the timing of migration at points along major migration routes, where birds of most migrant species neither breed nor winter, can be particularly instructive in this regard. The Texas Central Coast, located along the western shore of the Gulf of Mexico, is one such site. Cumulative records from this region show that migration really never stops; the latest spring migrants overlap with the earliest fall migrants in late June and early July while the latest fall migrants overlap with the earliest spring migrants in early January (Rappole and Blacklock 1983).

For the purpose of comparing the timing of migration among groups, we choose to set July 1 (non-leap-year Julian date 182) as the beginning of the fall migration period for temperate North America; December 31 (Julian date 365) as the end; and October 1 (Julian date 274) as the midpoint, recognizing that these dates vary considerably by latitude.

FACTORS AFFECTING FALL DEPARTURE DATE

North American migrant species fall into five major categories in terms of their timing of departure:

Very early (before August 15)
Early (August 15 to September 15)
Median (September 15 to October 15)
Late (October 15 to November 15)
Very late (after November 15)

Several hypotheses have been proposed for the observed timing of migrant fall departure from their breeding areas. In the following we discuss the principal ones based on broad generalizations derived from observed fall departure timing for selected migrants.

• *Breeding latitude or elevation.* Species or populations of species that breed at high latitudes or elevations tend to have earlier fall departure dates than similar species or populations breeding at lower latitudes or elevations, presumably due to the short period for frost-free days that occur in such regions. Examples include the Pectoral Sandpiper (*Calidris melanotos*) and Solitary Sandpiper (*Tringa solitaria*), both of which breed in the high Arctic regions from where they depart in July or early August.

• *Winter latitude.* Birds that winter at lower latitudes tend to leave earlier than species similar in other respects that winter at higher latitudes. The Veery (*Catharus fuscescens*) and Hermit Thrush (*Catharus guttatus*), for instance, both feed on invertebrates taken near the forest floor and have similar distributions in terms of breeding latitudes. However, the Veery departs much earlier on fall migration and winters much farther south, in central South America, than the Hermit Thrush, which winters in the southeastern United States and northern Middle America (Central America plus Mexico).

• *Diet.* Birds that feed on insects and other terrestrial invertebrates tend to be earlier fall migrants than those that feed on seeds. Also, terrestrial-feeding species tend to be earlier migrants than aquatic-feeding species. As an example, entire groups of high-latitude-breeding species that feed on arthropods (e.g., shorebirds) tend to be much earlier fall migrants than most waterfowl that feed on aquatic plants and animals.

• *Molt.* As discussed in chapter 3, timing of molt is an important factor in terms of fall departure. Both adult and juvenile Purple Martins depart for the wintering ground before completing the prebasic molt (Brown 1997), whereas in the Yellow-bellied Flycatcher (*Empidonax flaviventris*), adults migrate before they molt but juveniles migrate after completing a partial prebasic molt (body feathers only), resulting in a peak passage date along the Texas coast that is nearly a month later for juveniles than adults (Winker and Rappole 1992). In fact, although there are major patterns in terms of the timing of molt and migration, minor variations on these patterns are apparent for nearly every species, presumably based on species-specific life history characteristics and the period of time for which a population has been migratory (see chapter 3).

• *Sex and age.* Differences in the departure time and migration distances between sex and age groups from the same site or region have been reported for many species. In fact, although not thoroughly investigated, it is likely that some difference in mean timing of departure from a given region for the different sex and age groups exists for most migrants simply because of the different life history challenges faced by each during the reproductive period. However, there is one common situation that is of particular interest, that in which males become aggressive in fall and appear to compete with adult females and juveniles for territories or critical resources, perhaps stimulating migration in these groups earlier than might have occurred in the absence of apparent competition (Newton 2008:336, 425–456). Such competition has been reported in many species of partial or differential migrants (Gauthreaux 1978, 1982; Ketterson and Nolan 1983) and could represent a model for how the first migrants in a population initiate return to the region or site from which they originated; that is, adult females, accompanied by juveniles, precede adult males in migrating back to site of origin followed by adult males when (if?) food availability declines to the point where they can no longer persist at breeding sites. We propose that this pattern likely would be modified quickly by selection to produce the optimal departure time for each age and sex group so that those species in which the pattern persists are usually short-distance migrants that use resources (e.g., seeds) that do not disappear completely with the onset of winter. Presumably, adult males in these species balance reduced probability of survivorship against increased reproductive success associated with remaining on, or closer to, the breeding territory.

• *Differing fitness trade-offs between early completion of breeding, early fall departure, and raising of multiple broods.* Many North American migrants show a departure pattern that is the reverse of what would be predicted based on mean breeding latitude. For instance, several insectivorous wood warblers (Parulidae) that breed in the southeastern United States and winter in Middle America are early fall migrants (peak migration prior to September 15), yet congeners that breed at more northern latitudes show similar or later migration peaks along their routes, contrary to the latitudinal timing pattern discussed earlier (i.e., higher breeding latitudes = earlier fall migration) (Rappole et al. 1979). This pattern can be seen even within a species. Consider, for instance, the Prairie Warbler (*Dendroica discolor*), which breeds across eastern North America and winters in the Caribbean region (figure 4.2). In his study of this species, Nolan (1978:443) found that members of southern-breeding populations migrated earlier (peaking in late August) than northern breeding populations (which peak in early October). This difference may result from the fact that southern breeders of this species are single-brooded, completing nesting by mid-June (Burleigh 1958:540), whereas northern breeders mostly attempt to raise at least two broods, completing nesting by late July (Nolan 1978; chapter 3, this volume). This finding seems counterintuitive, at least based on the larger number of frost-free days available to southern breeders, which seemingly should allow them to raise at least as many broods as their northern counterparts. One possible explanation is the total amount of daylight hours available at the two latitudes. Daylength on June 16 at 30°N (southern Georgia) is 13.9 hours whereas that at 40°N (southern Indiana) is 14.8 hours (Baker and Baker 2012). Even though there are more frost-free days available in Georgia, there is less daylight during those days. This difference may result in nestlings taking longer to fledge and reach independence in Georgia than Indiana, which could explain why Georgia populations of the Prairie Warbler do not attempt to raise two broods whereas Indiana populations of the species generally do. Of course, there are many other possible explanations, somewhat different for each species, based on their ecology and life history. Analysis of variation in nesting season and brood production at different latitudes for the American Robin (*Turdus migratorius*) revealed that a combination of climatic variables (e.g., dry and wet bulb temperature) were the best predictors of initiation and duration of the reproductive period for this species (James and Shugart 1974).

Regardless of what factors cause differences in reproductive period length, the fact that southeastern-breeding Prairie Warblers (and several other species) are single-brooded whereas northeastern populations are double-brooded is not sufficient in itself to explain why southeastern birds should depart on fall migration earlier than northeastern birds. We suggest that early departure from the breeding area by southeastern birds may enhance probability of successful migration and location of suitable over-wintering sites (Rappole et al. 1979). For northeastern birds, fitness costs for later departure may be outweighed by benefits if successful rearing of second broods were likely, as appears to be the case.

Even among many insectivorous species that are mostly single-brooded regardless of breeding latitude, southern populations leave prior to northern populations. The Purple Martin (*Progne subis*) is such a species: Georgia populations (30°N to 34°N latitude) depart in July and August (Burleigh 1958:399), whereas New York populations (41°N to 44°N latitude) show peak migration in late August or early September (Bull 1974:389). Of course, Georgia birds arrive in spring in March whereas New York birds arrive in late April, so clearly southern birds can complete

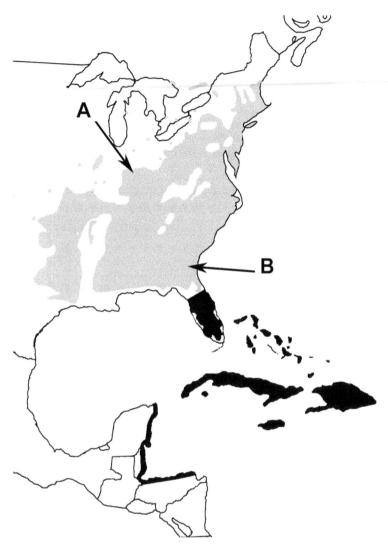

FIGURE 4.2   Breeding (*medium gray*), Florida resident (*light gray*), and winter (*black*) distribution of the Prairie Warbler (*Setophaga discolor*) in eastern North America showing location of Indiana (*A*) and Georgia (*B*) breeding populations mentioned in the text (based on Nolan and Mumford 1965; Nolan 1978).

breeding activities long before northern birds if both populations are mostly single-brooded, which seems to be the case (Brown 1997). However, this departure pattern raises the issue of whether or not increasing probability of cold weather is the key selective factor favoring fall departure, as is usually considered to be the case (Cohen 1967; Berthold 1999; Bell 2005). The threat of hard frost comes much later in Georgia than New York (U.S. National Oceanic and Atmospheric Administration 2008), so Georgia birds should remain on the breeding grounds longer, if onset of cold weather were the principal factor affecting departure.

We propose that the fact that southern populations of many migrants (e.g., Purple Martin [*Progne subis*]) leave months before the threat of cold weather is present is further evidence of other selective factors that might favor early southward movement (e.g., weather or predation along the migration route) (Rappole et al. 1979; Buskirk 1980) and/or the importance of early arrival at the wintering quarters to sequester sites or resources important for nonbreeding-season survival (Rappole et al. 1979; Rappole and Warner 1980; Rappole et al. 1989a; Winker and Rappole 1992). The effects of bad weather on food supply certainly must serve as a key factor in departure of new migratory populations, but we suggest that departure timing evolves quickly to balance optimal survival and offspring production, which, for many migrants, appears to involve departure from the breeding area, or at least preparation for departure, long before the move is necessary based on deteriorating weather.

Timing of Departure Within the Circadian Period

All migrants fall into one of three categories in terms of when during the 24-hour cycle they actually begin migratory flight: (1) daytime departure, (2) nighttime departure, or (3) either daytime or nighttime departure (Newton 2008:85). The reasons behind these differences are not well understood. Several authors have suggested that foraging ecology is a major determinant of when during the circadian cycle a bird should migrate (Brewster 1886; Palmgren 1949; Baker 1978:631; Alerstam 1990:310). Others have suggested vulnerability to predation as an important factor (Lincoln 1952; Mukhin et al. 2009). Nisbet (1955), Raynor (1956), Bellrose (1967), and Kerlinger and Moore (1989) cite the benefits of a nocturnal atmosphere for long-distance flight as perhaps the major factor favoring evolution of nocturnal migration. Biebach (1990) and others suggest that water conservation can play a major role in determining when during the circadian period birds choose to migrate, especially when transients must traverse hot, arid environments (e.g., the Sahara). We propose that the specific mix of nocturnal versus diurnal migratory activity seen in a given species or population represents a complex set of trade-offs involving morphology (body size, wing shape), ecology (prey distribution and activity), the flying environment (turbulence, winds aloft, visibility, thermals), vulnerability to predation, water conservation, and, presumably, many additional factors. In addition, length of evolutionary time for which a population has been migratory likely has a critical effect on this balance. Kerlinger (1995:93–94) suggests that the first migratory movement developing in a sedentary population must be during the normal activity period for the species (i.e., diurnal if the species forages in daylight hours; nocturnal if it forages at night). However, it may be that many of the same factors that favor nocturnal migratory movement (e.g., avoidance of predators or reduced turbulence in the air column) favor nocturnal dispersal by sedentary birds as well, and, if so, the first migratory movement by a sedentary diurnal species may be nocturnal. Mukhin et al. (2009) found that breeding individuals of a normally diurnal species when displaced from their nesting sites or whose nests were destroyed traveled long distances by night in attempting to return to nesting sites or when attempting to locate new sites.

As discussed earlier, light levels and some aspect of weather related to wind direction aloft appear to serve as important proximal cues for departure for many species of migrants. An additional cue may involve behavioral interaction with conspecifics in which initiation of migratory flight is stimulated by call notes of departing individuals or birds already in flight (Hamilton 1962, 1966; Larkin and Szafoni 2008).

MIGRATORY FLIGHT

Exaptations for migration include flight and, perhaps, migratory flight (*Zugstimmung*) as well. Three types of data support the latter hypothesis:

1. Some resident birds (few have been tested) undergo migratory restlessness (*Zugunruhe*), as well as migrants (Berthold 1988; Helm 2006; Helm and Gwinner 2006b).
2. Caged birds, when deprived of food, show *Zugunruhe* (Lofts et al. 1963), a presumably adaptive response for dispersing, resident offspring deprived of food by conspecifics.
3. Some individual migrants build up fat reserves and depart from wintering sites in the middle of the nonbreeding period (Rappole et al. 1989a).

Nevertheless, even if *Zugstimmung* is an exaptation for migration, long-distance migratory flight may impose a selective regime quite different from normal daily or dispersal movements. How much that regime differs from the day-to-day demands of foraging, predator avoidance, and travel to and from roosting, feeding, and bathing sites will depend upon many other aspects of the bird's life history (e.g., the foods that it eats and the habitat in which it lives). For birds that do most of their foraging on the wing, like terns (Laridae), swifts (Apodidae), and swallows (Hirundidnidae), the modifications required for migratory flight presumably are quite different from those that forage mostly on the ground, like grouse (Phasianidae), quail (Odontophoridae), and bustards (Otididae). The fact that these morphologically and ecologically distinct groups, as well as most other types of birds, have migratory representatives illustrates the potential fitness benefits of migratory movement. Nevertheless, consideration of the optimal adaptations for horizontal, powered, long-distance flight may provide some insight into how and why migration occurs in such disparate groups and provide information on how long a migratory habit has been followed within a given population or species.

We propose that regardless of the kind of bird involved, finding an optimal balance between the demands of day-to-day flight and migratory flight required modification of the anatomy and morphology, physiology, and behavior of the resident ancestors of migrants, which we discuss in the following.

ANATOMY AND MORPHOLOGY

We consider here only two of the most obvious anatomic and morphologic changes involved in shifting from a resident to a migratory lifestyle: muscle fiber type and wing shape.

Muscle Fiber Type The pectoralis and supracoracoideus are the principal muscles involved in flight, the former being responsible for the downward stroke of the wing and the latter for the upward stroke (George and Berger 1966) (although more recent work has shown the relationship to be more complex [Raikow 1985]). The pectoralis provides most of the power for flight and composes as much as 35 percent of the bird's mass (Greenewalt 1962). Three kinds of muscle fibers make up the pectoralis:

1. Fast oxidative glycolytic (FOG) fibers are relatively small and possess glycolytic enzymes that produce ATP rapidly over long periods of time, but they require large amounts of oxygen.
2. Fast glycolytic (FG) fibers are relatively large and possess glycolytic enzymes that produce ATP very rapidly anaerobically.
3. Slow oxidative (SO) fibers produce ATP through oxidative processes more slowly than the previous types (Butler and Woakes 1990).

The proportions of the different kinds of muscle fibers composing the pectoralis varies from species to species and group to group (Butler and Woakes 1990). FOG fibers compose the majority of pectoralis fibers in most migratory species (as well as many residents), providing most of the power for long-distance, powered horizontal flight. FG fibers are used mainly when bursts of muscle activity are required (e.g., during takeoff, landing, or directional change) and compose large portions of the pectoralis (50%) in species like the Ruffed Grouse (*Bonasa umbellus*) that rely on rapid takeoff to escape predators. SO fibers are thought to be involved in maintaining posture and, perhaps, in gliding and soaring, although their function is not entirely clear in all species.

Despite the fact that high FOG/FG ratios are more favorable for powered migration whereas lower ratios are much less favorable in terms of energy requirements, a broad range of ratios can be found among migrant species. In fact, there are many migrant species that have low ratios although average migration distance in these species tends to be shorter than for those with higher ratios. Studies have shown variation in the proportions of FOG to FG fibers in migrants that seem related to distances migrated (Lundgren and Kiessling 1988). Presumably similar kinds of variation could be found in comparing migrant versus resident populations of the same species that reflect the period of time for which a population has been migratory. We suggest that FOG/FG ratios along with capillary density and total wing surface area represent an aspect of migrant anatomy likely to be modified rapidly by natural selection (Lundgren and Kiessling 1988) (i.e., they are adaptations for migration).

Wing Shape Wing shape has a direct effect on power requirements for flight (Pennycuick 1975:13). As in the case of muscle fibers, the shape of a wing that is used primarily for rapid takeoff or short flights involving many directional changes differs dramatically from that required for long-distance, powered, horizontal flight. Many aspects of wing structure are important for migrants, but the two main features are the length of the wing relative to its width ("pointedness") and its width at the tip relative to width at the base ("convexity") (figure 4.3).

In general, the most efficient wing for powered, horizontal, migratory flight is a convex, pointed wing, as is found in terns and swallows, and the least efficient is a concave, rounded wing, as is found in grouse and quail (Lockwood et al. 1998). The most interesting factor regarding this relationship from the perspective of the evolution of the migrant wing is that migrant species are represented in all of the different wing type categories. In other words, it appears that whereas it is advantageous to have a convex pointed wing to be a migrant, it is not necessary; the fitness benefits derived from a migratory habit outweigh the costs, even for birds with concave, rounded wings. Nevertheless, one would predict that the longer a population of concave, rounded-wing birds is migratory, the more convex and pointed their wings are likely to be. If this hypothesis is correct, one would expect to find differences in concavity and pointedness between migrant and resident populations of the same or closely related species; for example, the wings of the temperate

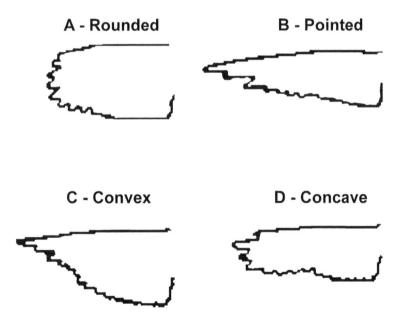

FIGURE 4.3   Basic wing shapes for flying birds (based on Lockwood et al. 1998).

migratory Northern House Wren (*Troglodytes aedon aedon*) would be expected to be more convex and pointed than the wings of its tropical resident relative, the Southern House Wren (*Troglodytes aedon musculus*).

The speed of horizontal, powered, migratory flight is, of course, affected markedly by the adaptations discussed earlier as well as by the mass of the bird: Typical groundspeeds for small songbirds are 20 to 30 km/h, whereas larger migrants (e.g., ducks) fly at speeds of 40 to 60 km/h, although there is a wide range of variation among species, even those of similar body size .

PHYSIOLOGY

Fat storage is an exaptation for migration. Most birds, whether resident or migratory, store energy as fat in evident preparation for predictable periods of food shortage (e.g., overnight survival, fledging, dispersal, egg laying, cold weather, and migratory flight) (King 1972; Ramenofsky 1990). The type of fat in which the energy is stored is usually triacylglycerol, which is formed by combining fatty acids in a reaction with glycerol (Allen 1976). This lipid is stored subcutaneously or between tissues in the body cavity without water and when oxidized during migratory flight produces roughly twice the energy per unit of weight of either protein or carbohydrate (Hochachka et al. 1977; Schmidt-Nielson 1983). It has long been known based on laboratory and field studies that migrants that use up their fat stores entirely will metabolize muscle tissue (protein) during efforts to initiate migratory flight (Schüz 1952; Lofts et al. 1963; Gwinner 1971). Recent studies now suggest that some migrants switch between fat and protein metabolism during migratory flight, depending on their water needs (Lindström et al. 2000; Klaassen and Biebach 2000). As mentioned earlier, fat is stored without water, which is an advantage if weight is the sole concern. However, if water stress is also a concern, then storage of some portion of needed energy reserves as muscle tissue, which is greater than 60 percent water, provides an option. Therefore, if water is limiting, as perhaps in long flights at high temperatures over ocean or desert, use of a higher proportion of energy stored as muscle may be advantageous (Klaassen and Biebach 2000). The physiologic, presumably hormonal, cues that signal switching from fat to protein metabolism and back during the course of a given migratory flight could represent exaptations for migration. However, it seems highly unlikely that the bird would store the optimal percentage of fat and protein likely to be required for a successful passage without considerable input from generations of natural selection.

BEHAVIOR

Optimal altitude for migratory flight varies according to the characteristics of the bird, the atmosphere, Earth's surface, and the season (table 4.2). When considering this situation, it is important to remember that the bird *chooses* that altitude at

TABLE 4.2   Factors Involved in Determination of the Optimal Altitude for Migratory Flight

CHARACTERISTICS OF THE ATMOSPHERE

Oxygen content. The fast oxidative glycolytic fibers that predominate in flight muscle require large amounts of oxygen to metabolize fat for the energy needed during flight. The amount of oxygen in the atmosphere declines with increasing altitude and at very high altitudes could serve as a limiting factor.

Temperature. Temperature declines at roughly 7°C per 1,000 m of altitude. Therefore, very high altitudes increase the danger of hypothermia. However, muscle activity involved with flight generates heat (Biebach 1990; Newton 2008:151–152). In addition, ambient temperatures at ground level can be very high over tropical deserts. When the flying migrant's body temperature becomes too high, the bird must make some adjustment to lower it, either by flying at a higher altitude or metabolizing protein to produce water that can be used for evaporative cooling. Thus, temperature can play a role in influencing selection of optimal altitude for migratory flight (Biebach 1990).

Air density. Air density decreases with altitude, reducing drag, and decreasing energy required for flight per unit of time. The effect is greater for smaller birds (Pennycuick 1975).

Direction of air movement (wind). By flying in the direction of air movement (i.e., downwind), the migrant can increase its groundspeed.

CHARACTERISTICS OF THE BIRD

Size. The surface area of a bird varies according to roughly two-thirds power of its mass; thus, smaller birds have a larger surface area to mass ratio than that of larger birds, requiring relatively more power per unit of mass to overcome the drag of the body (Pennycuick 1975:7) and favoring a higher altitude for migratory flight—all else being equal.

Shape. Wing span and shape have a direct effect on power and energy requirements for flight (Pennycuick 1975:13; Lockwood et al. 1998).

Form of energy storage. Birds can store the energy required for long-distance flight either as fat or muscle with different trade-offs that result depending on the ratio that is used. A greater amount of energy is produced per unit of fat metabolized compared with that produced per unit of muscle metabolized. However, muscle is two-thirds water, which may be important for cooling purposes in flights over tropical desert environments (Biebach 1990).

Mass. At the beginning of a migratory flight, a considerable portion of the bird's mass is fat or muscle (as much as 50 percent in some migrants), which will be used as energy during the course of the flight. As this energy is used, mass declines, affecting optimal altitude for flight. All else being equal, a higher altitude is favored for individuals of lower mass (Pennycuick 1975), so the bird should gradually move to a higher altitude as mass declines.

(continued)

TABLE 4.2   (*continued*)

CHARACTERISTICS OF EARTH'S SURFACE

Different ground environments have different effects on the atmosphere above them, which in turn can affect the optimal altitude for migratory flight. For instance, the high ambient daytime temperatures over tropical deserts could favor higher altitudes to reduce evaporative water loss.

TIME OF DAY

Variation in ambient temperature over the circadian cycle may favor higher altitudes for diurnal migration compared with nocturnal migration for birds transiting tropical deserts.

SEASON

Seasonal changes—for example, direction of prevailing winds at various altitudes, frequency of storms, and ambient temperatures—likely affect the optimal altitude for any particular time during the year.

---

which it will fly during migration. We suggest that this choice could be, in part, an exaptation for migration, as dispersal could require similar ability to respond to the environment in choosing altitude in ways that are most suitable for long-distance flight. Nevertheless, it seems probable that generations of natural selection on members of a migratory population affect this choice (i.e., that the longer the population has been migratory, the better the choice of any individual in that population is likely to be). Although dispersal involves some of the same aspects as migratory movement, the optimal altitude for migration will depend to some extent on the specific conditions likely to be met along the routes followed in moving between the breeding and nonbreeding portions of the range. Therefore, selection of optimal altitude for flight along a particular migratory route is likely to be an evolved adaptation to a migratory habit.

Liechti et al. (2000) used radar data for migrants flying over Israel to examine the relative effects of energy needs versus water constraints on migrant flight altitude and concluded that wind direction (increased flight speed = energy savings) appeared to be the best predictor.

The foregoing discussion further illustrates the complex nature of the adaptations required for optimal migration via powered flight (as opposed to soaring [Newton 2008:163–192]). Adaptations for long-distance movement may be required for dispersing individuals of many resident species. Nevertheless, the balance between the kinds of adaptations in terms of body shape and size, wing shape, energy storage adapted to a specific altitude required for maximizing flight efficiency (unit of energy expended per kilometer of distance traveled) is likely to be quite different for a dispersing bird as opposed to a long-distance migrant. Thus, selection of optimal altitude for long-distance flight and evolution of the various adaptations that go along with it are likely to represent clear reflections of how long a species has been a migrant.

Migration Route

Experimental Studies

Migration route (distance, direction, and timing of duration of migratory flight) is a statistical phenomenon representing the median choice of path followed by all individuals of a population in travel between the principal breeding and non-breeding sites. These routes are often referred to as "flyways" when applied to the migrations of waterfowl and some other groups. That migration routes might have genetic components was long surmised (Lowery 1945; Wolfson 1948), and extensive laboratory work over the past half century appear to confirm this hypothesis (Kramer 1952; Gwinner 1968, 1971; Berthold 1996; Pulido 2007). Most of this research has been based on measurements of the duration and direction of *Zugunruhe* or migratory restlessness, a behavior seen in putative migrants when they are kept in captivity during the migratory period (Gwinner 1968; Berthold 1973). Investigators of the phenomenon of migration have measured *Zugunruhe* in birds kept in circular cages using various techniques, including an observer recording direction and activity (Kramer 1952); perches located around the perimeter of the cage equipped with microswitches attached to event recorders (Berthold 1973); and an ink pad located at the bottom of the cage with white blotting paper around the walls (Emlen 1975) (figure 4.4).

Laboratory experiments with *Zugunruhe* have had a profound influence on migration studies. Nevertheless, when considering the results of these experiments, it is important to keep some aspects of experimental design in mind:

1. The birds used in the experiments usually are hand-reared, a practice that can alter the behavior of individuals in ways that are not always obvious unless compared with that of parent-reared birds.
2. The experimental subjects are kept in a controlled environment, assumed to be isolated from environmental conditions that are known to affect migratory flight (e.g., wind, weather, and directional light sources). Thus conditions are abnormal, and the resulting behaviors may be abnormal as well in ways difficult to understand without some comparison with free-flying birds. In addition, the range of factors known to affect orientation and navigation includes some that are not controlled for in the experimental design (e.g., light levels and magnetic fields), which could affect results.

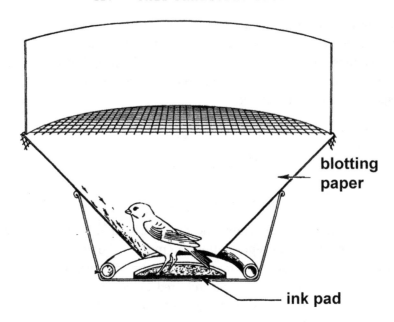

blotting paper

ink pad

FIGURE 4.4   Emlen cage for measuring *Zugunruhe* direction (based on Emlen and Emlen 1966).

3. Experimental subjects usually are isolated from either caged or free-flying members of their population. Hamilton (1966) demonstrates that social interaction, in terms of audiovisual cues, are likely important to various aspects of migratory flight as has been suggested by Larkin and Szafoni (2008). This possibility is bolstered by the results of some laboratory experiments, which have shown that juvenile birds demonstrate different *Zugunruhe* patterns when they are kept alone compared with when they can see or hear birds in neighboring cages (Pulido and Berthold 2003:57). As discussed by Helm et al. (2006), social environment may play an important role in migration decisions.

4. It is important to remember that *Zugunruhe* (migratory restlessness) is not *Zugstimmung* (migratory flight); it is a behavior that is assumed to result from the frustration of the bird's attempt to initiate migratory flight.

With these caveats in mind, we consider some of the findings of *Zugunruhe* studies.

The results of these experiments have yielded extraordinary information about factors that affect timing of initiation, direction, and duration of *Zugunruhe* (Gwinner 1968; Emlen 1975; Berthold 1988) including the fact that responses to these factors are heritable (Helbig 1991; Pulido and Berthold 2003; Pulido 2007). For instance, Berthold and his colleagues and students have demonstrated that both direction and duration of *Zugunruhe* differ for juvenile birds derived from populations that breed and winter in different areas (Helbig 1996) (figure 4.5)

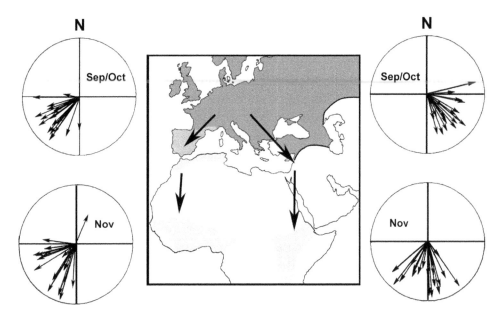

FIGURE 4.5   Map of Blackcap (*Sylvia atricapilla*) breeding (*dark gray*), resident (*medium gray*), and wintering (*light gray*) areas for eastern and western European populations. Arrows show direction of major migration routes for the two populations. Circular diagrams show individual mean vectors for *Zugunruhe* orientation of hand-raised birds derived from populations from west (*left*) and east (*right*) of the central European migratory divide for different time periods during fall migration (based on Helbig 1996).

and that cross-breeding of birds derived from these different populations produces *Zugunruhe* in offspring that is intermediate in terms of direction and duration (Helbig 1991, 1996). Experiments involving artificial selection have shown that these behaviors can be changed in a relatively few generations (Berthold et al. 1990; Pulido 2007).

HYPOTHESES FOR MIGRATION ROUTE

The work on *Zugunruhe* in young caged birds along with examination of existing ranges of species and subspecies of migrants provide the bases for hypotheses suggesting that the behavioral traits governing the tracing of an actual migration route between breeding and wintering areas are genetically fixed in the individuals of any given migrant population (Berthold 1996, 2001; Pulido 2007). "Genetically fixed" here can have at least two meanings:

- "Fixed" in terms of the inability of an individual to modify direction and distance of migratory flight to the wintering ground according to environmental circumstances
- "Fixed" in terms of evolutionary stasis in genetic factors controlling the route lasting thousands of generations in populations of migrants

Pulido (2007:167) summarizes the individual aspect of the meaning of "genetically fixed": "[The] migration program is largely insensitive to most environmental perturbations, whether wind, unfavorable weather conditions, or food scarcity, for which birds do not seem to compensate (Berthold 1996)."

The key elements of this "migration route" hypothesis (also known as "clock-and-compass," "bearing-and-distance," or "vector" hypothesis) may be summarized as follows: Juvenile migratory birds are able to locate the wintering ground for their population due to a genetic program for migration route. This program functions by providing genetic instructions concerning the directions the bird must follow along the route and the period of time that it must follow each direction to arrive at the proper place. Rabøl's (1978) "goal area" hypothesis is similar to the "migration route" hypothesis except that it posits navigation towards a goal area which in the course of the season moves down the migratory route.

The "migration route" hypothesis provides the basis for a considerable amount of related recent work on the phylogeny of migrants in which it is assumed that the migration route is a phylogenetic trait of individual migrants; a trait whose evolution dates back thousands of generations (Baker 2002; Ruegg and Smith 2002; Irwin and Irwin 2005; Brelsford and Irwin 2009). Irwin and Irwin (2005:34) summarize the concept for a specific group of Eurasian migrants:

> The many subspecies and species boundaries in central Siberia suggest that many taxa have similar histories of expansion into Siberia along two pathways. Ancestral migration routes appear to have been conserved during these expansions, such that species or subspecies that expanded from central Asia into Siberia still migrate through central Asia to India, whereas those that expanded from eastern China into Siberia still migrate through eastern China to Southeast Asia.

They provide support for this idea using data on known breeding and wintering ranges for several Eurasian migrants, which show that birds that breed in western Siberia appear to migrate around the western side of the Himalayas to wintering areas in India, Pakistan, and Iran, whereas those that breed in eastern Siberia migrate around the eastern side of the Himalayas to wintering areas in Southeast Asia and southern China (figure 4.6).

Ruegg et al. (2006) present a similar hypothesis for a Western Hemisphere migrant, the Swainson's Thrush (*Catharus ustulatus*). Populations of this species that breed in peninsular Alaska follow a long migration route to South America, a route whose evolution, they propose, must date back to the Pleistocene (figure 4.7). Their reasoning is that if the route were not a Pleistocene relic, the Alaskan-breeding birds would follow a shorter route: a faster and presumably safer route similar to that of neighboring conspecifics to wintering grounds in coastal California or Middle America.

FIGURE 4.6   Assumed migration routes (*arrows*) for eastern-breeding (*dark gray*) and western-breeding (*light gray*) populations of the Greenish Warbler (*Phylloscopus trochiloides*) (based on Irwin and Irwin 2005): *medium gray* = overlapping breeding range; *black* = winter range for both populations.

Brelsford and Irwin (2009) further argue that the reason hybrids between members of populations with different migration routes are apparently unsuccessful (assumed because known interbreeding has not subsumed population differences) is that the traits governing migration route might be completely inappropriate, resulting in very low fitness for hybrids. We do not question their conclusion (low hybrid fitness), but we do question whether or not mixing of migration route programs is the cause. As we have discussed in previous chapters, there are many other genetically programmed differences in life history (e.g., timing of breeding and molt) that could explain low hybrid fitness, and there are important questions concerning the validity of the migration route program hypothesis, at least as currently formulated.

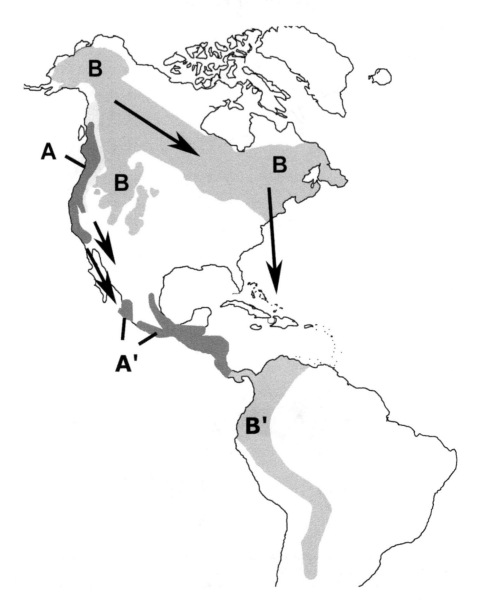

FIGURE 4.7   Migration route (*arrows*) followed by coastal-breeding (*A*; *dark gray*) Swainson's Thrushes (*Catharus ustulatus*) to coastal and Middle American (Mexico plus Central America) wintering grounds (*A'*; *dark gray*) and that followed by continental-breeding birds (*B*; *medium gray*) to South American wintering grounds (*B'*; *medium gray*) based on band-recovery data (based on Ruegg et al. 2006): *light gray* = areas of possible overlap in breeding population.

The fundamental assumption underlying all of the experimental and theoretical work on migration route is that ability to follow a particular migration route is the behavioral trait that is being considered. The strongest support for this assumption comes from two types of studies:

- Laboratory experiments in which *Zugunruhe* direction is tested for individuals derived from populations that winter in different localities (figure 4.4)
- Displacement experiments in which young individuals are moved to points outside the normal migration route and whose subsequent movements indicate an inability to correct for this displacement to reach the proper winter quarters (Perdek 1958; Chernetsov 2004; Thorup et al. 2007)

However, there are many types of data that do not support the migration route hypothesis (table 4.3; figure 4.8).

Based on the information in table 4.3, it appears that the "migration route" hypothesis has serious deficiencies in terms of explaining the orientation behavior of juveniles. This finding, however, begs the question of how naive migrants are able to find their way to appropriate winter quarters. This question is especially problematic from the perspective of our concept of migration as a form of dispersal (see chapter 1). If the offspring of the first generation of migrants are dependent upon exaptations for their movement capabilities, how could they find their way to their winter (i.e., ancestral breeding) areas? One way would be for them to accompany adults (e.g., their parents or other conspecifics) as suggested by Rappole and Tipton (1992:52). However, there is another possibility, which we term the "destination" hypothesis (also known as the "locality-fixation" hypothesis of Williams 1958:58). According to this hypothesis, juvenile migratory birds have a genetic program enabling them to home to the ancestral area from which the parent population was derived (often the wintering area). No genetic program for route (i.e., direction and distance to be migrated) is required, only an ability to use environmental cues to navigate. As has been pointed out by a reviewer of this book in manuscript form, this hypothesis has strong Lamarkian overtones (i.e., the genetic changes required seem to precede the need). This idea presumes that a genetic "destination" program would have no value for a sedentary species, which, of course, we do not know. Perhaps the best way to approach this question would be to test for evidence of the existence of such a program in resident species. One way to do this might involve the hatching and raising of offspring at a site distant from where their parents lived; then either release and follow the birds via radio tracking or place them in experimental cages and record the direction of their *Zugunruhe*. The second type of experiment has been performed, as mentioned in table 4.3, using offspring derived from a resident population of Blackcaps (*Sylvia atricapilla*) from the Canary Islands (Berthold 1988). These naive juveniles, housed in Germany, showed *Zugunruhe* in the general direction of their parents' tropical home. Berthold (1988) explained these results as indicating that the resident

TABLE 4.3   Summary of Observations and Studies That Raise Questions Regarding the Sufficiency of the "Migration Route" Hypothesis as an Explanation for Fall Migration Movements of Juveniles

OBSERVATION	COMMENTS
Juveniles experimentally displaced outside the normal migration route show orientation appropriate to the wintering area's actual location.	More than 80 studies have reported this behavior (Thorup and Rabøl 2007; Thorup et al. 2011). Explaining these findings using the "migration route" hypothesis requires its expansion to include an additional hypothesis; namely, "compensation" (i.e., birds are capable of computing deviation from their normal route and making navigational corrections) (Thorup and Rabøl 2007). There are no data to test this second hypothesis except the data that were used to derive it.
Juveniles naturally displaced (i.e., by wind, weather, or obstacles) migrate to appropriate winter quarters.	Juvenile migratory birds often follow different routes than those of adults during migration to the wintering ground; in particular, juveniles tend to collect in numbers that are orders of magnitude larger than those of adults along areas bordering major obstacles (e.g., the ocean) (Murray 1966). This phenomenon, often referred to as the "coastal effect" (Ralph 1978; Rappole et al. 1979; Rappole 1995:87–88), has been reported from both North American and Eurasian migration systems in a large number of migrant species (Ralph 1978; Chernetsov 2006; Ydenberg et al. 2007). These young birds have the same genetic makeup as adults (with regard to navigation and orientation) and yet are able to arrive at the same wintering areas as adults despite taking routes that differ by hundreds of kilometers from those taken by adults (e.g., Blackpoll Warbler [*Dendroica striata*] [Nisbet et al. 1963; Rappole et al. 1979]) (Thorup and Rabøl 2001).
Migration can appear and disappear in a few generations (Rappole et al. 1983:47; Able and Belthoff 1998; chapter 8, this volume).	The "migration route" hypothesis seemingly requires some evolutionary period in which to develop the intricate timing between direction and distance that would be involved in precise migration from a particular breeding area to a particular wintering area. No time period is required for the "destination" hypothesis to work—the resident bird has the program in place when it leaves on dispersal/migration for the first time, and its offspring will have the same program.

Observation	Comments
Correct homing to winter range by juveniles despite recent range expansion to breeding areas thousands of kilometers from the original breeding range and in different hemispheres (e.g., Northern Wheatear [*Oenanthe oenanthe*] [Kren and Zoerb 1997] [figure 4.8] and Pectoral Sandpiper [*Calidris melanotos*] [Holmes and Pitelka 1998]).	A southbound migration route for offspring raised in a new breeding range in a new hemisphere obviously would take them to a new wintering area. Newton (2008:641) states that the wheatear migration from the New World to African wintering quarters results from the migration program being fixed. However, genetic fixing of the "migration route" program (i.e., a specific direction and duration for migration) would not result in return to Africa for juveniles—it would result in travel to a completely new wintering area in the New World tropics.
Zugunruhe in displaced tropical resident birds (e.g., Blackcaps [*Sylvia atricapilla*] from the Canary Islands [Berthold 1988]).	To explain this using the "migration route" hypothesis, a second hypothesis must be included; namely, that the behavior is a "remnant" from when the population was migratory. There are no data to test this "remnant behavior" hypothesis except those used to derive it.
In some species of migrants (e.g., Pectoral Sandpiper [*Calidris melanotos*] and Broad-winged Hawk [*Buteo platypterus*]), a portion of the population winters in one area as juveniles and in another area thousands of kilometers away as adults (Tabb 1979; Holmes and Pitelka 1998).	If the migration route is what is programmed in a migrant, then presumably the same route would be followed each year, regardless of the individual's age.

Canary Island population was derived from an ancestral German migratory population and that the *Zugunruhe* behavior they demonstrated was a genetic remnant of the migratory route program evolved when the population was migratory. However, there is no way of knowing which population is derived from which, and the behavior of these juveniles could as easily be explained by a genetic program for destination as for route. In any event, further experimental work with other resident species could resolve the issue.

A second approach to the question of whether or not a "destination" program might exist would be to look for evidence of it in other taxa. In this regard, Monarch Butterflies (*Danaus plexippus*) perhaps provide the strongest evidence for such a program. These butterflies migrate to a single, small wintering area in the mountains of central Mexico to which they have never been, and to which they

FIGURE 4.8   Breeding (*light gray*) and wintering (*black*) distribution of the Northern Wheatear (*Oenanthe oenanthe*) showing fall migration routes (*arrows*) for Old World and New World breeding populations.

will not return, although their offspring will (Brower and Malcolm 1991; Brower 1995) (figure 4.9). This work demonstrates that an ancestral site can be genetically programmed. Similarly, although less certainly, there is evidence that catadromous eels (*Anguilla* sp.) can home both to their oceanic breeding and spawning area and their freshwater living areas (Aoyama 2009).

The "destination" hypothesis provides theoretical support for some general concepts important to understanding the initiation and subsequent modification of migration over evolutionary time. We will address these ideas in greater detail in chapter 8. In proposing that many migrant birds possess a genetic program for winter destination location, we do not intend to suggest that such a program excludes existence of a "migration route" program. In fact, we argue in chapter 6 ("Spring Transient Period") that differences between fall and spring routes (elliptical or loop migration) are best explained by a "migration route" hypothesis. Indeed, it seems probable to us that many migrants possess both types of programs. However, the "destination" program appears to us to provide the most parsimonious explanation for the majority of field data concerning how juveniles are able to migrate to the appropriate wintering area (figure 4.8).

APPARENT "OFF-ROUTE" MIGRATION

At times during the course of a migratory journey, birds of many species have been observed to migrate in a direction that is not along the expected route toward their destination or to land in places distant from their breeding or wintering range

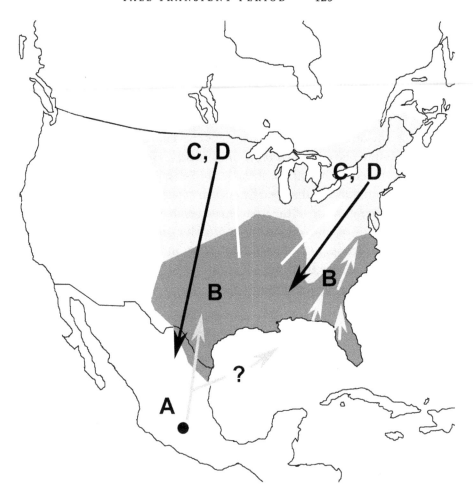

FIGURE 4.9   Monarch Butterfly (*Danaus plexippus*) migration in eastern North America. Generation *A* migrates from the wintering site in Mexico (*black circle*) in April and May to breeding sites in the southeastern United States (*dark gray*); females of generation *A* die after laying eggs for generation *B*; generation *B* migrates northward in June to summer breeding areas (*light gray*); two or more generations (*C* and *D*) are produced over the summer. By September, individuals of generation *D* are migrating southward to Mexico, where they will overwinter and serve as the new generation *A* for the following spring season (based on Brower and Malcolm 1991; Brower 1996).

(Newton 2008:267–299). Often such occurrences are the obvious result of wind or weather, which can force birds far from their normal range or in different directions from those normally taken. There are, however, occurrences of apparent systematic travel by large numbers of individuals, usually juveniles, in directions considered to be incorrect or, at least, abnormal. Among various hypotheses to explain these deviations are some that propose that they result from genetic or neurologic mistakes resulting in misorientation. Among these are "mirror-image migration," in which the bird apparently confuses right from left, and hence migrates in a direction 90 degrees off from what is correct (DeSante 1983), and "reverse migration," in which the bird migrates in a direction that is 180 degrees off (Rabøl 1969). For each of these hypotheses, there are no data other than descriptions of the occurrences to support them. Patton and Marantz (1996) take a more fruitful approach while examining some of the same observations used by DeSante (1983) to posit mirror-image misorientation, finding that there are adaptive behaviors that could explain the phenomenon.

There is no one hypothesis that will explain all instances of evident misorientation or any other apparently systematic, deviant migration movement. The data for each must be evaluated on its own merits. However, we do not consider explanations based on hypothetical classes of behaviors to be useful. Our view is that close examination of the specific circumstances, meteorological conditions, and genetic programming of the individuals involved will provide a logical explanation. We realize, of course, that our approach cannot explain all displacements (e.g., flamingos dropping from the winter sky in Siberia [Krulwich and Block 2011]), but then perhaps there are some things that are best left unexplained for lack of full knowledge of the circumstances.

ORIENTATION AND NAVIGATION

Hamilton (1966:59) states:

> In line with the general assumption that the role of [a] sophisticated orientation mechanism is confined to long-distance travelers, it might be concluded that celestial orientation is irrelevant to the dispersal of birds from a local roost where local landmarks are likely to be supplemental or primary bases for the orientation of locomotion. The fallacy of this assumption is established by the sharp decline in the accuracy of orientation of birds dispersing from such roosts under overcast skies. (Hamilton and Gilbert, manuscript)

We concur. The ability to be able to move from one place to another and return has extraordinary adaptive value whether or not an organism is migratory or whether or not it is a bird (Storm 1966). Therefore, we consider the range of abilities involved in use of the many cues that have been discovered as exaptations for migration. In addition to Hamilton's observation, this hypothesis is supported by the fact that much of the work on avian ability to detect sensory cues has been done on resident birds (e.g., Rock Pigeon [*Columba livia*]). The assumption that ability to use most cues required for orientation and navigation are exaptations for migration forms the basis for the following discussion.

Genetic Programming

As discussed, and regardless of whether or not residents possess the same adaptations for orientation and navigation as migrants, it is evident that the foundation of migrant movement is genetically programmed. All the orientation and navigation adaptations (exaptations?) listed in table 4.4 are focused on following this program, at least for naive juveniles.

SENSORY DETECTION OF ENVIRONMENTAL CUES

The senses usually are listed as five in number: touch, taste, smell, sight, and hearing. Birds and many other kinds of organisms possess more senses than these (e.g., ability to detect and respond to differences in Earth's magnetic field and in wind direction) and are able to use them to know where they are going, where they have been, and how to travel between them (table 4.4). There are many summaries of the mechanisms used by birds in orientation and navigation, including Storm (1966), Emlen (1975), Gauthreaux (1980), Åkesson (2003), Wiltschko and Wiltschko (2003), and Newton (2008:241–257).

SOCIAL CUES

The importance of social interactions in terms of assisting migrant orientation and navigation has not been studied to the degree that sensory cues have been but is recognized as being important to life history event timing (Helm et al. 2006). Although laboratory experiments have established that juvenile experimental subjects have different responses in terms of *Zugunruhe* direction and distance when kept in isolation as opposed to having neighbors (Pulido and Berthold 2003:57), most information on the importance of social interactions to migrants is based largely on inference. Nevertheless, the importance of this cue is obvious for species that normally forage in family groups or flocks during the nonbreeding period (e.g., geese and swallows) because many also are easily visible migrating in family groups or flocks during the daytime at low altitudes (Newton 2008:257). Indeed, Ward and Zahavi (1973) proposed this function for flocks; that is, that they served as "information centers" where individuals could benefit from knowledge of resources by accompanying experienced conspecifics. In the case of migrants, the "resource" would be information on the appropriate direction for migration toward suitable sites for stopover or wintering. Hamilton (1966) proposed that behavioral interaction is an extremely important cue for high-flying, nocturnal migrants that are typically solitary when foraging or resting during the day; Larkin and Szafoni (2008) have suggested a similar function for behavioral interaction among migrating social species as well. The clearest support for Hamilton's contention is that such migrants almost always migrate as members of loose flocks as confirmed by several different types of data: television-tower kills (Kemper 1996); ceilometer observations (Avery et al. 1975); moon-crossing observations (Lowery and Newman 1966); radar studies (Richardson 1990; Liechti et al. 2000); and audiovisual observation of flocks of these birds forced to abnormally low flying levels by bad weather (Hamilton 1966). In addition, migrating members of these loose flocks usually

TABLE 4.4   Sensory Detection of Environmental Cues Used by Birds for Orientation and Navigation

SENSE INVOLVED	ENVIRONMENTAL CUE	DESCRIPTION	CITATIONS
Vision	Landmarks	Birds learn through visual observation during flight and use memorization to identify features on the ground that they can subsequently refer to for orientation.	Mettke-Hofmann and Gwinner 2003; Mouritsen 2003; Newton 2008:241
Vision	Sun's location in the sky	Many different kinds of organisms, including birds, can use the location of the Sun in the sky as an orientation mechanism, although perhaps mainly in combination with other orientation information.	Kramer 1957; Able 1980:290, 1982a; Moore 1982, 1987
Vision	Polarized light	Animals able to see the plane of polarized sunlight (e-vector) can use this information for orientation purposes, and this ability has been found to be present in birds.	Muheim et al. 2006; Castelvecchi 2012
Vision	Star patterns	Early (1940s) experiments with caged birds demonstrated that proper orientation appeared to depend upon a clear view of the night sky. Subsequent work of increasing sophistication has demonstrated that star patterns serve as important cues for many migrants, usually in combination with other factors (e.g., magnetic field and polarized light).	Kramer 1952; Emlen 1975; Mouritsen and Larsen 2001; Cochran et al. 2004
Unknown	Earth's magnetic field	Able (1980:306) states that the idea that organisms could orient using Earth's magnetic field, just as we use a compass, dates back more than a century, but it was not until the latter half of the twentieth century that extensive experimental work documented this capability in many species of birds (including nonmigratory Rock Pigeons) as well as other kinds of organisms.	Yeagley 1947; Keeton et al. 1974; Wiltschko and Wiltschko 2003; Fischer et al. 2003

Sense Involved	Environmental Cue	Description	Citations
Hearing	Sounds	Rock Pigeons, and perhaps other birds, can detect sounds that are beyond human hearing range (infrasound). Certain features of Earth's surface (e.g., wind through mountain passes or waves pounding a shoreline) produce site-specific infrasound signatures that could be used for orientation purposes by those species able to detect them.	Kreithen 1978; Able 1980:325–327; Keeton 1980
Olfaction	Smells	Every site on Earth's surface has its own characteristic odor (i.e., trace molecules in air or water), which can be used by some organisms during migration to locate that site (e.g., Pacific salmon [*Oncorhynchus*] spp. [Dittman and Quinn 1996]). Few birds are known to possess the olfactory capabilities necessary to allow this type of orientation, but there is evidence that at least some species or groups of species do (e.g., Rock Pigeon [*Columba livia*]).	Papi 1989; Wallraff 2004
Unknown	Wind direction	Migratory birds, insects, and perhaps other organisms can use wind direction in combination with other factors (e.g., polarized light) to orient during migratory flight—or by choosing when and when not to fly based on ambient wind.	Able 1980:322–323

vocalize at regular intervals, presumably to maintain contact with flock members during nocturnal migratory flight (Hamilton 1962, 1966, 1967; Nolan 1978:452; Able 1980:326; Newton 2008:89, 257).

EXPERIENCE

It is our contention that both adult and juvenile migrants possess the ability, in terms of genetic programming, to return to the region in which the population from which they are derived evolved (Williams 1958:58). In addition, of course, adults will have had the experience of at least one round-trip journey, not only to a particular region but also to specific sites within that region. Extensive evidence in the form of band returns documents that many adults of many different kinds of birds are capable of migrating point-to-point (i.e., from a specific breeding site to a specific wintering site), as will be discussed in greater detail in chapter 5.

Stopover

Some individuals of some species are capable of migrating essentially nonstop from breeding to wintering area; for example, a Bar-tailed Godwit that flew more than 11,500 km within 8 days from its breeding grounds in Alaska to its winter quarters in New Zealand (Gill et al. 2009). This pattern, however, is not common. Most individuals of most migrant species make one or more stops along the route between breeding and wintering areas (Biebach et al. 1986; Moore et al. 2005). Studies indicate that these stops can be of two different types: (1) temporary pauses in migratory flight, usually lasting less than 12 hours and (2) stopovers lasting for days or weeks. The cause of temporary pauses seems to be one of two things: (1) the onset of daylight for nocturnal migrants or nightfall for diurnal migrants as long as they are not over water or desert; or (2) development of weather conditions that are not favorable for migration (e.g., winds in the wrong direction or heavy rain). The cause of stopovers seems to be the need to rebuild fat reserves or to continue or complete molt. Behavior of individuals of the same species at short-term versus long-term stopover can be quite different. Migrants at temporary stopover sites often show little regard for habitat choice, are gregarious, even if normally solitary, and spend an hour or more resting immediately prior to resuming migratory flight (Rappole and Warner 1976). Members of the longer-term stopover group tend to be more selective with regard to habitat, can be aggressive toward conspecifics in competing for resources, and undergo normal foraging activities almost throughout the duration of the stopover up until the last few hours of the day of departure when, like members of the temporary group, they become quiet and inactive. Rappole and Warner (1976) proposed that individuals in these two groups were in different physiologic states as well as behavioral states. They suggested that birds in the temporary group remained in a "flying" state (*Zugstimmung*) during their short stay at a site, whereas those that stayed for longer periods were in a "feeding" state (*Zugdisposition*) while they rebuilt their fat reserves. They further suggested that a normal fall migration, at least for most long-distance, terrestrial migrants, might involve several alternations between feeding and flying states before the winter destination was reached. It now seems likely that there are at least four physiologic/behavioral states associated with migration (see chapter 6).

The alternation between *Zugdisposition* and *Zugstimmung* (or, rather, its experimental stand-in, *Zugunruhe*) is well documented based on laboratory experiments

(Farner 1955; Lofts et al. 1963; King et al. 1965; King 1972). There is tantalizing but much less complete information from the field (Mueller and Berger 1966; Robl 1972; Rappole and Warner 1976; Biebach et al. 1986; Moore 2000; Wikelski et al. 2003). The chief difficulty lies with the fact that the precise hormonal controls governing the different physiologic/behavioral states of migration are not known (Holberton et al. 1996). Documenting a relationship between a particular set of behaviors observed in free-flying birds with a particular hormonal state has not yet been accomplished to our knowledge.

CHAPTER 5

WINTERING PERIOD

T HE "WINTERING period" is the nonbreeding portion of the life cycle in which the migrant spends the majority of its time, and the "winter range" is the place where the majority of its population spends that period. For many species of migrants (e.g., Red-eyed Vireo [*Vireo olivaceus*]), this classification is straightforward, with a wintering period spent in northern South America that lasts from the end of October to the beginning of April (figure 5.1). For some other species of migrants—for example, the Garden Warbler (*Sylvia borin*)—the movement to the wintering ground is more complex, involving at least one prolonged stopover of 6 to 8 weeks in sub-Saharan savannah before continuation on to wintering grounds in trans-equatorial Africa (Jones 1995) (figure 5.2). We will discuss the basic pattern of the migrant wintering period as well as its principal variations in this chapter.

ARRIVAL

Birds can migrate from a specific breeding area to a specific wintering location across thousands of kilometers once they have been to a site (Rappole 1995:59–61; Newton 2008:753). Exactly how they are able to accomplish this without a GPS or

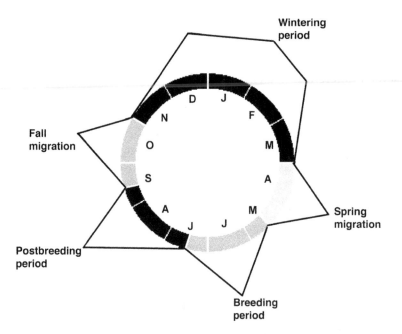

FIGURE 5.1   Annual cycle for the Red-eyed Vireo (*Vireo olivaceus*).

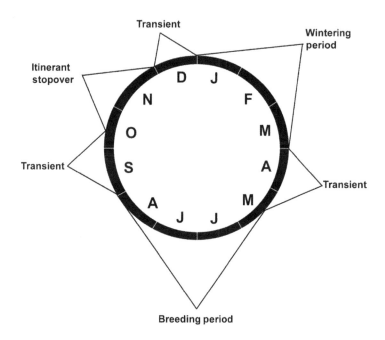

FIGURE 5.2   Annual cycle for the Garden Warbler (*Sylvia borin*).

a sextant and chronometer is not well understood. Nevertheless, the fact that they possess this ability is incontrovertible as documented by numerous band recovery studies for hundreds of migrant species representing most major migrant taxa from several different migration systems (Moreau 1972; McClure 1974; Rappole 1995:59–61; Newton 2008:501–508). As an example, in one study of Hooded Warblers (*Setophaga citrina*) wintering in the Veracruz rainforest, six of 10 birds banded on their territories in December 1973 were recaptured on the same territories in November 1974 (Rappole and Warner 1980). As part of this same study, a Kentucky Warbler (*Geothlypis formosa*) banded as an adult (skull pneumatized) on December 28, 1973, was recaptured 60 m northeast from its original capture point on December 5, 1980, presumably having made seven round-trips between its eastern United States breeding area and southern Mexican wintering area in the interim. Holmes and Sherry (1992) reported higher return rates to wintering sites in Jamaica for adult American Redstarts (*Setophaga ruticilla*) and Black-throated Blue Warblers (*Setophaga caerulescens*) than to breeding sites in New Hampshire.

DURATION OF THE MIGRATORY JOURNEY

Factors controlling duration of the migratory journey are something that we know very little about. Adult birds of many migrant species evidently recognize when they have arrived at their wintering quarters because they have been there before. Juveniles have no experience on which to draw and must therefore find their way by obtaining information from experienced conspecifics, a genetic program, or some combination of the two. We have proposed that both adults and juvenile migrants may possess the ability to return to the point of origin for the ancestral population from which they derived, which, in most cases, will be the wintering area (see chapter 4). If our hypothesis were correct, the first offspring of the first migratory generation of any population thus would have the capacity to migrate to the wintering area without assistance from experienced conspecifics, although social cues (e.g., vocalizations of conspecifics in migratory flight) could enhance orientation (Hamilton et al. 1967; Hamilton and Gilbert 1969).

Berthold (1988:228) and his students (e.g., Pulido 2007) propose an alternative mechanism to explain how juvenile migrants could accomplish migration to their population's wintering area; namely, by possessing a genetic program for the correct direction and duration of the migratory journey. According to this theory, the naive, juvenile bird sets out in a genetically programmed direction and continues for a genetically programmed period until arriving at the wintering region. The nature of *Zugunruhe* experiments with juvenile caged birds raises some questions, however, concerning their meaning. *Zugunruhe* is not migratory flight; it is a behavior observed in captives that are not allowed to migrate. Also, initiation of *Zugunruhe*, as a stand-in for the beginning of migratory flight, is not an event that is isolated from other life history events. In fact, it may depend not only on distance of the wintering site but also on differential timing of molt in the different populations (Berthold and Querner 1982; Helm and Gwinner 2006b). Thus, a key part of duration (i.e., initiation) depends not just on distance of the wintering area but also on timing of molt. Although timing of these two life history events, molt and migration, may in fact be related to location of the wintering area, that relationship is likely to be complex, not just a simple matter of distance. This difficulty is illustrated by the fact that timing of both molt and migration can differ for different members of the same population (e.g., Yellow-bellied Flycatcher [*Empidonax flaviventris*]). In this species, most adults migrate before molting and most juveniles migrate after molting, which results in different mean timing of the fall transient period for members of the two groups. At a stopover site located along the western Gulf Coast in southern Texas, the mean median Julian date for peak passage was 234 for adults versus 253 for juveniles (Winker and Rappole 1992). Thus, differences in timing of molt alone could explain part of the reason for observed differences in *Zugunruhe* between different populations (i.e., different initiation of migration dates resulting from different molt completion dates).

Despite the concerns regarding their meaning, there is no question that *Zugunruhe* experiments have established that there is a genetic basis for timing of life history events, including initiation of migration (Gwinner 1968; Berthold 1973; Pulido 2007). Cross-breeding experiments between individuals of southeast- and southwest-migrating populations of Blackcaps (*Sylvia atricapilla*) have found that F1 offspring demonstrated *Zugunruhe* orientation that was "intermediate between and significantly different from, that of both parental populations" (Helbig 1991). In addition, Berthold (1988) found that not only direction but also duration of *Zugunruhe* differs between populations that winter at different latitudes (figure 5.3).

The argument that a genetic program for duration and distance of migratory flight could take a naive juvenile to the correct wintering area makes theoretical sense based on experimental data of the type displayed earlier, but what relationship do the experiments on which the theory is based have to what actual migrants do? Can distance between the breeding and wintering areas be equated in some way with duration of *Zugunruhe*? Consider, for example, the *Zugunruhe* durations shown in figure 5.3A: greater than 50 days for the nonmigratory Canary Island birds; 100 days for French birds; and greater than 150 days for German and Finnish birds. If we allow for 10 hours of flying time per night at a mean ground speed of 22 km/h, this amounts to a migratory journey of greater than 5,500 km for the sedentary, resident population; greater than 11,000 km for French migrants; and greater than 16,500 km for German and Finish migrants, even if we assume that they spend half their time refueling. The longest actual migration journey, that for the Finnish birds to their West African winter quarters, is 6,500 km. Available information on duration of fall migration indicates that most birds complete the movement within 60 days, during the months of September and October (Aymí and Gargallo 2006), not the more than 100 days indicated by *Zugunruhe* experiments. In addition, the average number of days spent in migratory flight by birds

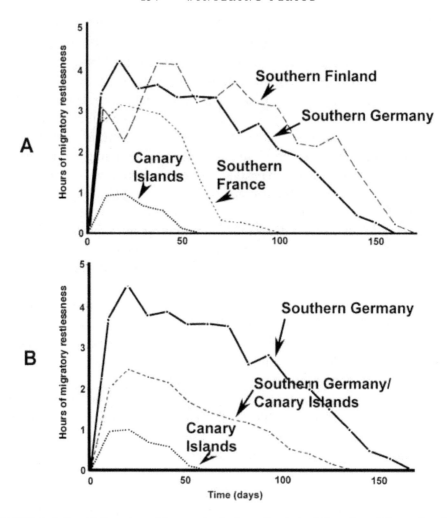

FIGURE 5.3   (A) Duration of *Zugunruhe* for juveniles derived from four different breeding populations of Blackcaps (*Sylvia atricapilla*): Canary Islands (tropical resident); southern France; southern Germany; and southern Finland. (B) Duration of Blackcap *Zugunruhe* for parental stocks from the Canary Islands (resident) and southern Germany (migrant) and offspring derived from making genetic crosses between these parental stocks (southern Germany/Canary Islands) (based on Berthold 1988:229).

resident in the Canary Islands is zero so far as we know, not 50. (Note that the Canary Island Blackcap is not the only resident species known to exhibit *Zugunruhe* [Mewaldt et al. 1968].)

These observations demonstrate that duration of *Zugunruhe* is not a good stand-in for duration of an actual migratory journey. Nevertheless, one might argue, the experimental differences found in duration of *Zugunruhe* must mean something. We agree with this argument but suggest that the relationship between *Zugunruhe* and an actual migratory journey are not well understood. As shown earlier, assumptions concerning any sort of direct relationship are not supported by available information in terms of what we know about the duration of *Zugunruhe* versus the actual time required to migrate from breeding to wintering sites. Certainly, the hypothesis that *Zugunruhe* experiments document the existence of a genetically fixed program for duration of migration, in Blackcaps or any other migrant, are not supported by available data. In fact, there are very few field data on how the migration journey terminates.

We have proposed that migrants do not possess a genetic program for route (i.e., direction and distance) (see chapter 4); rather, we suggest that they possess a genetic program for ability to locate the wintering area (ancestral resident population range). In terms of duration of migration, we further propose that not only is it not programmed, but also that it will last as long as it takes for the bird to find an appropriate wintering site (i.e., one where it can obtain resources necessary for survival throughout the wintering period). This shift in perspective would mean that a bird could find a site that was suitable for a portion of the wintering period but that might later become unsuitable, and that alternation between *Zugdisposition* and *Zugstimmung* could continue even after the bird has arrived at the appropriate wintering area for its population. We propose that if the individual cannot obtain the appropriate resources at a site because of inability to find or defend appropriate habitat, then it should continue to move. Also, nonmigratory birds could use *Zugstimmung* for dispersal or the kinds of intratropical movements reported by Beebe (1947) in Venezuela and McClure (1974) in the Philippines by individuals of species presumed to be resident.

There are some data from studies of migrants on their wintering grounds to indicate that migratory flight (*Zugstimmung*) and the migratory journey between breeding and wintering areas are not equivalent. This hypothesis is supported by the fact that juvenile, resident birds undergo *Zugdisposition* and *Zugunruhe* (Helm and Gwinner 2006b). These behaviors have been characterized as "remnant" from when ancestral populations were migratory. However, they could represent adaptive behavior, both for juvenile birds of resident species forced to disperse from their natal areas and for juvenile migrants unable to find sites with adequate resources on first arrival at wintering areas. In both cases, a state of *Zugstimmung* may be appropriate. There are some data indicative of at least the second category (i.e., continuing migratory movement even after arrival within the wintering region). Rappole and Warner (1980) found that whereas individuals of many migrant species were typically sedentary, territorial, and carried low fat reserves (sufficient for overnight needs) during their stay at wintering sites; nonterritorial individuals were not sedentary and occasionally were found with moderate or heavy fat reserves, perhaps indicative of capacity for initiating migratory flight to locate more suitable wintering sites. Further studies of wintering migrant Wood Thrushes (*Hylocichla mustelina*) using radio tracking provided additional data in support of this hypothesis (Rappole et al. 1989a).

These findings provide no conclusive evidence of the precise nature of when and where *Zugstimmung* or *Zugunruhe* occur in natural populations, but they do provide further evidence of the complexity of the relationship between them and migratory journeys.

THE MEANING OF WINTERING GROUND "ARRIVAL"

"Arrival" at the wintering area can mean very different things for different species, as well as for different age and sex groups within species. For some types of migrants (e.g., pelagic species like the Wandering Albatross [*Diomedea exulans*]), "wintering area" can include vast parts of the world's oceans. The only sites to which they show fidelity so far as is known are their breeding islands. For others, like the Brown-chested Jungle Flycatcher (*Rhinomyias brunneatus*), the wintering area appears to be well defined (figure 5.4). In between these extremes, there exists a variety of wintering strategies that differ among species. The Palearctic–African migration demonstrates several of these strategies, as nicely summarized by Newton (2008:708) based on the work of Moreau (1972) and Jones (1985, 1995, 1999) (table 5.1). Thus, in some species of trans-Saharan migrants, the "wintering area" can include two or more different regions occupied at different times during the wintering period.

Age also has an effect on what "arrival" means. Consider that roughly half of any given population of migrants "arriving" in the appropriate region and habitat for the wintering period is likely to be composed of naive juveniles that have never

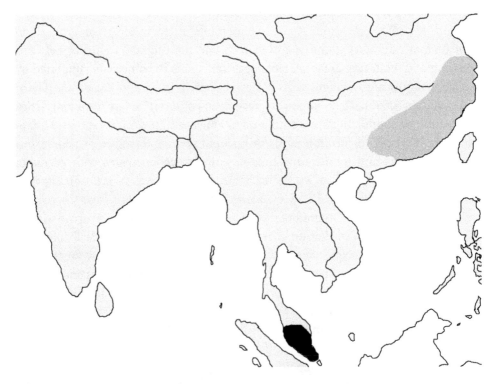

FIGURE 5.4   Brown-chested Jungle Flycatcher (*Rhinomyias brunneatus*) breeding distribution (*medium gray*) in eastern China and wintering distribution (*black*) in Malaysia.

TABLE 5.1   Wintering Strategies of Palearctic Migrants to Sub-Saharan Africa

WINTERING STRATEGY	EXAMPLES
Entire period spent in dry season conditions of the northern Sahel	Greater Whitethroat (*Sylvia comunis*), Short-toed Snake Eagle (*Circaetus gallicus*), Black-tailed Godwit (*Limosa limosa*)
Initially in dry-season conditions of Sahel and Sudan; then moving south to dry-season conditions of Guinea	Woodchat Shrike (*Lanius senator*), Common Nightingale (*Luscina megarhynchos*), Garden Warbler (*Sylvia borin*)
Initially in wet season of the northern tropics; then moving across the equator into the southern tropics	Tree Pipit (*Anthus trivialis*), Thrush Nightingale (*Luscina luscina*), Willow Warbler (*Phylloscopus trochilus*)
Entirely in wet-season conditions of the southern tropics	Common Cuckoo (*Cuculus canorus*), Hobby (*Falco subbuteo*), Eurasian Golden Oriole (*Oriolus oriolus*)

been there. Regardless of whether or not they are following a "migration route" program or a "destination" program to determine arrival, no specific wintering site awaits them because most suitable sites will be occupied by experienced adults (as discussed in the section "Social Structure During the Nonbreeding Period"). Therefore, presumably, they must continue to move until they are able to locate such a site, which may take much of the wintering period (Rappole and Warner 1980; Rappole et al. 1989a).

SOCIAL STRUCTURE DURING THE NONBREEDING PERIOD

FITNESS ASPECTS OF NONBREEDING PERIOD SOCIALITY

Sociality in terms of intraspecific interactions during the nonbreeding period are potentially shaped by at least two distinct aspects of fitness, but to different degrees for different species of migrants: reproduction and survival.

Reproduction Breeding pairs of some species of migrants (e.g., geese and cranes) keep their young of the year with them throughout most or all of the nonbreeding period, often traveling thousands of kilometers between breeding, molting, and wintering areas as family groups (Tacha 1988; Gill et al. 1996). This behavior presumably enhances fitness for long-lived, *K*-selected species wherein increased parental investment increases the likelihood of offspring survival (Trivers 1972; Maynard Smith 1977). In other migrants, certain aspects of behavior during the nonbreeding period appear to be related to preparation for enhancing reproductive success for the next year's breeding season. This preparation is most obvious among many species of waterfowl in which birds begin courtship in attempts to form pair bonds with potential mates during fall or winter (Bellrose 1976); pairs then fly together to their breeding area, which often for many species is the site at which the female bred (or was raised in) the previous year (Greenwood 1980).

Territoriality by adult males that remain on and defend the breeding area while juveniles and adult females migrate is a behavior observed in several species of partial migrants (e.g., Song Sparrow [*Melospiza melodia*], European Robin [*Erithacus rubecula*], Blackbird [*Turdus merula*], Skylark [*Alauda arvensis*], and Blue Tit [*Cyanistes caeruleus*]) (Nice 1937; Lack 1943, 1944a, 1944b; Schwabl et al. 1984; Smith and Nilsson 1987; Adriaensen and Dondt 1990; Newton 2008:597–599; Hegemann et al. 2010). "Behavioral dominance" is one possible explanation for this behavior, in which dominant individuals sequester the local food resources for the nonbreeding period forcing subordinate individuals to migrate (Gauthreaux 1978b, 1982; Lynch et al. 1985; Parrish and Sherry 1994). This argument is based on two assumptions:

1. Adult male behavior (territoriality) is predicated solely on defense of food resources.
2. Adult female and juvenile behavior (migration) is dictated by adult male behavior; that is, they would remain on the breeding area if not forced to leave.

Few data support either assumption for any species, let alone for partial migrants as a group (Rappole 1995:17). An alternative explanation is that the movement and defense strategies observed in the different sex and age groups reflect the different alternatives for maximizing fitness confronting each group at a particular point in the life cycle (Fretwell 1972; Baker 1978; Myers 1981). This argument has been developed by Greenwood (1980) in which he contrasts avian breeding systems (mainly monogamous) with mammalian breeding systems (mainly polygamous) (table 5.2). As delineated in table 5.2, male fitness in monogamous species theoretically can be enhanced by holding on to breeding territory (resource defence) despite the potential of lowered survivorship because of the possibility of enhancing reproductive success, whereas no such benefit accrues to females (or to juvenile males who have no breeding territories).

A similar argument can be made for other types of sexual distributional differences that occur between migrants (e.g., differential migration). Differential migration occurs when all members of a population migrate, but one age or sex group migrates a different mean distance from the breeding area than another (Ketterson and Nolan 1983). It is similar to partial migration in that it is usually adult males that tend to winter closer to the breeding area, apparently displacing subordinate groups, and it has been similarly explained as resulting from social dominance in which adult males sequester the best food resources, forcing adult females and juveniles to migrate farther and occupy wintering areas with

TABLE 5.2   Monogamous (Avian) Versus Polygamous (Mammalian) Resource Defense Systems

MONOGAMOUS BIRDS (RESOURCE DEFENSE)	POLYGAMOUS MAMMALS (MATE DEFENSE)
High male investment in resources in presence or absence of mate(s)	Low male investment in resources particularly in absence of mate(s)
Low female investment in resources	High female investment in resources
Inter-male competition for resources	Inter-male competition for mates
Mainly monogamous (but include leks?)	Mainly polygamous
Male philopatry High cost to male dispersal?	Female philopatry High cost to female dispersal?
Greater female natal and breeding dispersal: (1) Reproductive enhancement: female choice of male resources (2) Inbreeding avoidance	Greater male natal and breeding dispersal: (1) Reproductive ehhancement: increase access to females (2) Inbreeding avoidance
Evolution of patrilineal social organization	Evolution of matrilineal social organization

Source: Greenwood (1980).

poorer resources (Gauthreaux 1982:117). However, the same alternative arguments expressed for partial migration can be used to explain differential migration as well; that is, that adult males have different reproductive roles and face different selective pressures than adult females and juveniles that require adult males to make fitness trade-offs between survival and earliest-possible territory occupancy (i.e., females and juveniles migrate to areas optimal for overwinter survival whereas adult males migrate to areas optimal for balancing survival and future reproductive success).

Survival Regardless of the possible role played by reproduction in terms of nonbreeding-season behavior, clearly survivorship is the central focus for most migrants. Therefore, it is food resource use that determines much of what migrants do during this period. Because many migrants join communities during the wintering period when food resources are at their lowest ebb (Rappole 1995:58), intraspecific competition for those resources may be important and intense. In practice, nonbreeding-season sociality varies with the kinds of resources being harvested and their abundance and distribution in space and time. Some kinds of foods in certain habitats are more economically and safely harvested by the individual foraging alone, whereas other types are optimally harvested in groups. Whether or not migrants exploit foods alone or in groups depends on a balance between costs and benefits involving at least four factors:

1. *Resource distribution.* If the food is distributed in such a way as to be economically defendable—that is, if more energy is gained in harvesting the resource than is lost defending it—then solitary harvest behavior may be favored, whereas if more energy is required to defend it than can be gained in consuming it, then individuals should tolerate conspecifics (Brown 1964; Kaufmann 1983).
2. *Vulnerability to predation during harvest.* In general, solitary individuals appear more vulnerable to predation than those foraging in groups (Miller 1922; Powell 1985; King and Rappole 2000, 2001a).
3. *Resource location.* Group members may serve as sources of information regarding possible food locations for other group members (Ward and Zahavi 1973) or even as food sources via commensalism or kleptoparasitism (King and Rappole 2001b, 2002).
4. *Resource harvest.* Some kinds of resources (e.g., schools of fish or grain fields) are more efficiently harvested by groups than by solitary individuals (Taylor 1984).

PATTERNS OF NONBREEDING-PERIOD SOCIALITY

Five major patterns of migrant nonbreeding-season sociality derive from the different kinds of food resource patterns.

Solitary, Sedentary, and Territorial A territory is any defended area (Kaufmann 1983). Two types are seen in nonbreeding-season migrants: short-term, in which an ephemeral resource in a confined space is defended for a few minutes or hours (e.g., a flower bed, a section of beach, or a fruiting tree) (Armitage 1955; Kale 1967; Recher and Recher 1969; Emlen 1973); and long-term, in which relatively stable resources dispersed over a wider area are defended for days, weeks, or months (e.g., the arboreal insect resources in a forest canopy or understory arthropods) (Eaton 1953; Schwartz 1964; Rappole and Warner 1980). There is published evidence of territoriality in at least 75 species of migrants, although the number is probably far higher than this figure because social behavior of the majority of migrants has not been investigated during the nonbreeding period (see table in Rappole 1995:35–37). Taxonomic and geographic biases in occurrence of winter territoriality (e.g., 28% of those found to be territorial are Parulidae and 83% are from the neotropics) probably result mainly from biases in where and on which groups most work on the topic has been done rather than actual differences in migrant social behavior between regions or major taxa. Almost no work has been done on wintering migrant sociality in Asia or on intratropical migrants anywhere, and most investigation of Palearctic migrant sociality in Africa has been focused on interspecific interactions between migrants and residents (e.g., Leisler 1990).

Note that in migrant species known to be territorial during the nonbreeding period, both males and females defend individual territories, with females using the same acoustic and visual displays to exclude both male and female conspecifics (including song) as males (Rappole and Warner 1980; Rappole 1988).

Solitary, Nonterritorial Wanderer Classic studies of wandering migrants have been performed in steppe habitats of the Sahel region of sub-Saharan Africa where entire populations of several species of Palearctic migrants are wanderers, at least during the early part of the winter season (Moreau 1952; Morel and Bourlière 1962). These birds appear to be capitalizing on seasonal resource superabundance before migrating farther south to more stable wintering habitats (Jones 1985, 1995, 1999; Jones et al. 1996).

Prior to intensive work on wintering migrant communities of the past few decades, most observers assigned terrestrial migrants to this behavioral class (Buechner and Buechner 1970:93; Tramer 1974). In part, this classification resulted from the fact that apparently wandering migrants are indeed very common, as mentioned earlier. However, the classification also resulted from untested assumptions regarding the individual's residency status, the problem being that it cannot be determined whether or not an individual is "wandering" unless its movements and site tenacity can be documented, which, at present, can only be done as part of a long-term study using individually marked birds. Even as of today, few such studies have been done. Of those that have been done, several have shown that while a portion of any given population of wintering migrants may be composed of wanderers, another portion may demonstrate a different type of social behavior (e.g., territoriality) and may have a relatively restricted home range (Rappole and Warner 1980; Rappole et al. 1989a, Rappole 1995:35–37). Furthermore, some studies have shown that nonbreeding-season migrant social behavior can change for any particular individual over the course of a season or even daily, depending on how resources are distributed (Recher and Recher 1969; Rappole 1995:33–48).

Conspecific Flocks Many species of migrants occur in flocks of conspecifics during the nonbreeding season. We discussed earlier the various reasons given for flocking, but we suggest that there are two main reasons for the behavior in wintering migrants: food resource distribution and vulnerability to predation. If critical food is distributed in dispersed clumps, then a flock may be the most economical and safest means of harvesting it (Taylor 1984). Thus, seed-and-fruit eaters (e.g., Red-winged Blackbirds [*Agelaius phoeniceus*], Cedar Waxwings [*Bombycilla cedrorum*], and Great-tailed Grackles [*Quiscalus mexicanus*]) commonly occur in flocks of conspecifics during this period, feeding mostly on seeds and fruits (Rappole et al. 1989b). In contrast with mixed-species flocks discussed later, single-species flocks of migrants tend to range over a large area during the nonbreeding period, sometimes even involving major geographic shifts (Witmer et al. 1997).

Observation of large flocks of any type gives the impression that such groups represent entirely stochastic associations. However, although larger flocks certainly demonstrate characteristics of randomness, the size of all foraging flocks likely bears a direct relation to resource abundance and distribution (Pulliam and Caraco 1984).

Solitary Representative of a Particular Migrant Species in Mixed (Interspecific or Multispecies) Foraging Flocks Mixed-species foraging flocks are well-known aspects of nonbreeding-period avian communities from many different habitat types and parts of the world (Bates 1863; Winterbottom 1943; Morse 1970; Powell 1985; Terborgh 1990; King and Rappole 2001a). Such flocks range in temporal and spatial organization from a few individuals of a few species in apparent ad hoc, temporary associations lasting only a few hours (Rappole 1995:43) to well-organized, long-term associations lasting several months composed of many species foraging over a relatively small, well-defined home range that is defended from other flocks (Powell 1980; Munn 1985) (figure 5.5). Several explanations have been given for these associations (table 5.3), but it seems likely that the principal reasons for their occurrence are the same as for conspecific flocks: foraging efficiency and safety.

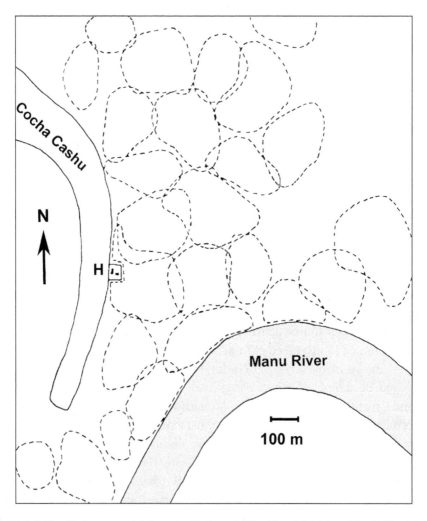

FIGURE 5.5   Understory mixed-species flock territories (*dotted lines*) in 1.8 km² of lowland forest in the Peruvian Amazon region. "Cocha Cashu" is a village; H = two houses in a clearing (based on Munn 1985).

TABLE 5.3   Suggested Benefits to Members of Mixed-Species Flocks

CATEGORY	BENEFIT	DESCRIPTION	CITATIONS
Foraging enhancement	Commensal prey flushing by flock members	Normal movements and foraging activities of some flock members inadvertently flushes prey items of use to other flock members.	King and Rappole 2002
Foraging enhancement	Information center	Experience and knowledge of prey location by some flock members may be used by other flock members.	Ward and Zahavi 1973
Foraging enhancement	Niche overlap reduction	A theoretical argument based on the idea that individuals can better adjust their niche breadth in terms of foraging behavior when associating with and observing potential competitors.	Morse 1970
Foraging enhancement	Increased possibility for kleptoparasitism	Some flock members may observe foraging activities of others and steal prey whenever possible.	King and Rappole 2001b
Predator avoidance	Increased surveillance	Experimental and field data document the effectiveness of predator detection and avoidance by groups compared with that by individuals.	Powell 1985:724
Predator avoidance	Decreased predator attack efficiency	Large, tight flocks may make individual prey selection and attack more difficult.	Miller 1922; Tinbergen 1946
Predator avoidance	Decreased predator encounter probability	A theoretical argument based on the idea that flocking increases predator search time.	Olson 1964
Predator avoidance	Decreased individual exposure	A theoretical argument based on the idea that a group member, especially of large groups, has less chance of being eaten than an individual solely because of group membership.	Brock and Riffenburgh 1960

Source: Powell (1985).

Individual wintering migrants show four main types of long-term relationship with these kinds of stable, mixed-species foraging flocks:

1. An individual remains with the flock throughout the flock's home range, essentially defending the flock itself as a moving territory by preventing other conspecifics from joining (Morton 1980b:447; Rappole and Warner 1980; Rappole et al. 1983:35; King and Rappole 2000, 2001a).
2. An individual joins the flock only while the flock is on its territory (Rappole and Warner 1980).
3. An individual associates with more than one flock moving on or through its territory (Chandler and King 2011).
4. Two or more unrelated individuals associate with the flock throughout the wintering period (King and Rappole 2000).

These behaviors differ from that of most resident flock attendants, many of which attend flocks as mated pairs or family groups (Moynihan 1962; Powell 1985).

Different species of migrants that typically associate with mixed-species flocks during the wintering period demonstrate characteristic differences in terms of conspecific tolerance. Some species normally are the sole member of their species when in attendance at flocks, whereas others may usually be represented by several individuals (King and Rappole 2000). Differences in sociality probably have to do with several factors including the species of the migrant, resource distribution and density, the species composition of the flock (often varies by habitat and region), and the experiences and competitive ability of the individual (can vary by age or sex) (Rappole 1995:45; King and Rappole 2000). As for conspecific flocks, the number of conspecifics in a given multispecies flock probably is not stochastic, but rather is determined by the amount of available resources (Mewaldt 1964; Davis 1973; Pulliam and Caraco 1984; King and Rappole 2000). In those few migrant species in which attendance at mixed-species flocks has been documented by banding, individuals demonstrate both within- and between-season fidelity to the flock or flock vicinity (Buskirk et al. 1972; Powell 1980; Rappole and Warner 1980; R. B. Chandler 2011).

Mixed Wintering Behavioral Strategies. Individuals of the same species of wintering migrants can display different types of social behavior at different times or places (table 5.4). We suggest that these different behaviors result principally from variations in resource abundance or distribution.

PLUMAGES

Many closely related species pairs in which one is migratory and one is resident differ in terms of their molts and plumages. In particular, one of the most common differences between migrant and resident congeners is the existence of a prealternate molt for migrants into an alternate (breeding) plumage, which usually takes place during the wintering period. The question is what purpose does the plumage

TABLE 5.4   Reported Social Behavior of Nearctic Migrants in Different Neotropical Habitats, Localities, or Time Periods

Species	Habitat	Locality	Social System[1]	Citations
Wood Thrush (*Hylocichla mustelina*)	Rainforest	Mexico	T	Rappole and Warner 1980; Rappole et al. 1989a
Wood Thrush (*Hylocichla mustelina*)	Rainforest	Mexico	A	Rappole and Warner 1980; Rappole et al. 1989a
Yellow Warbler (*Setophaga petechia*)	Pasture, hedgerow	Mexico	T	Greenberg and Salgado Ortiz 1994
Yellow Warbler (*Setophaga petechia*)	Pasture, hedgerow	Costa Rica	T	Skutch in Bent 1953
Yellow Warbler (*Setophaga petechia*)	Forest edge	Panama	F	Moynihan 1962
Bay-breasted Warbler (*Setophaga castanea*)	Old forest	Panama	A, S	Morton 1980a; Greenberg 1984
Bay-breasted Warbler (*Setophaga castanea*)	Young forest	Panama	A, S	Morton 1980a; Greenberg 1984
Chestnut-sided Warbler (*Setophaga pensylvanica*)	Rainforest	Panama	F	Greenberg 1984
Chestnut-sided Warbler (*Setophaga pensylvanica*)	Rainforest	Costa Rica	F	Powell et al. 1992
Chestnut-sided Warbler (*Setophaga pensylvanica*)	Pasture, hedgerow	Costa Rica	S, T?	Powell et al. 1992
Black-throated Blue Warbler (*Setophaga caerulescens*)	Low scrub	Puerto Rico	A, C, M	Staicer 1992
Black-throated Blue Warbler (*Setophaga caerulescens*)	Forest	Jamaica	T	Holmes et al. 1989
Black-throated Green Warbler (*Setophaga virens*)	Rainforest	Mexico	S, T?	Rappole and Warner 1980
Black-throated Green Warbler (*Setophaga virens*)	Rainforest	El Salvador	C	Dickey and van Rossem 1938

(*continued*)

TABLE 5.4 *(continued)*

Species	Habitat	Locality	Social System[1]	Citations
Black-throated Green Warbler (*Setophaga virens*)	Montane forest	Panama	F	Buskirk et al. 1972
Black-throated Green Warbler (*Setophaga virens*)	Highland pine-oak forest	Honduras, Guatemala, Mexico	C	King and Rappole 2000
American Redstart (*Setophaga ruticilla*)	Pasture	Panama	A	Morton 1980b
American Redstart (*Setophaga ruticilla*)	Low scrub	Puerto Rico	A	Staicer 1992
American Redstart (*Setophaga ruticilla*)	Forest	Jamaica	T	Holmes et al. 1989
Black-and-white Warbler (*Mniotilta varia*)	Rainforest	Mexico	F	Rappole and Warner 1980
Black-and-white Warbler (*Mniotilta varia*)	Rainforest	Belize	F	J. H. Rappole, personal observation
Black-and-white Warbler (*Mniotilta varia*)	Pasture, hedgerow	Mexico	S	J. H. Rappole, personal observation
Black-and-white Warbler (*Mniotilta varia*)	Low scrub	Puerto Rico	A, C, M	Staicer 1992
Black-and-white Warbler (*Mniotilta varia*)	Second growth	Belize	M	J. H. Rappole, personal observation
Black-and-white Warbler (*Mniotilta varia*)	Montane forest	Panama	J	Buskirk et al. 1972
Black-and-white Warbler (*Mniotilta varia*)	Deciduous forest	Mexico	F	Hutto 1994
Wilson's Warbler (*Wilsonia pusilla*)	Rainforest	Mexico	T	Rappole and Warner 1980
Wilson's Warbler (*Wilsonia pusilla*)	Montane forest	Panama	F	Moynihan 1962

Species	Habitat	Locality	Social System[1]	Citations
Summer Tanager (*Piranga rubra*)	Rainforest edge	Panama	F	Moynihan 1962
Summer Tanager (*Piranga rubra*)	Rainforest	Mexico	T	Rappole and Warner 1980
Summer Tanager (*Piranga rubra*)	Deciduous forest	Mexico	F	Hutto 1994

[1]F = attendant at cohesive interspecific flocks, defended flock against conspecifics; J = joiner of cohesive interspecific flocks, attending for short periods; A = took part in multispecific feeding aggregation at temporarily abundant food source; C = foraged with flock of conspecifics; M = foraged with flock of individuals of mixed species; T = territorial; S = solitary individual, status unknown.

produced serve for a migrant that is not necessary for a resident? We see three possible purposes for coloration of any plumage:

- Camouflage to make the individual as inconspicuous as possible to potential predators
- Status signaling to same-sex conspecifics in competition for mates or food
- Status signaling to opposite-sex conspecifics in competition for mates or food (Ewald and Rohwer 1980; Rohwer and Butcher 1988; Stutchbury and Morton 2001)

The alternate (breeding) plumage presumably is no different, resulting from a balance of the various selection factors created by the different needs confronting migrants. We propose that the wide range of intersexual variation in plumages found in many migrants is indicative of this balancing selection resulting from intersexual competition for resources (Rappole 1988). In about two thirds of Nearctic avian migrants that have wintering populations in the neotropics, male and female plumages are identical although in many of these there are sexual differences in other morphologic characters (e.g., wing length, bill length, or body size) (Rohwer and Butcher 1988). Among the one third of species in which males and females differ in plumage, the differences for many are not "dimorphic" with a brightly colored male and a dull-colored female; rather, the differences are between a brightly colored male plumage and a wide range of female plumages (i.e., polymorphic rather than dimorphic) varying from dull-colored to male-like (Oberholser 1974; Rappole 1983, 1988). We suggest that these plumage variations result from balancing selection on both males and females and present the following hypothesis from Rappole (1988:2314–2315):

Information from the study of wintering migrants suggests [a] possible explanation for the evolution of female ornamentation in these species: "andromimesis" or female mimicry of males. This hypothesis is based on Darwin's theory of sexual selection and observed intraspecific competition during the non-breeding season. The theory is as follows:

(a) Competition among males of monogamous as well as polygynous species is intense, favoring evolution of morphological and/or behavioral characters to enhance success in this competition for females. Emlen and Oring (1977) and Mayr (1976) argued that sexual selection in monogamous species is negligible. However, sexual selection should occur in monogamous species as well (Fisher 1930). The fact that males of many monogamous species do not breed because of failure to obtain a territory and/or a mate (Rappole et al. 1977) indicates that there is intermale competition. Bright male coloration can be easily explained as a result of such competition.

(b) Many of the brilliant plumage characteristics evolved by males are used in agonistic displays exclusively or in addition to courtship displays. Lack (1943) reported this phenomenon for the English Robin (*Erithacus rubecula*). There are similar observations on the American Redstart (*Setophaga ruticilla*) (Ficken and Ficken 1962) and the Red-winged Blackbird (*Agelaius phoeniceus*) (Peek 1972).

(c) The same displays as used by males in agonistic encounters during the breeding season are used by both sexes in defense of winter territory during the non-breeding season (Rappole and Warner 1980).

(d) Displays given by females that lack the plumage characters of the males are less visually effective. Therefore, selection should favor evolution of male-like characters that are important in agonistic displays by females that compete with males for resources during the nonbreeding season. An example is the Hooded Warbler, most females of which do not have the black cowl possessed by males [figure 5.6]. The cowl is a key part of agonistic displays on the wintering ground [figure 5.7] (Rappole and Warner 1980).

Suggested costs and benefits of bright plumages are summarized in table 5.5, illustrating how andromimesis could form a key part of how balancing selection might result in both the similarities and differences seen between male and female migrant plumages. This theory is a part of a larger hypothesis, presented in the following, to explain observed intersexual structural differences as well as microgeographic and macrogeographic intersexual differences in geographic distribution (Rappole 1995:19–21):

1. We assume that, in the absence of significant sexual selection (Payne 1984), the favored plumage for a bird is cryptic (Rohwer 1975).

FIGURE 5.6   Female (*left*) and male (*right*) Hooded Warblers (*Setophaga citrina*).

FIGURE 5.7   Displays used by both male and female Hooded Warblers (*Setophaga citrina*) in defense of winter territory (drawing by C. P. Barkan, based on Rappole and Warner 1980).

2. In the absence of significant balancing factors, with both sexes competing for the same limiting resources during the nonbreeding season, selection should force them toward equivalent morphology (Rohwer 1975).

3. In many species in the which the male and female are identically cryptic, the sexes differ morphologically (usually in wing length) (Rappole 1988).

4. Because nonbreeding intersexual competition for resources should force equivalent morphology and appearance, the presence of differences could be the result of breeding ground factors (Williamson 1971; Kodric-Brown and Brown 1978; Rappole 1988). One such factor could be a form of "ecological release," in which the male and female members of a mated pair, forced by the nature of their cooperative rearing of young to share a foraging space in which resources are relatively abundant, evolve structural differences to minimize intrapair competition for food and maximize efficiency in feeding of young (Snyder and Wiley 1976; Temeles 1985).

TABLE 5.5   Exemplary Costs and Benefits Imposed by Selection Forces on Individuals Wearing Bright Rather Than Cryptic Plumage

COSTS OF BRIGHT PATTERN

Increased probability of predation for the individual

Increased probability of predation for offspring (brightly colored parent bringing attention to nest location)

Increased aggressive response from territorial males during the breeding season

Energetic demands of additional molts (for males that molt from a more cryptic basic plumage to a more bright alternate (breeding) plumage or for females to molt from a cryptic alternate (breeding) plumage to a brighter basic (nonbreeding) plumage to enhance success in competition with males for nonbreeding season resources

BENEFITS OF BRIGHT PATTERN

Enhanced attractiveness to females during breeding season (males only)

Enhanced effectiveness in competition with other males for and defense of breeding territory (males only)

Enhanced effectiveness in competition for and defense of nonbreeding-season food resources (both sexes)

Source: Rappole (1995:20).

5. The presence of intersexual morphological differences favors the development of differences in foraging habitats (Kodric-Brown and Brown 1978). Such differences could be so great that, under certain circumstances of resource dispersion male-female pairs could coexist on the same nonbreeding site by using different resources or different microhabitats (Zahavi 1971; Leck 1972; Greenberg and Gradwohl 1980; Wunderle 1992; Rappole et al. 2000a; Chandler and King 2011).
6. For species in which the male is brightly patterned, and males compete with females and other males for nonbreeding resources, females should mimic males in nonbreeding appearance (andromimesis) (Rappole 1988).
7. Costs for malelike plumage (e.g., additional molts or higher predation rates) and macro- or microgeographic sexual differences in habitat preferences cause different balances to be struck for different species in terms of morphological differences between the sexes.

Differences in plumages also occur between adult and subadult males in many species of migrants (Oberholser 1974; Ginn and Melville 1983; Pyle 1997).

Rohwer and Butcher (1988) summarize data on subadult male plumage variation in North American passerine migrants in which adult males on average are brighter in color than females, showing five major classes of variation in subadult male molts and plumages (table 5.6). Subadult male differences in plumage from adult males are attributed mainly to breeding-ground factors, for example, status signaling (reducing competition with adult males for breeding territories while retaining some potential for attracting mates) and predator avoidance (Montgomerie and Lyon 1986), but may result from balancing selection involving both breeding-period factors and nonbreeding-period crypsis requirements and competition for resources (Rohwer and Butcher 1988). In any event, in addition to the variety of subadult male plumage patterns described by Rohwer and Butcher (1988), similar kinds of polymorphism as described for females has been found in subadult male plumages (Rappole 1983), indicating that the kinds of balancing selection discussed earlier for females also affect subadult males but in ways that are different than for females.

TABLE 5.6   Categories of Yearling (Subadult) Male Plumages in North American Migrant Passerines Having Sexually Dichromatic Adults in the Winter and in the Summer

CATEGORY	EXAMPLES
Winter, subadult male plumage; summer, adult plumage; spring molt, extensive body molt in all known cases	Bluethroat (*Luscina svecica*), Yellow Warbler (*Setophaga petechia*), Dickcissel (*Spiza americana*)
Winter, subadult male plumage; summer, subadult male plumage; spring molt, partial body molt (in all cases yearlings are more like adult males)	Vermillion Flycatcher (*Pyrocephalus rubinus*), Purple Martin (*Progne subis*), Black-capped Vireo (*Vireo atricapillus*)
Winter, subadult male plumage; summer, subadult male plumage; spring molt, none (little seasonal change in appearance of adults and subadults)	Painted Bunting (*Passerina cirus*), Yellow-headed Blackbird (*Xanthocephalus xanthocephalus*), Purple Finch (*Carpodacus purpureus*)
Winter, adult plumage; summer, adult plumage; spring molt, partial (a few species) or extensive (most species) spring body molt (yearling and adult males more conspicuous for summer)	Blue-gray Gnatcatcher (*Polioptila caerulea*), Blackburnian Warbler (*Setophaga fusca*), Snow Bunting (*Plectrophenax nivalis*)
Winter, adult plumage; summer, adult plumage; spring molt, none (little seasonal change in appearance of adults and subadults)	Golden-crowned Kinglet (*Regulus satrapa*), Red Crossbill (*Loxia curvirostra*), Black-throated Blue Warbler (*Setophaga caerulescens*)

Source: Based on Rohwer and Butcher (1988).

The variation in molts and plumages among roughly one-third of North American passerine migrants is not typical of tropical residents. Although subadult male plumages exist in tropical species, they appear to have different functions from those seen in migrants (e.g., subordinate status signaling to adults among species in which the young remain on the adult territories for the first year of life) (Stutchbury and Morton 2001:100). Thus, in many cases, migrant molts and plumages appear to be adaptations to a migratory lifestyle and, when considered in comparison with molts and plumages of tropical resident congeners, may be informative with regard to the length of time for which a population has been migratory. In particular, differences in mating systems that occur between even closely related migrants may be a fruitful area of research in determining the reasons for differences in plumages and molts between migrants and residents (Stutchbury and Morton 2001).

NONBREEDING/WINTER DISTRIBUTION

The factors responsible for nonbreeding distribution patterns of migrants are extremely complex. Previously, the prevailing fundamental assumption governing the study of distribution was that interspecific competition constituted the main organizing principle for animal communities (MacArthur and Wilson 1967; MacArthur 1972; Diamond 1976; Greenberg 1986). During this period, it was believed that interspecific competition was the most important aspect governing both avian community organization and migrant distribution. MacArthur (1972) proposed that the niches filled by migrants in Temperate Zone and Boreal Zone communities were filled by "ecological counterparts" beginning at the southern end of their breeding ranges. These ecological counterparts prevented migrants from participating as members in the stable communities of the subtropics and tropics, which resulted in a "fugitive species" strategy in which migrants passed the nonbreeding period subsisting on superabundant resources that actual community members could not completely harvest (MacArthur and Wilson 1967:82; Dingle 1980:63). As a result, a great deal of field work in the 1960s through the 1980s was designed to measure the ways in which organisms competed to fill the niches available in a given ecological community (Karr 1971; Cody 1974), including how migrants fit into the communities in which they spent the winter period (Willis 1966; Cox 1968; Karr 1976; Greenberg 1986). However, there is considerable evidence that whereas competition plays a role, other aspects are involved in community structure (Verner 1977; Wiens 1977, 1983). The approach taken in our treatment is that it is the environment as a whole that dictates a migrant species' nonbreeding range, including geography, climate, habitat, food, interspecific competition, predators, diseases, sex, age, and evolutionary history. In the following discussion, each of these, which together form the ecology of the wintering migrant, is discussed and considered from the perspective of how it may contribute to shaping the nonbreeding range.

GEOGRAPHY

The specific features of Earth's surface (e.g., oceans, shorelines, rivers, plains, and mountains) set the basic outlines for a species' winter range (or ranges) both by what is available (e.g., land masses at particular latitudes for terrestrial species) and by what is reachable, based on various obstacles or barriers (e.g., oceans and mountains).

CLIMATE

Climate is the prevailing weather (temperature, air pressure, humidity, precipitation, sunshine, cloudiness, and winds) of a region throughout the year averaged over a series of years. Climate dictates many important aspects of the migrant environment, especially habitats, foods, and diseases. Within-year variation (seasonality) is a key aspect of climate for migrants, although not necessarily the seasonality of the winter range. In general, the winter range is far less seasonal than the breeding range, although the terrestrial environments of sub-Saharan Africa are an exception.

The most intriguing aspect of African climate is that timing and severity of the dry season varies sharply across the continent during the Palearctic "winter" when Eurasian migrants are present. As a result, steppe habitats in the Sahel that are suitable for many migrants in September are no longer suitable by October or November, at which point members of several species move farther south (table 5.1). For instance, birds following strategy 2 as listed in table 5.1 migrate from Eurasia to the steppe and savanna regions south of the Sahara Desert in Palearctic fall (mainly September) as the rainy season is ending. They remain there for 1 to 2 months, with members of some migrant species undergoing molt (Pearson 1973, 1975, 1990; Pearson and Backhouse 1976, 1983; Pearson et al. 1988), then depart with onset of the dry season for less parched habitats to the savanna or forest habitats to the south (Jones 1985, 1995, 1999). An example is a Nightingale (*Luscina megarhynchos*), shown by geolocator data to arrive in Senegal in November, where it remained for 1½ months before traveling to Guinea in December, where it stayed presumably for the remainder of the wintering period (British Trust for Ornithology 2011b).

Similar kinds of itinerant migration by Holarctic migrants based on intratropical climatic seasonality are known to occur in other migration systems, but they are not as well documented as the African situation. For instance, the Red-breasted Flycatcher (*Ficedula parva*) arrives in north-central Burmese dipterocarp forest in September from Eurasian breeding grounds just as the rainy season is ending (figure 5.8). They remain there as solitary (territorial?) winter residents until the dry season begins in mid-February when most dipterocarp forest trees lose their leaves. Where they go to is not known, although it is known that they do not return to their Palearctic breeding areas until April (Rappole et al. 2011a).

The movements reported for some species that breed in western North America and migrate to "monsoon" areas of the American southwest to molt before

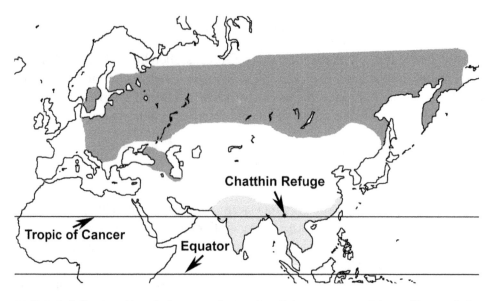

FIGURE 5.8   Breeding (*dark gray*) and wintering (*light gray*) range of the Red-breasted Fly-catcher (*Ficedula parva*). The Chatthin Refuge is the Burmese dipterocarp forest site from which wintering members of this species disappear by mid-February, as discussed in the text.

proceeding on to their wintering areas in Mexico and Central America may fall into this category of itinerant movement related to seasonality as well (Butler et al. 2002), as may the movements of Eastern Kingbirds and Swainson's Thrushes in Panama (Morton 1980).

HABITAT

Habitat is the sum of the physical characteristics of the place where an organism lives (e.g., vegetation and topography) (Odum 1971:234). For most kinds of organisms, determination of habitat use is a straightforward proposition: If the organism is known to live there, then that is its habitat. This approach works fairly well for migrants on their breeding grounds: If the migrant is known to have raised offspring there, then that is its breeding habitat. However, determination of nonbreeding-season habitat for migrants is more problematic. They can occur in almost any habitat type, especially when in transit, so simple occurrence may not be enough to identify habitat preference. Therefore, number of occurrences often is cited as an additional indicator. But numbers alone can also be misleading. Sudden changes in weather can cause the appearance of hundreds of migrants coming seemingly from nowhere and landing in inappropriate habitats (e.g., beaches or shipboard) (James 1956; Newton 2008:805–821). In addition, temporary resource concentrations can change an unsuitable habitat into a suitable one very quickly, attracting significant numbers for a short period. As a result, a third type of information is often required to identify a suitable nonbreeding habitat: persistence.

That is, if many individuals of a given species are observed in a given habitat over a long period of time (days or weeks), then the habitat is assumed to be a preferred nonbreeding-season habitat.

Based on occurrence, number, and persistence, migrants occupy a much broader array of non-breeding habitats than their resident relatives or than they themselves on the breeding grounds. Nevertheless, the meaning of this finding is not totally clear. For one thing, most observations of extreme nonbreeding-season habitat breadth occur for species that are forest-related on the breeding grounds but are observed to use a wide variety of second growth, scrub, steppe, and even agricultural or residential environments on the wintering grounds (Leisler 1990). There are very few comparable observations for the two thirds of migrant species that breed in aquatic or open environments, most of which use similar habitats during the nonbreeding period.

Four hypotheses have been presented to explain the extensive data that migrants that are forest-related during the breeding period use a much wider array of, or at least different, habitats in winter:

• *Niche breadth* (Petit et al. 1995). A "generalist" species in ecological terms is one whose niche dimensions appear to be broad relative to those of another species ("specialist"). Thus, by definition, many migrant species are habitat generalists. Unfortunately, this categorization does not advance our understanding of *why* they are habitat generalists. The implication is that they can exploit a wider variety of resources than residents and are therefore able to use a larger number of habitats (Herrera 1978; Tramer and Kemp 1980; Petit et al. 1995). Certainly, one can think of examples that seem to fit this idea. For instance, several tropical resident species (e.g., Spotted Antbird [*Hylophylax naevoides*]) seem to be obligate army ant (e.g., *Eciton burchelli* and *Labidus praedator*) swarm foragers, dependent on swarm activity to expose hidden prey. The niche of this species appears to be quite narrow in comparison with the somewhat similar, but migratory, Hooded Warbler (*Setophaga citrina*) (Willis 1966). This migrant is an occasional ant-swarm follower when wintering in Panamanian rainforest but can survive equally well in tropical forest and second-growth understory in the absence of ants or in temperate forests of the southeastern United States. Similarly, tropical frugivore and nectarivore species are poorly represented among migrants—at least those that travel to breed in temperate or boreal regions. But is this because these species have narrower niches than migrants or because their niches simply do not exist outside certain tropical habitats? If so, the explanation for migrant ability may depend on something in addition to, or other than, niche breadth.

• *Niche ubiquity* (Rappole 1995). This idea seems similar to that of the "habitat generalist" concept but differs in that it is not the fact that migrant niches are broader than those of residents, which allows them to exploit more environments, but that their niches are common to a greater variety of habitats.

For example, consider the Red-eyed Vireo (*Vireo olivaceus*), which breeds across much of North America and winters (and breeds) in South America in several forest and second-growth habitats. The Red-eyed Vireo forages by peering into cracks, crevices, and crannies on the outer branches, twigs, and leaves of broad-leaved trees, a niche that does not appear especially broad. Nevertheless, it is a niche that seems to be present in a large number of different habitats. Niche ubiquity rather than niche breadth seems especially pertinent for explaining why many aquatic and open-country migrants are able to exploit seasonal temperate and boreal habitats.

• *Resource superabundance.* A species' niche is what it does better than any other in a given ecological community (Hutchinson 1957; Emlen 1973:210). Thus, in a stable community, the niche space occupied by each member appears well defined. But communities, even tropical ones, are seldom stable. Resource concentrations occur at ant swarms, fruiting trees, blow-downs, or, on a longer timescale, over entire regions as a result of sharp seasonal changes. When such instability occurs, more resources may be available for a period than what the species especially adapted to exploit them can consume, changing an unsuitable habitat into a suitable one for invading species from outside the community (e.g., migrants). Something like this may explain why so many migrants that winter in African steppe, scrub, and savanna breed in Palearctic forest habitats (44 species); that is, there are few competitors during summer in Palearctic forests for the seasonally abundant arthropod resources that occur there.

• *Juvenile inexperience.* Juvenile migrants occur in disproportionately larger numbers in a broader array of habitats than adults (Murray 1966; Ralph 1978; Rappole et al. 1979; Chernetsov 2006; Ydenberg et al. 2007). This observation is not surprising considering that when juveniles first arrive on their wintering range, they may have to spend some time searching for suitable habitat, including the possibility that they may have to find a site that they have to defend from conspecifics (Rappole and Warner 1980; Rappole et al. 1989a; Rappole 1995:34–37).

Each of these hypotheses explaining migrant nonbreeding-season habitat breadth appears to have some validity for at least some migrant species under some circumstances. However, perhaps none applies to "migratory birds" as a class. With this caveat in mind, migrant winter habitat use can be summarized in general terms. One or more species of migratory birds occupy nearly every major habitat type on Earth's surface that is livable (i.e., with sufficient food, water, and shelter) during the wintering period. Nevertheless, there are certainly some habitats that have more migratory species occupying them than others. A breakdown of use for 15 major habitat types by number of migrant species that winter in the neotropics is shown in figure 5.9.

A somewhat cruder comparison of habitat use by migrant species for three of the world's major migration systems is provided in the appendices in Rappole (1995:183–197) and Rappole et al. (1983) (table 5.7).

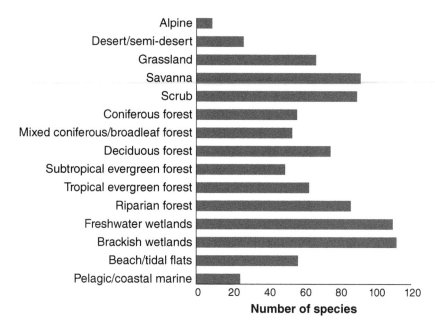

FIGURE 5.9   Wintering migrant habitat use in the neotropics (based on Rappole 1995:10).

FOOD RESOURCES

That distribution of food resources has a major effect on migrant nonbreeding distribution is perhaps best illustrated by the extraordinary itinerant migrations in sub-Saharan Africa described by Moreau (1952) and elucidated by Jones (Jones 1985, 1995; Jones et al. 1996) in which large numbers of migrant species spend 1 to 2 months in the Sahel region of sub-Saharan Africa using superabundant,

TABLE 5.7   Number of Migratory Species Inhabiting One of Three Major Habitat Types by Wintering Region

WINTERING REGION	AQUATIC[1]	FOREST[2]	OPEN[3]
Middle and South America	105	113	120
Africa	74	13	98
South Asia, Pacific Islands, and Australia	139	107	92

Source: Based on data from Rappole et al. (1983:7) and Rappole (1995).

[1]Lakes, ponds, rivers, and shorelines. Excludes pelagic species.

[2]Any habitat in which trees present a closed canopy including thorn forest.

[3]Includes grassland, cropland, desert, savanna, steppe, and scrub.

post–rainy season resources there until they disappear, and then heading south for the remainder of the wintering season. Similar kinds of itinerancy, involving both small-scale and large-scale intratropical movements, have been described for some Nearctic migrants as well. Morton (1980) describes understory-foraging Swainson's Thrushes (*Catharus ustulatus*) in Panama as abundant early in the wintering period (November) at the end of the wet season, but departing, evidently for South America, by mid-December, whereas the largely frugivorous (in winter) Eastern Kingbird (*Tyrannus tyrannus*) apparently travels initially to northern South America in fall and does not appear in Panama until the dry season has begun in January.

A summary of food use for 338 species of Nearctic migrants during their winter sojourn in the neotropics is shown in figure 5.10, illustrating that migrants exploit nearly every major food category available. Nevertheless, it is obvious that certain categories (e.g., arboreal invertebrates) are favored, at least by long-distance migrants, whereas other categories (e.g., nectar) are rare (Morse 1971). Generalizations concerning migrants have been made with regard to their winter food use habits compared with taxonomically similar residents including the following:

1. They are more opportunistic than residents (Leisler 1990).
2. They use a broader range of foraging behaviors than residents (Herrera 1978; Tramer and Kemp 1980).
3. They are less able to use mobile or hidden prey (Thiollay 1988).
4. They have morphology more suitable for long-distance flight (smaller size, attenuated wings) than for foraging (Leisler 1990).

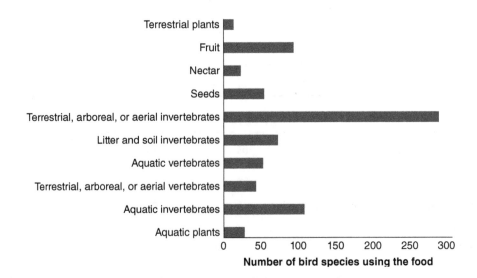

FIGURE 5.10   Major categories of foods used by Nearctic migrants wintering in the neotropics (based on Rappole 1995:29).

5. Their foraging structures (e.g., bill size and shape) are better suited for exploiting a variety of foods (generalist) rather than a particular type of food (specialist) (Cox 1968; Herrera 1978; Greenberg 1981; Leisler 1990).
6. They have less fear of feeding on new foods or approaching new situations ("neophobia") than residents (Greenberg 1990; Leisler 1990).

Each of these claims may have some basis in fact for some migrant species in some situations, but there are numerous exceptions to each generalization (Rappole 1995:49–74). In addition, it is important to remember that many migrants that winter in tropical communities and breed in temperate or boreal regions also have populations that are tropical residents: 23 percent of Palearctic–African migrants; 31 percent of Palearctic–Asian migrants; and 48 percent of Nearctic–neotropical migrants. Therefore, any generalizations concerning structural or behavioral differences between migrant and residents as classes must take into account that large percentages of migrants are very close to being exactly the same as a large number of residents.

As an illustration of this latter point, consider that whereas only 13 species of nectarivorus long-distance migrants breed in the Nearctic region, 145 additional nectarivore species undertake migrations within the tropics (Rappole and Schuchmann 2003). When intratropical migrants such as these are included in the comparison, it can be seen that many of the foraging and flight structure adaptations considered to be characteristic of migrants are adaptations specifically to a particular kind of migration—namely, long distance. Most long-distance migrants to the Holarctic are insectivores presumably because that is the superabundant food resource that becomes seasonally available in those regions. Essentially, arboreal insectivores are preadapted for exploiting the seasonal resources of the Holarctic, but neither insectivory nor any other particular food-use category is a prerequisite for migration.

COMPETITION

MacArthur (1972:21) defines competition as follows: "two species are competing if an increase in either one harms the other." Unfortunately, what constitutes "harm" to a species can be quite difficult to document. For instance, if species A consumes the same food as species B, is species A harming species B if there is no observable effect on populations of species B (or vice versa)? In practice, two kinds of data are used to infer competition: (1) observations (e.g., if members of two species are observed using the same resources at the same time) and (2) distribution (i.e., if when species B is present, species A is absent).

Despite operational difficulties in establishing when competition is present and important (i.e., in terms of population effects) and whether or not it is responsible for the absence of any given species, competition theory has dominated the way in which we understand migrant biology for the past half century, as discussed in the following sections on competition among migrants and residents and between members of congeneric species of migrants.

Migrant–Resident Competition The role played by competition in terms of shaping the winter range of migrant species has long been of major interest to researchers. In particular, how millions of migrants could make a semi-annual invasion of complex tropical communities filled with hundreds of highly specialized members was a key question shaping much of the research on migratory birds in the mid- to late twentieth century and remains an intriguing question. This conundrum plays out in full view to even the casual observer in sub-Saharan Africa each fall when millions of migrants of at least 183 terrestrial species invade the region.

Moreau (1952:264) first addressed the issue of migrant accommodation in tropical wintering habitats based on his observations in African tropical steppe and savanna habitats, concluding that migrants "fit in" in the following ways:

- As full members of tropical communities, exploiting niches not filled by the resident avifauna (e.g., migrant Yellow Wagtails [*Motacilla flava*], which capture insects disturbed by the passage of grazing ungulates in African grasslands)
- By wandering across the landscape and exploiting temporarily superabundant resources wherever they can find them (e.g., Barn Swallow [*Hirundo rustica*])
- By itinerancy, that is, moving from one region of seasonal resource superabundance to another (e.g., Marsh Warbler [*Acrocephalus palustris*])

In this catalog of migrant wintering strategies, Moreau makes an important distinction between temporary versus seasonal resource superabundance. He does not define these terms, but from the context he appears to mean that "temporary resource superabundance" applies to situations in which more resources occur at a site than can be exploited by African resident species over a period of hours or days, whereas "seasonal resource superabundance" refers to situations in which more resources occur across an entire region than can be exploited by African resident species over a period of weeks or months. Examples of temporary resource superabundance include fruiting trees, army-ant raids, locust swarms, termite emergence, and wildfires. Seasonal resource superabundance, affecting entire regions, occurs for certain kinds of resources at the end of the Sahel rainy season but also during summer in Palearctic temperate and boreal habitats.

Morel and Bourlière (1962) documented this phenomenon of migrant use of superabundant resources as well, finding that migrants arrived at the end of the rainy season in the Sahel region of sub-Saharan Africa at a time when insect resources were abundant. Members of these Palearctic migrant species exploited these resources in mobile groups, referred to as "floating populations," in contrast to many of the tropical resident species, which appeared to be sedentary. They concluded that whereas the sedentary residents were members of the African ecological community, the migrants were simply skimming the excess as temporary interlopers. Subsequently, many field studies have documented migrant exploitation of superabundant resources in both the Old World and New World tropics (e.g., Leck 1972; Karr 1976; Leisler 1990). Observations of this type of migrant resource-use can be summarized as follows:

1. Migrants, especially landbird migrants that are forest-related on their breeding grounds, tend to occur in a wider variety of habitats on their wintering grounds than on their breeding grounds (Petit et al. 1995).
2. Unlike resident members of these communities, many migrants appear to stay at sites for short periods of time (hours or days).
3. Migrants do not appear to be members, in terms of occupying a specific niche, in the communities that they invade during the nonbreeding period; rather, they seem to use temporarily superabundant resources.

As mentioned, these observations led the great theoretical ecologist Robert MacArthur (1972) to propose that migrants effectively were fugitive species whose strategy during the nonbreeding period, forced upon them by seasonal environmental deterioration on their breeding grounds, was to move southward colonizing warmer environments where they could subsist on temporary resource surpluses, such as insect blooms, fruiting trees, or grain fields. This concept had a number of predictions that could be tested for in the field, including the resident ecological counterpart, which would prevent migrants from occupying stable niches over long periods of time (weeks or months) in tropical communities, except under conditions in which resources were superabundant. More than three decades of research on wintering migrants have now shown that certain aspects of migrant wintering biology are not in accord with the fugitive species concept, at least as originally formulated:

1. Long-term banding studies have shown that many forest-related migrants live in a small, well-defined part of the tropical avian community throughout the wintering period and that they return to their wintering sites within these communities year after year at rates comparable to, or even exceeding, return rates to their breeding sites (Rappole and Warner 1980; Holmes and Sherry 1992).
2. Members of some migrant species of both sexes defend territories within these communities against conspecifics (Rappole and Warner 1980); others join mixed-species flocks, either as single representatives of their species or as groups of conspecifics (Rappole et al. 1983:35). When single members are present, they defend the flock space as if it were a territory, excluding conspecifics (Rappole et al. 1983:35; Powell 1985; King and Rappole 2001, 2002).

3. No evidence has been found of obvious tropical resident ecological counterparts for migrants that were functioning as long-term (throughout the winter period) members of tropical communities (Rappole and Warner 1980).

4. Contrary to superabundant resource dependency that had been proposed as an explanation for migrant persistence in tropical communities, many migrants were found to join these communities during periods when insect, nectar, and fruit resources often were at their lowest ebb for the year (Rappole 1995:29–30).

5. Investigation of the systematics of these migrant species shows that 23 percent of Palearctic migrants that winter in the African tropics have conspecific populations that are actually resident in the tropics along with 31 percent of Palearctic migrants that winter in the Asian tropics and 48 percent of Nearctic migrants (Rappole 1995:129–130).

These data document that the fugitive species hypothesis does not apply to wintering migrants as a group. Many species of migrants can, and do, serve as competitive members of their wintering communities. For these migrant species, there is no evidence that resident species directly or indirectly prevent them from exploiting stable resources over long periods in tropical communities. Nevertheless, many of the data on which the fugitive species hypothesis was based (e.g., migrant exploitation of superabundant resources) have validity for explaining migrant movements at certain sites and during certain periods of the year, requiring careful further consideration.

Community competition theory is based on the idea that each ecosystem possesses a certain volume of resources that can be divided only in so many ways (MacArthur 1957; MacArthur and Wilson 1967). This assumption may be true in a completely stable environment, but no such environment exists on Earth. Indeed, most terrestrial environments experience both short-term (e.g. hourly or daily) and long-term (seasonal, annual) variability in weather and climate. This variability in the physical environment produces resource variability, such that no population of organisms can match exploitation with production instantaneously ("irregularity principle" of Willis 1966:221). Indeed, prey species often appear to exploit this inability of predators to track resource variation by concentrating reproductive output into periodic bursts, making resource tracking just that much more difficult (Taylor 1984). Successful dispersal (i.e., movement away from a home area and occupation of a new area) by definition requires the presence of resources along the route and at the new area in excess of those that can be exploited by the organisms already living there.

Migration, as a strategy, does not depend on any particular kind of food resource or habitat type; rather, it depends on two kinds of resource superabundance: temporary and seasonal. Certain kinds of migration depend on certain kinds of resources, but migration as a movement strategy to maximize fitness can be used by any kind of bird, so long as seasonal resource superabundances exist

(Mayr and Meise 1930). In this scenario, the more stable community, usually the wintering area, is that from which the migrant ancestral population is derived—it is that place in which it can compete for the stable resources of the community (Williams 1958). The areas through which it moves and which serve as its destination must be areas with excess resources—temporary or seasonal. Thus, migrants avoid competition with residents in two ways:

1. Because the populations from which they are derived evolved in the communities in which they winter, these migrants are capable of competing with residents because, functionally, they are residents.
2. They are able to exploit temporary or seasonally superabundant resources that cannot be used or defended by resident species in the communities through which they pass along the migration route and which they join on their breeding grounds.

This second aspect of migrant behavior is precisely what MacArthur (1972) proposed, except that it provides a better description of their transient and breeding behavior than their wintering behavior, at least for many migrant species.

Congeneric Competition The role of congeneric competition in shaping migrant distribution during the wintering period is not well understood (Greenberg 1986). Nevertheless, it is obvious from congeneric winter range patterns that some effect exists (Lack 1944b; Salomonsen 1955; Mengel 1964; Rappole 1995). Cox (1968) hypothesized two scenarios for interspecific competition shaping summer and winter distribution of migrant species pairs (figure 5.11). Both scenarios begin with two species of partial migrants allopatric to each other throughout the annual cycle. In scenario I, species 2 colonizes and outcompetes species 1 on the resident portion of species 1's range, whereas species 1 colonizes and outcompetes species 2 in the migrant (i.e., breeding and winter) portions of species 1's' range. In scenario II, species 1 colonizes and outcompetes species 2 on the breeding portion of species 2's range; species 2 colonizes and outcompetes species 1 as a breeder in the resident portion of species 1's range; species 1 outcompetes species 2 as a winter resident in the resident portion of species 2's range; and species 2 colonizes and outcompetes species 1 on the winter resident portion of species 1's range. The proposed mechanism on which the competition is based is bill structure, which is suspect for a number of reasons (Rappole 1995:55–56), not the least of which is that the mechanisms affecting interspecific competition that result in allopatry may be quite different during the wintering period, when mates are irrelevant and food is critical, as opposed to the breeding period, when food resources may be superabundant but mates, territories, or nesting sites may be limited (see chapter 7). In addition, competition may have little to do with factors favoring allopatry, especially for closely related congeners. If the two separate populations are subspecies rather than species, a third complicating factor affecting

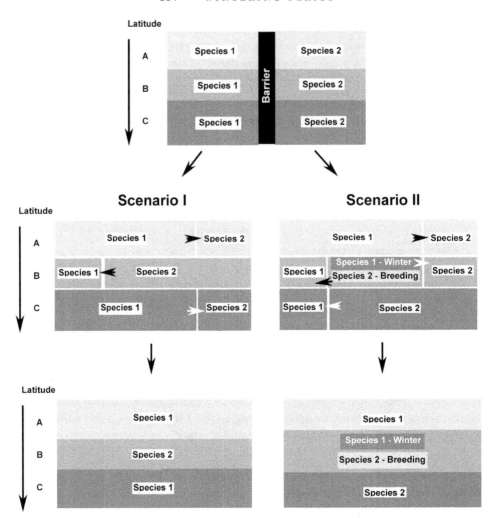

FIGURE 5.11   Evolutionary scenarios explaining breeding and winter distribution of two closely related species resulting from interspecific competition for food based on bill size, as explained in the text (based on Cox 1968): *light gray* = breeding-season distribution only; *medium gray* = breeding and winter distribution; *dark gray* = winter distribution only.

distribution of the two populations could include incompatible timing of major life history events (e.g., molt) or genetic programs for migration route or wintering area location (see chapters 3, 4, and 9).

PREDATORS

The effect of predators on migrant nonbreeding distribution rests mostly on inference: Predators are observed preying on migrants, which appears to affect their physiology and/or behavior; therefore, if the predation levels are high enough, it is assumed that such high levels must affect distribution as well (although see Ydenberg et al. 2007). There are some data that actually appear to document this connection (e.g., studies demonstrating a relationship between effects of hunting on location of major staging areas for waterfowl) (Madsen 1995). However, most documentation comes from studies of birds in transit where it has been demonstrated that predators can have a marked effect on migrant populations at stopover areas, reducing numbers by as much as 10 percent (Metcalfe and Furness 1984; Lindström 1989, 1990), and affecting fat reserves (Krapu and Johnson 1990; Ydenberg et al. 2002). If there are few data on the effects of predators on migrant behavior en route, there are even fewer for the wintering period. One of these found that nomadic Wood Thrushes (*Hylocichla mustelina*) wintering in second-growth scrub in southern Veracruz suffered higher predation rates (mostly avian predators, e.g., the Feruginous Pygmy Owl [*Glaucidium brasilianum*]) than sedentary, territorial birds in neighboring rainforest (Rappole et al. 1989a). Separating the effects of habitat, experience, and predators on distribution is probably impossible, but the results of this work demonstrate that predators have a potential role.

DISEASES

As in the case of predators, little is known regarding how diseases might shape winter ecology and range in migratory birds. Nevertheless, the findings of Beadell et al. (2006) demonstrating that avian malaria has a profound influence on the elevational distribution of a number of Hawaiian bird species is at least indicative of potential.

SEX

Whether or not an individual is male or female has a profound effect on many aspects of life history, including nonbreeding distribution for many migrant species (Gauthreaux 1978b, 1982). As discussed in the section "Social Structure During the Nonbreeding Period," adult male migrants of a large number of species spend the nonbreeding period on or near the breeding territory (partial migration) or at least closer on average to the breeding ground than adult females (differential migration) (Ketterson and Nolan 1983). Differences in factors affecting male reproductive success could account for some of the observed differences in migration distance (Fretwell 1972; Myers 1981).

Gauthreaux (1982:117) hypothesized that sexual differences in winter distribution resulted from dominance by one sex over the other, usually the male, in which members of the dominant sex prevent members of the subordinate sex from using preferred habitat. Perhaps the single most convincing study of the role that intersexual dominance during the wintering period might play in the life history of migrants comes from Studds and Marra (2005). They studied American Redstarts (*Setophaga ruticilla*) wintering in two different habitat types in Jamaica: mangrove and scrub. Mangrove has a higher density of potential prey items for redstarts (small flying insects) than scrub. It is possible to distinguish adult from juvenile males by plumage. Adult and juvenile females are identical in plumage, but juveniles can be identified by examining for incomplete ossification (pneumatization) at least until about mid-November by which time the skulls of many juveniles are completely ossified (Grant and Quay 1970). Adult males occur at disproportionately higher rates than females and juvenile males in mangrove, whereas the reverse is true in scrub. Individuals defend winter feeding territories in both habitats, which they occupy throughout the winter season, and often return to in subsequent years. Within-season survivorship is high and does not differ for individuals wintering in the two habitats, but between-season return rates to the mangrove habitat are higher (based on recapture or resighting of marked individuals). Redstarts wintering in mangrove habitat develop a different stable carbon isotope signature than that of those wintering in scrub, presumably based on differences in the stable carbon isotope signature of the vegetation that their prey items are eating. Between January 15 and February 20 in both 2002 and 2003, they permanently removed a total of 28 birds wintering in the mangrove habitat (15 adult males, 9 juvenile males, and 4 females). During the same period, 42 birds were captured in the scrub habitat (8 adult males, 2 juvenile males, and 32 females), bled (for carbon isotope analysis), weighed, measured, color-marked, and released. Studds and Marra (2005:2381) actually split the female samples into adults and juveniles, although they do not provide information on how this was done—members of these two different age groups are identical in plumage and cannot be differentiated reliably after mid-November by degree of skull ossification (i.e., an increasing percentage of juveniles look the same as adults). These 42 scrub-wintering birds are referred to as the "control" group. Subsequent to removal of redstarts from the mangrove habitat, at least 23 new birds moved in (2 adult males, 6 juvenile males, and 15 females) (again, the authors split the female sample into adults and juveniles although they do not explain how this would be possible). These birds are referred to as "upgraded." At least 14 of these upgraded birds (sex and age not given) evidently were captured within the January 15 to February 20 sampling period, bled, weighed, measured, color-marked, and released. These birds had stable carbon isotope signatures typical of open habitat. A month later (March 20 to April 15) they began attempts to recapture, bleed, weigh, and release marked birds in the two different habitats, obtaining samples from 11 birds recaptured in scrub habitat and 14 birds from mangrove habitat (sex and age not given). The birds recaptured from mangrove habitat, which previously showed stable carbon isotope signatures typical of scrub now showed signatures typical of mangrove. In addition, they maintained body mass over the two capture dates whereas those from scrub habitat lost an average of 8 percent of body mass. Intensive monitoring of birds in the two habitats showed that individuals wintering in the mangrove habitat departed earlier than those wintering in scrub.

On the basis of these data, the authors drew the following conclusion: "Findings here demonstrate that winter habitat occupancy can be an important determinant of individual performance in migratory birds. Restricted access to food-rich winter habitats may limit survival of females and immature males, an outcome that could be an important driver of population structure and dynamics," (Studds and Marra 2005:2380) or, to put it more plainly in the context of our current discussion, male winter-season dominance has a significant negative effect on female fitness. This conclusion rests on three pieces of evidence:

1. Redstarts in mangrove habitat (that came from scrub habitat) maintained "condition" (presumably body fat) better than those from scrub habitat.
2. The same birds departed earlier on migration (than scrub-wintering birds).
3. They returned at a higher rate the following winter.

We examine each of these results in line with other information.

• *Superior "condition."* Use of the word "condition" in reference to some morphologic or physiologic characteristic (e.g., body mass, fat reserves, hormone levels) assumes a relationship between the characteristic measured and fitness (e.g., a bird with high fat reserves is thought to have greater fitness than an individual with low fat reserves). For instance, Studds and Marra (2005:2383) state, "All upgraded and control redstarts [i.e., birds wintering in scrub] survived through the wintering period, suggesting that experimentally induced differences in physical condition did not become limiting until late spring." Lacking data on actual survivorship differences between mangrove-wintering versus scrub-wintering birds, they assume that "condition" affects departure timing from the wintering ground, which affects arrival timing on the breeding ground, which affects breeding territory quality, which affects reproductive fitness (and, perhaps, survivorship (Runge and Marra 2005). There are no data to support this train of assumptions.

• *Early wintering ground departure.* A fundamental basis for the conclusions reached by Studds and Marra is their argument that timing of departure from the wintering ground derives principally from an individual's physiologic condition (i.e., redstarts in "better condition" depart earlier than those in poorer condition). Assuming that "condition" = fat reserves = body mass, they found that birds that had moved into mangrove habitat were indeed in better condition and departed earlier than birds in scrub habitat. Certainly, ability to accumulate sufficient resources to build fat reserves during the premigratory period must have some effect on this timing, but there is no obvious measure to demonstrate this relationship at present that does not involve the confounding issues of sex and age. Indeed, there are powerful arguments for why adults should precede juveniles and why adult males should precede adult females that have nothing to do with physiologic condition (see chapter 2). One can see how condition might affect timing within a given sex or age grouping, but it is especially difficult to understand a claim that an entire sex (females) is disadvantaged in terms of reproductive fitness by later departure when the female plays such a huge role in determining what the male's fitness will be (through her choice of mate). Certainly, a great deal more evidence *directly* linking departure timing differences with fitness will have to be presented before these inferences can be evaluated properly.

• *Subsequent season return rates.* The most direct link between wintering habitat and fitness that Studds and Marra provide is a significant difference in subsequent wintering season return rates: 59 percent for mangrove-wintering birds versus 33 percent for scrub-wintering birds. There are at least two problems, however, in using these data to imply fitness differences. First, as mentioned by the authors, return rates are not equivalent to survivorship. We do not know what happened to birds that were not recaptured or resighted. Second, and perhaps more tellingly, other studies using larger sample sizes of redstart return rates of the same Jamaican population of redstarts found no significant differences in return rates for birds wintering in scrub versus mangrove habitats (e.g., Holmes and Sherry 1992).

There are many species of migrants in which, as in the American Redstart, male and female wintering distance from breeding habitat is similar, yet microgeographic and macrogeographic differences in distribution between the sexes are evident (Nisbet and Medway 1972; Rappole and Warner 1980; Rappole 1988; Lopez Ornat and Greenberg 1990; Wunderle 1992; Parrish and Sherry 1994; Marra and Holberton 1998; Marra et al. 1998; Sillet and Holmes 2002; Chandler and King 2011). The importance of the study by Studds and Marra (2005) is that they have attempted to demonstrate negative fitness consequences for females resulting from competition with males for quality winter habitat. We question both the factual and theoretical reasoning on which their conclusions are based. Because natural selection is a population-based phenomenon (Darwin 1859), one must question exactly how perpetuation of such a system of intersexual competition in which the fitness of one sex is evidently damaged relative to that of another can work in an evolutionary sense. We suggest that, in fact, it cannot without having corresponding effects on many other aspects of sexual morphology, ecology, and distribution. In examining this situation for those migrants in which it exists, we begin with the assumption that there is only one valid measure of evolutionary success and that is contribution of offspring to the next generation. If competition with males for winter food resources is important for females from a fitness perspective, then natural selection should favor evolution of females that are exactly like males, all other aspects being equal. The fact that most female migrants differ morphologically in terms of plumage, wing length, or other morphological factors (Rappole 1988) indicates that fitness for females is determined by more than intersexual competition for winter food resources (i.e., that there are other factors involved that balance out effects of competition). Presumably, the same is true for males.

AGE

As in the case of sex, age must have an important effect on migrant nonbreeding distribution. Such effects have already been discussed with regard to fall migration route (see chapter 4). Gauthreaux (1982) in fact considered sex and age as comparable factors affecting nonbreeding distribution, maintaining that adult males secure the best parts of the winter range or habitats, pushing females and juvenile males into poorer (in a survival sense) parts of the winter range or wintering habitats. Sex and age, however, are not equivalent in terms of the role that natural selection can play or with regard to evolutionary expectations for adaptive response. As discussed earlier, balancing selection can modify the appearance and structure of males and females to produce fitness optima in both male and female plumages, foraging structures, and habitat selection (Slatkin 1984). However, selection cannot alter two fundamental competitive liabilities for young birds: (1) lack of experience and (2) the shorter period for which selection has acted on them relative to adults so that on average, young birds represent less well-adapted individuals than older birds. Thus, young birds should be expected to suffer in competition with adults and to avoid such competition whenever and wherever possible unless resources are superabundant. We suggest that avoidance should result in different microgeographic or macrogeographic wintering distribution, on average, between adults and juveniles of many migrant species, as well as in different wintering strategies (e.g., perhaps including greater dependence on temporary resource superabundance and mobility in juveniles compared with adults). There is considerable evidence that such differences exist, at least among partial and differential migrants (Ketterson and Nolan 1983), but detailed comparison for most long-distance migrant species is lacking for at least two reasons: lack of investigation and difficulty in distinguishing adults from juveniles in many species.

EVOLUTIONARY HISTORY

Many species of migrants have broad breeding distributions and relatively restricted winter distributions (del Hoyo et al. 1992–2012; Poole 2010). The Barred Warbler (*Sylvia nisoria*), for instance, breeds across 7,000 km of longitude and 2,000 km of latitude in Eurasia whereas it winters across 800 km of longitude and latitude in tropical Africa (figure 5.12). We have proposed that most migrant populations initiate a migratory habit as dispersing individuals by moving from a natal area of greater stability to a new breeding area of seasonal resource superabundance. As applied to the Barred Warbler, this hypothesis presumes that it is the African wintering area where the species evolved and from which migratory populations to Eurasia were derived. If this hypothesis is correct, then obviously evolutionary history plays a major role, if not the most important role, in migrant winter distribution.

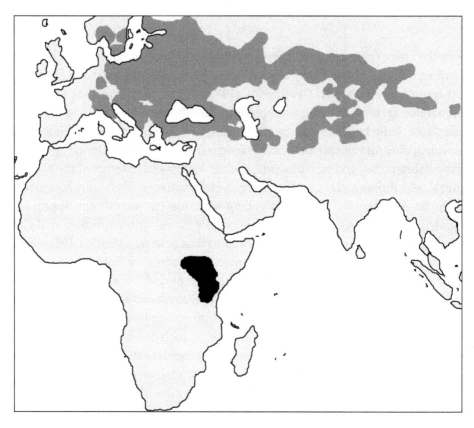

FIGURE 5.12   Barred Warbler (*Sylvia nisoria*) breeding (*medium gray*) and wintering (*black*) distribution.

CHAPTER 6

SPRING TRANSIENT PERIOD

T HE SPRING transient period begins with departure from the wintering area and ends with arrival at the breeding area. As in the case of fall migration, the ultimate cause for the behavioral and physiologic changes associated with departure for the breeding area seem clear (i.e., to place the individual in the optimal environment for survival and/or reproduction) (Mayr and Meise 1930; Williams 1958; Rappole and Jones 2002; Rappole et al. 2003). The context, however, is different from fall migration in that, at least for most adult migrants, the emphasis is on the "reproduction" aspect of the fitness equation rather than the "survival" aspect. As stated by Newton (2008:352), "Birds leave their wintering areas so as to reach their nesting areas in time to breed at the most favorable season." This shift in emphasis from survival to reproduction has far-reaching effects on migrant adaptation for movement from nonbreeding to breeding areas.

PREPARATION FOR DEPARTURE

The principal evidence of preparation for spring departure presumably is hyperphagia accompanied by weight gain beginning a few days prior to actual migratory flight, at least based on experimental studies with caged birds (Gwinner 1972; King 1972). However, little work has been done with free-flying birds to confirm these preparations. Timing is probably based on an endogenous program, influenced to a greater or lesser extent by local conditions depending on whether or not the species is a short-distance or long-distance migrant, respectively. Internal, physiologic controls over a migratory bird's responses in preparation for departure are discussed in chapter 4. As in the case of fall departure, the characteristic behaviors and physiologic changes associated with premigratory hyperphagia begin suddenly and, at least in those small passerines that have been studied, last for 6 to 12 days prior to reaching a plateau in terms of percentage of body mass when the bird is seemingly ready to depart (King and Farner 1959; King 1972).

CUES FOR DEPARTURE

CUES FOR THE FIRST MIGRANTS

The "timing of departure" problem is somewhat different in spring from that confronted in fall. Consider these departure issues from the perspective of the first generation of migrants. They know when to leave in fall because either resources are becoming scarce or intraspecific competition for food resources (or breeding territories for the next year) is intense (see chapter 4). But how does this first generation of migrants know whether or when to move toward the breeding ground in spring? Why not simply remain on the wintering ground and breed there like their resident relatives? We propose that two interacting factors trigger this initial return migration: (1) endogenous timers governing initiation of preparation for reproduction, presumably common to all members of the ancestral population, migratory and resident alike, and (2) intraspecific competition for breeding territories and mates. In other words, intraspecific competition for breeding sites could serve as the driving force favoring the second and subsequent spring migrations, just as it did for the first. This scenario presumes that timing of seasons favorable for reproduction in both the wintering area (home to the ancestral population of migrants) and breeding area (for the migrant population alone) must be at least somewhat compatible. If this theory provides a feasible scenario for how spring departure on migration begins, then some predictions concerning the structure of departure timing can be made:

• The fewer the number of generations separating current migrants from first migrants, the greater the likelihood of overlapping resident and migratory portions of a population on the wintering ground.
• The fewer the number of generations separating current migrants from first migrants, the greater the similarity in terms of timing of major life history events (e.g., breeding and molt) between resident and migratory portions of populations

(i.e., natural selection acting to "fine tune" timing is likely to result in considerable differences between migrant and resident populations).

• The fewer the number of generations separating current migrants from first migrants, the greater the overlap in timing of departure for the breeding grounds between the different age and sex groups of the migratory population (i.e., migration tends to place adult males, adult females, and juveniles in very different selection regimes from those encountered by their resident counterparts).

• The greater the number of generations separating current migrants from first migrants, the greater the degree of endogenous control over departure (i.e., selection should favor a departure timing system based on breeding-ground suitability rather than wintering ground competition). A corollary of this prediction is that whereas distance separating breeding and wintering area is likely to increase selection pressure to produce a more precise endogenous timer, some degree of endogenous control will likely be favored regardless of distance because proximal environmental cues on the wintering ground are unlikely to provide more exact information regarding the seasonal suitability of the breeding area than a genetically programmed, endogenous timer (in combination with information gathered en route during actual approach to the breeding area).

The last prediction assumes that there are both costs and benefits involved with early departure. Costs include probability of encountering poor food resources or inclement weather on the migration route or at the breeding area; benefits include the probability that early arrival on the breeding area may result in obtaining the highest-quality territory (males) or mate (females), allow for production of multiple broods, or at least additional time for successful re-nesting after predation. For a migrant optimally to balance costs and benefits, we suggest that whereas departure should be principally under endogenous control (commensurate with availability of resources for fattening), speed of return should be subject to modification according to the specific conditions encountered along the route (van Noordwijk 2003).

ENDOGENOUS FACTORS AND DISTAL CUES GOVERNING TIMING OF SPRING DEPARTURE

Inferential evidence of endogenous control over timing of spring departure is extensive based on observation of departure times for equatorial migrants (Gwinner 1972; Piersma et al. 1990). Experimental investigation of the interplay between endogenous program, environmental calibrator (usually photoperiod), and proximal environmental cues governing timing of departure from the wintering ground have not been as well studied as the fall departure from the breeding ground. Nevertheless, documentation of endogenous timing of spring departure has been obtained for a few species of both short-distance and long-distance migrants (Merkel 1963, 1966; Zimmerman 1966; Gwinner 1968; King 1968; King 1972:212).

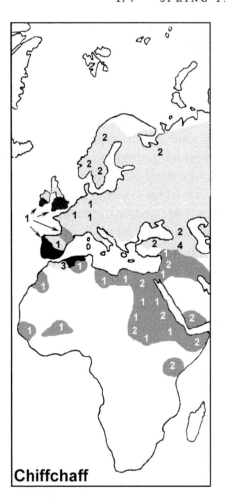

FIGURE 6.1   Breeding (*light gray*) and winter (*dark gray*) distribution for the Chiffchaff (*Phylloscopus collybita*) and Willow Warbler (*Phylloscopus trochilus*) in western Eurasia and Africa. Numbers represent the different subspecies for each species. Note that breeding areas for some subspecies occur in Eurasia east of the map boundaries (based on Gwinner 1972).

An example is the seminal work by Gwinner (1972), who tested endogenous control over timing of departure from the wintering area for two different breeding populations of a long-distance migrant (Willow Warbler [*Phylloscopus trochilus*]) and a short-distance migrant (Chiffchaff [*Phylloscopus collybita*]) (figure 6.1). He measured onset of *Zugunruhe*, weight changes, and molt in hand-reared, juvenile birds kept in special cages under uniform (12-hour light, 12-hour dark) light conditions and found considerable evidence of endogenous control in the long-distance migrant populations with less evidence in the short-distance migrant. He concluded: "The Willow Warbler with its tight schedule doubtless profits from a rigorous endogenous control of its annual cycle. The Chiffchaff, on the other hand, may derive advantage from a higher dependence on external factors, enabling higher adaptability to changing environmental conditions" (Gwinner 1972:233). The data, however, did not necessarily demonstrate differences in relative amounts of endogenous control over timing of spring departure between the two species, but rather differences in the role of photoperiod in recalibrating the circannual rhythm: Willow Warblers maintained a nearly annual rhythm (±1 or 2 months) in timing of major life history events even after being maintained for years on a uniform photoperiod, whereas Chiffchaff timing precision deteriorated after 13 months.

Further documentation of the endogenous nature of spring departure timing for short-distance migrants is provided by work on captive (but kept outside) White-crowned Sparrows (*Zonotrichia leucophrys gambelii*) wintering in Pullman, Washington (King and Farner 1959, 1965; King 1972). Standard error in a year-to-year comparison of date of onset of vernal, premigratory fattening among individuals and years (over an 8-year period) was ±1.0 day (King 1972:207), despite considerable variation in local weather conditions (King and Farner 1959). Several other experimental studies lend support to the hypothesis of endogenous control over timing of spring movement, although work has been done on a relatively small number of species (see reviews in King 1972; Newton 2008:352–359).

Timing of Departure Within the Spring Transient Period

Ultimate Factors Affecting Spring Departure Date

North American migrant species fall into five major categories in terms of their timing of departure:

> Very early (before February 15)
> Early (February 15 to March 15)
> Median (March 15 to April 15)
> Late (April 15 to May 15)
> Very late (after May 15)

Assuming that timing of departure is based on genetic programming, whether triggered endogenously, by environmental cues, or some combination thereof, several evolutionary reasons have been suggested for precisely when migrants depart from their wintering areas. In the following we discuss the principal ones based on broad generalizations derived from observed spring departure timing for selected migrants.

• *Breeding latitude.* Breeding area latitude is well known to have an influence on presumably endogenous spring departure programs in both short-distance and long-distance migrants (Blanchard 1941; Curry-Lindahl 1958, 1963; Fry et al. 1972;

Gwinner 1972; Piersma et al. 1990; Wood 1992). For example, several populations of the Yellow Wagtail (*Motacilla flava*) winter in the same region of central Africa, but southern-breeding populations depart on spring migration before northern populations (Curry-Lindahl 1963). Similar patterns are seen in some waders (Piersma et al. 1990) and the White-crowned Sparrow (Blanchard 1941).

• *Winter latitude.* Short-distance or partial migrants that winter in temperate latitudes tend to be earlier migrants on the whole than similar species that winter in the tropics (Tryjanowski et al. 2005), although this factor often is found to co-vary with other factors (e.g., diet or breeding latitude), so that it is difficult to assign cause.

• *Diet.* Migrants that feed on arboreal insects tend to be later migrants than those that feed on seeds or aquatic plants and invertebrates (Tryjanowski 2005). As an example, peak passage dates for waterfowl along the subtropical Texas Central Coast (28°N latitude) occur in February and March, whereas peak passage dates for insectivorous passerines occur in April and May (Rappole and Blacklock 1983).

• *Molt.* Dugger (1997) found that timing of the early-winter prebasic molt affected spring departure time for female Mallards, with early-molting birds departing earlier than later-molting birds wintering at the same latitude. He found no relationship between age and timing of molt or departure. However, he did not measure another potential factor affecting endogenous timing of both molt and spring departure: breeding latitude. Mallards wintering in Arkansas, where the study was done, could derive from a wide range of breeding populations (Bellrose 1976:229–243).

• *Sex and age.* Sex and age have large effects on timing of departure in many migrants. The usual pattern in those species in which timing differences occur is for adult males to depart earliest from the wintering area followed after a delay of some days or even weeks by adult females (Marra and Holberton 1998), although the pattern is reversed in some polyandrous species (Oring and Lank 1982). Immature or subadult birds (2, 3, or even 4 or more years after hatching depending on species) may delay departure, delay migration along the route, or remain on the wintering grounds throughout the next and even subsequent breeding season until adulthood is reached (Gauthreaux 1978b; McNeil et al. 1994; Pierotti and Good 1994; Holmes and Pitelka 1998; Poole et al. 2002).

The principal argument advanced to explain early departure is that it is the prerogative of the most fit individuals (in a physiologic sense) because early-arriving individuals on the breeding ground (usually adult males) generally appear to be in the best physiologic condition; that is, departure, travel speed, and arrival differences are best explained as resulting from a straightforward competition for food resources (Gauthreaux 1978b, 1982; Moore et al. 2005). This argument has two major weaknesses: It assumes that (1) early departure has no costs and (2) the fitness value of early departure is equivalent for different age and sex groups. We have presented alternative hypotheses that take these considerations into account in chapter 2, at least from the perspective of intersexual differences in breeding-ground arrival date (and wintering-ground departure date). Here, we suggest that intergenerational differences in departure time also have more complex origins than a simple "competition for food resources" argument.

Consider, for instance, that whereas a competition-based hypothesis could explain differences in physiologic condition between different age and sex groups occupying the same wintering ground site, it cannot explain carryover of such differences to the breeding area. Once adults have left the common wintering area, food resources should not be a problem for most subadult birds because their putative conspecific competitors have departed, and, often, resources in the wintering region are actually increasing and resident birds are beginning or in the middle of breeding (Skutch 1950, 1954, 1960, 1972; del Hoyo et al. 1992–2011; Urban et al. 1997; Fry et al. 2000, 2004; Rasmussen and Anderton 2005b). Indeed, subadults of several migrant species do remain at the wintering area throughout the coming breeding season. Nor are food resources likely to be a problem on the migration route given that (1) the adults will not be there as competitors if subadults delay departure and (2) resources along the route are likely to be increasing as well. In addition, weather is likely to be better later in the season so that subadult travel is likely to be less hazardous than that of earlier-migrating adults (Buskirk 1980).

Thus, competition for food resources can only provide a partial explanation for delay in subadult movement and no explanation at all for the reported poor physiologic condition on arrival on the breeding ground. Nor can this food limitation hypothesis explain why subadults of many migrant species remain on the wintering ground or along the migration route rather than completing travel to the breeding areas, as is seen in a number of migrants (McNeil et al. 1994). We suggest that it is competition with adults for reproductive resources that affects subadult wintering-ground departure timing and other spring migration movement strategies. We propose that the key determinant regarding whether, or for how long, to delay wintering-ground departure for subadult migrants is probability of successful reproduction. In short-lived birds like many passerines in which adult survivorship is less than 60 percent per year (Roberts 1971; Rappole and Warner 1980; Holmes and Sherry 1992), making the trip north to the breeding ground is probably worth the risks, as many adults will not be returning and the subadult has a relatively low chance of survival to the next breeding season. However, for a long-lived migrant like the Herring Gull in which annual adult survivorship is greater than 90 percent (Pierotti and Good 1994), travel to the breeding area by subadults likely represents a wasted journey if breeding-site resources (e.g., mates or nesting areas) are limiting.

Resolution of questions regarding the role played by competition in timing of departure for American Redstarts (*Setophaga ruticilla*) (Studds and Marra 2005) or any other long-distance migrant probably cannot be accomplished through field experiments alone. The problem is that whereas it is assumed that competition is what is being tested, in reality it is impossible to control for other major factors that could affect departure timing (see chapters 2 and 5). Perhaps experiments of the type designed by Mewaldt, King, and Gwinner might help to elucidate this question (e.g., Mewaldt et al. 1968; Gwinner 1972; King 1972). For instance, if redstarts,

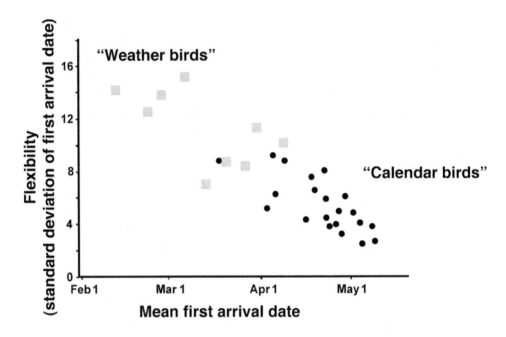

FIGURE 6.2   Mean and variation in first arrival date for selected species of facultative migrants ("weather birds"; *gray squares*) versus "calendar birds" (*black circles*) in Poland, 1983–2003 (based on Tryjanowski et al. 2005).

or any other long-distance migrant, were held in cages and fed *ad libitum* during the transitional period from winter to spring migration, we might learn whether or not there were programmed differences in migration timing by observing the physiologic and behavioral changes (e.g., *Zugunruhe*) that took place in the different age and sex groups in the absence of competition.

• *Facultative versus calendar migrants.* Facultative migrants whose movements appear to occur solely in response to local weather conditions tend to begin spring migration earlier than those whose timing is mainly under endogenous control (calendar migrants) (figure 6.2) (Tryjanowski et al. 2005). However, this observation does not mean that facultative migrants do not experience endogenous control over their departure timing (Berthold 1999). We suggest that, in fact, endogenous cues control the individual facultative migrant's responsiveness to weather cues. This hypothesis may explain why Mallards wintering in the same region undergo winter (prealternate molt) and spring migration at different times although exposed to the exact same weather systems (Dugger 1997).

• *Balancing of fitness trade-offs.* This factor is assumed rather than measured but is probably true for all departure times for which some degree of endogenous control is involved, which we suggest is probably all migrants. For each migratory population, the timing of major life history events will be different from all others based on evolutionary history and ecology in addition to the factors listed earlier. Thus, optimal balance in departure timing will be unique, not only for members of a given population but also for the different age and sex groupings within that population.

PROXIMATE FACTORS AFFECTING SPRING DEPARTURE DATE

• *Photoperiod.* Experimental evidence for a few species of short-distance migrants wintering in temperate areas has confirmed that spring daylength plays a role in timing of departure (King 1972).

• *Physiologic status.* Timing of initiation of premigratory fattening is under tight endogenous control among those migrant species studied (King 1972; Gwinner 1972). However, if an individual is unable to attain proper fat levels within the normal period (6 to 12 days for passerines tested), then presumably they must delay departure until they are able to achieve the correct levels or go elsewhere in an attempt to improve food resource availability (Rappole and Warner 1976).

• *Local food availability in the wintering area.* Departure time can fluctuate year-to-year for the same individual, even for long-distance migrants whose initiation of preparation for departure (*Zugdisposition*) presumably is under endogenous control. For instance, Berthold et al. (2002) reported variation in departure date of nearly a month (March 22 in 1997; February 26 in 1998) for an adult male White Stork wintering in central Africa, and movements within the wintering area were quite extensive and not the same from one year to the next. Whether local environmental cues or the bird's personal physiologic state mediate these adjustments is not known.

• *Weather.* Proximal cues for actual initiation of fall migration in temperate regions appear to be mainly related to weather (see chapter 4). For timing of initiation of spring migration, the same is certainly true for many facultative migrants and probably true for at least some species of calendar migrants that winter in temperate regions (Tryjanowski 2005). However, for species that winter in tropical regions, local weather may not serve well as an important cue for initiation of migratory flight, except perhaps in a negative way (i.e., if local weather is especially bad, it may cause an individual to delay initiation). Thus, the bird's own physiologic state (in the sense of fat reserves) may serve as a critical cue for actual initiation of migratory flight (Cochran and Wikelski 2005), perhaps in addition to auditory cues indicative of migratory flight in progress for conspecifics (Hamilton 1962, 1966). For instance, Piersma et al. (1990:123) suggested that timing of departure on spring migration for several thousand waders of 13 species wintering in coastal Mauritania was largely under internal rather than local environmental control.

TIMING OF DEPARTURE WITHIN THE CIRCADIAN PERIOD

All migrants fall into one of three categories in terms of when during the 24-hour cycle they actually begin migratory flight: (1) daytime departure, (2) nighttime departure, or (3) either daytime or nighttime departure (Newton 2008:85). As far as is known, species that belong to a particular category in fall occur in the same category in spring.

Migratory Flight (Zugstimmung)

The exaptations and adaptations in terms of anatomy, morphology, physiology, and behavior exhibited by migrants for the purposes of migratory flight are discussed in detail in chapter 4.

Spring and fall migration presumably do not differ in terms of the basic requirements for migratory flight, at least in terms of anatomy and morphology. However, physiologic demands may be different during spring migration. For instance, migrating individuals appear to be able to adjust energy storage between fat and protein to meet balance needs for water along the route (Lindström et al. 2000; Klaassen and Biebach 2000). These requirements may differ between spring and fall. As an example, trans-Saharan migrants in fall confront lower mean temperatures than in spring and may use different proportions of protein (which contains water) and fat (which does not) during the different migratory periods. Obviously, such balancing of energy storage forms could only result from natural selection, although whether such adjustments occur as genetic changes or relatively rapid adjustments within reaction norms is not known.

Migration Route

Conclusions regarding the structure of routes, pathways, or corridors for migrants between their breeding and wintering areas in the past have been based largely on the assumption that most follow the most direct route possible between breeding and wintering area except when confronted with topographical features assumed to function as barriers (e.g., mountain ranges or oceans) and follow the same route on both journeys (Lincoln 1950; Fisher 1979). Actual documentation of routes has been based on accumulated observations analyzed on the assumption of a bell curve distribution pattern (i.e., the largest number of observations for a given species are assumed to occur along the main route at any given point in time). For some species, especially short-distance, intracontinental migrants (e.g., waterfowl), documentation has verified assumptions. But for the majority of migrants, data have been insufficient for accurate determination and comparison of fall and spring route.

Lack of knowledge or equivocal data on fall and spring routes have generated intense debate in some cases. For instance, it has long been known that several shorebirds that breed in the North American High Arctic follow an elliptical or loop migration route between breeding and wintering area: south over the western

North Atlantic Ocean to southern South America in fall and a much more westerly path northward in spring (Cooke 1915:12). Nisbet et al. (1963) suggested a similar route for the Blackpoll Warbler (*Setophaga striata*) based on observed flight direction, records at sea, and estimated physiologic capabilities. However, Murray (1965, 1976) opined that a more parsimonious explanation for Blackpoll occurrences at sea was wind drift and that the most likely southward route for the species was along the eastern coast of North America and across the Caribbean to South America.

The data presented by Nisbet et al. (1963), along with recent findings regarding the long-distance flight capabilities of passerine migrants (e.g., Stutchbury et al. 2009; Bächler et al. 2010), suggest that the western North Atlantic route is the likely fall migration path followed by adult Blackpolls and by at least some portion of other populations of passerine migrants that breed in northeastern North America and winter in eastern or southern South America. Not only do these birds have the physiologic capability for the 90-hour flight, but also the route represents the quickest and safest for fall passage to eastern South America, as long as they depart with sufficient fat reserves and a tailwind. As in the case of the shorebirds, the spring migration north likely follows a far more westerly route because prevailing winds do not favor a western North Atlantic path. Nevertheless, the western North Atlantic route for passerines, like most other migration routes, remains a hypothesis. Presumably, geolocators will be able to resolve this issue.

Similar questions have arisen concerning the routes followed by long-distance migrants that breed in eastern North America and winter in Middle America or western South America. Cook (1915) described these species as passing southward in fall over Florida and the Gulf of Mexico to their tropical wintering grounds and returning northward in spring by the same pathway. Williams (1945) presented data on migration from the western coast of the Gulf of Mexico that were orders of magnitude greater in spring than fall for many species, indicative of a westward bend in the spring route (circum-Gulf). Lowery did not accept Williams's data, and a spirited debate developed in the literature (Lowery 1945, 1946, 1951, 1955; Williams 1947, 1950, 1951), subsequently carried on by some of Lowery's students, and seemingly resolved in favor of a spring trans-Gulf route more or less identical to that followed south in fall (Stevenson 1957; Gauthreaux 1971). Rappole et al. (1979) reopened the debate, citing additional seasonal distribution data from the Texas Gulf coast implicating a westward shift in spring for many trans-Gulf migrants (figure 6.3).

No one can question the huge volume of songbird migration that crosses the Gulf of Mexico during spring migration, a fact that has been documented by several different kinds of observational data (e.g., Able 1972; Gauthreaux and Belser 1999; Barrow et al. 2000a, 2000b). Thus, the question is not whether or not spring trans-Gulf migration exists, but rather its volume and trans-Gulf track *relative to fall migration* for any given species. Rappole and Ramos (1994) presented several

FIGURE 6.3   Hypothetical major fall and spring migration routes (*arrows*) across the Gulf of Mexico region of North America for the Chestnut-sided Warbler (*Setophaga pensylvanica*) (based on Rappole et al. 1979): *light gray* = breeding range; *dark gray* = winter range.

reasons for why a westward shift in migration route might occur in spring compared with the migration route in fall:

1. Frontal systems usually move in a northwest to southeast direction in the region.
2. North winds generally follow the passage of fronts, providing favorable tailwinds for trans-Gulf flight in fall (figure 6.4B).
3. In addition, after passage of fronts in fall, the probability of confronting turbulence over the Gulf during migratory passage is low, at least for the early part of the season (August to October).
4. In spring, birds do not follow fronts on northward passage from the neotropics. The fronts still come down from the north and northwest, but a bird departing on a trans-Gulf flight from the north coast of Yucatán or Honduras cannot anticipate whether or not it will encounter a front, with contrary headwinds, while over the Gulf.

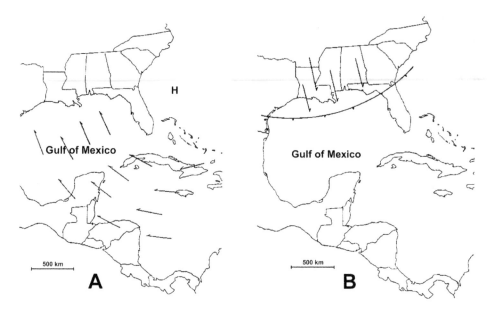

FIGURE 6.4   Direction of prevailing winds (*A*) and storm-front winds (*B*) over the Gulf of Mexico (based on Rappole and Ramos 1994): *H* (part A) = normal location of the dominant weather system of the region known as the "Bermuda High."

5. Prevailing winds over the Gulf in the absence of fronts are from the southeast (figure 6.4A).
6. Migrants traveling north in spring take advantage of southeasterly tailwinds to follow a route with a more westerly swing than is followed in fall that also reduces the possibility of exposure to strong headwinds over the Gulf. However, the later in spring that a bird attempts movement across the Gulf of Mexico, the less likely it is to confront unfavorable winds, and the safer a more direct route across the Gulf becomes (Buskirk 1980).

A few recent geolocator data for two trans-Gulf migrants seem to lend some support to this hypothesis, and more such data are likely to provide a final resolution. In the study by Stutchbury et al. (2009), two Wood Thrushes (*Hylocichla mustelina*) and two Purple Martins (*Progne subis*) were tracked using geolocators from their breeding area in northwestern Pennsylvania to their wintering areas in the neotropics and back. The spring migration routes across the Gulf region ranged from 600 to 1,300 km west of their fall routes (although it is important to remember that geolocator data have a standard error of ±200 km).

Given the differences in prevailing winds and weather as well as likely seasonal differences in location of suitable refueling sites, we propose that the greater the distance separating the breeding and wintering range and the earlier in spring their departure, the less likely that members of a population will follow the same route north as they followed south; and the more generations for which a population has followed a migratory habit, the more distinct the differences between south-bound and northbound routes are likely to be. Studies from other parts of the world document fall and spring differences in migration route for several species (e.g., Bächler et al. 2010).

The major issue for such a hypothesis is to what degree are the fall and spring differences in route that have been documented programmed versus simply dictated by obstacles, fat storage capabilities, and prevailing wind direction? We do not know the answer to this question for any species, let alone for migrants as a class. Routes such as those followed by several North American and European migrants summarized by Newton (2008:654) seem likely to have a genetic component, as they involve dramatic departures from the most direct route beginning from the moment the bird initiates either fall or spring movement. However, it could be argued that prevailing wind direction over a water obstacle is the principal route-shaping force in these cases.

Gauthreaux et al. (2006) argue that winds aloft have little effect on migration routes, at least for passerines crossing the Gulf of Mexico on spring migration. They used data from 10 weather surveillance radars, located at fairly regular geographic intervals around the northern Gulf coast from Brownsville, Texas, on the west to Key West, Florida, in the east. Using these radars, they measured volumes, altitudes, and landing longitude for flocks of millions of migrants representing 100 or more species arriving along the northern rim of the Gulf after completing presumptive trans-Gulf flights between March 15 and May 15 during the 2001–2004 spring migration seasons. They compared these landing data with information on direction of winds aloft over the Gulf at three different altitudes (500 m, 1,000 m, and 2,500 m) during the time periods when the birds were presumed to be crossing the Gulf. They found the following:

1. Peak landing longitude for the entire region over the time of the study was located at roughly 29°N, 94°W.
2. Longitude for peak landing volumes of migrants during a given time period did not vary much from year to year.
3. Peak longitude did not appear to be affected significantly by the direction of winds aloft over the Gulf.

They concluded that spring trans-Gulf migration routes are genetically fixed (i.e., migrants are following a genetic program for route across the Gulf that is affected very little by actual wind conditions encountered during the crossing). They further argue that these genetic programs likely date from the peak of the last glacial cycle (15,000 ybp) when physiographic conditions in the region were quite different. However, there is no need to hypothesize the existence of a genetic program for route or, indeed, any genetic program at all to explain these results. All the birds heading north in spring are going back to a place where they have already been—the site where they bred or were raised the previous year.

Banding data demonstrate that birds have the capability of homing to such sites over vast distances, even when displaced (see chapter 2). Thus, landing site along the Gulf is as likely to be determined by where along the Gulf the birds departed from (i.e., not all migrants departed the previous day from the southern shore of the Gulf; many come from the western shore) and where the birds' ultimate destinations are relative to that departure point (i.e., a breeding site in the eastern United States or Canada) as it is to be determined by a genetically programmed route. In addition, direction of arrival relative to winds aloft is relatively meaningless because winds aloft usually present at least 60 degrees in variation to choose from depending on the altitude chosen (Gauthreaux et al. 2006), and it is well known that birds can adjust flying altitude to fit the direction in which they wish to fly (see chapter 4).

The basic assumption concerning routes according to a hypothesis of route flexibility is that, all else being equal, a migrant will follow the same route on its outbound journey from breeding to wintering area as it follows on its return, as appears to be the case for at least some White Storks. The second assumption of the "route flexibility " hypothesis is that if routes have genetic components, they can be shaped by a very wide range of selection factors in addition to those already mentioned (wind direction, weather, shortest distance)—for example, food availability at stopover areas, predation, excessive heat (over deserts), excessive cold (over mountains), and competitors. Future work with radio tracking, geolocators, and perhaps other devices and techniques are likely to provide the data necessary to test these assumptions. At present, there are at least some routes that seem best explained by this "route flexibility" hypothesis—for example, the Rufous Hummingbird (*Selasphorus rufus*) (figure 6.5), which appears to follow different fall and spring routes based on food availability (Healy and Calder 2006).

Orientation and Navigation

This topic was discussed in chapter 5 with regard to fall migration, much of which applies here. There is, however, one major difference in which navigation of migrants in spring differs from that in fall: the role played by experience. In fall, half or more of the birds involved in movement between breeding and wintering areas have never been to the wintering area before, whereas all of the birds involved in movement between wintering and breeding areas have been to the breeding area. Nevertheless, the paradigm for returning second-year songbird migrants is that they do not home to natal area (Weatherhead and Forbes 1994). We have challenged this view (see chapter 2), maintaining that this conclusion is based on probable sampling error, suggesting that most do home to natal area but remain for such a short period that capture or observation is unlikely. In any event, whether or not second-year birds actually do home to natal area is probably immaterial from a navigation perspective; even the small number of captures recorded

FIGURE 6.5   Fall and spring migration routes (*arrows*) of the Rufous Hummingbird (*Selas-phorus rufus*) in western North America between breeding area (*light gray*) and wintering area (*dark gray*) (based on Healy and Calder 2006).

for those species sampled (1 to 2% usually) (Weatherhead and Forbes 1994) demonstrates that they have the capacity.

STOPOVER

The term "stopover" refers to any period of time spent on Earth's surface (land or water) during a bird's migratory journey. In our discussion of the phenomenon, we will address three main aspects: (1) the ultimate factors governing stopover occurrence, (2) the physiologic and behavioral states associated with stopover, and (3) the proximate factors governing stopover initiation and duration.

ULTIMATE FACTORS

Our fundamental assumption governing stopover as an aspect of migration is that if refueling is necessary for a bird to move successfully from wintering to breeding area, then sufficient stopover sites with superabundant food resources in appropriate habitat must occur along both the fall and spring routes or migration will not occur or persist. We further suggest that many aspects of stopover timing, physiology, and behavior may represent exaptations for migration, in the sense that these same factors could be requisite for successful dispersal. Nevertheless, as information presented in the following will indicate, certain aspects pertaining to particular within-season or between-season timing of stopover must represent adaptations evolved subsequent to development of a migratory habit.

PHYSIOLOGIC AND BEHAVIORAL STATES

Any given individual observed during stopover must be in at least one of two, and perhaps four, physiologic/behavioral states that differ from those of birds that are not in transit. Groebbels identified two of these: *Zugdisposition*, in which the bird prepares for a migratory flight by eating intensively and laying down fat reserves; and *Zugstimmung*, in which the bird actually undertakes migratory flight. Both of these states have been well documented by field and laboratory studies (Bairlein and Gwinner 1994; Lindström 2005), although the endogenous controls governing shifts between states are not known (Holberton and Dufty 2005). Jenni and Jenni-Eiermann (1992) have suggested a third physiologic state during stopover, a "fasting" state when plasma levels of key metabolites differ from those of birds in either *Zugstimmung* or *Zugdisposition*. On the basis of observations of behavior of migrants in-transit at stopover sites, we suggest that there may be a fourth physiologic/behavioral state associated with migration: a "transit" state (table 6.1 and figure 6.6).

Recent work with the physiologic/behavioral states associated with migration further emphasizes several key differences between *Zugstimmung* (migratory flight) and *Zugunruhe* (migratory restlessness of caged birds). For instance, investigations have now been done on the behavior and physiology of *Zugstimmung*, both with free-flying birds and captive birds in wind tunnels, providing information on wingbeat rate, heart rate, energy consumption, orientation, and responses to various environmental stimuli (Cochran and Wikelski 2005). These data demonstrate that *Zugunruhe* is not comparable to *Zugstimmung* in most aspects of behavior or physiology, suggesting that perhaps we have learned about as much as we could from this experimental methodology. Further work using it may risk providing data that are distorted in ways we don't understand.

TABLE 6.1   Physiologic and Behavioral States of Migrants in Transit in the Normal Sequence in Which They Are Hypothesized to Occur Once Migration Has Begun

STATE	BEHAVIORAL CHARACTERISTICS	PHYSIOLOGIC CHARACTERISTICS
"Flying" state (*Zugstimmung*)	Bird is actually in migratory flight.	Gut is completely emptied of food and waste. Bird metabolizes fat and/or protein stores to fuel flight. Heart rate is high relative to normal foraging or resting (Cochran and Wikelski 2005).
"Transit" state	Birds coming to the end of a migratory flight deviate from the migratory azimuth in evident search for appropriate habitat (Moore et al. 1995:133; Barrow et al. 2000a; Cochran and Wikelski 2005). Birds that normally forage alone may occur in large, loose flocks (Rappole 1995). Habitat may or may not be appropriate for the species (U.S. National Park Service 2011). Flocks may move in the direction of normal migration or a different direction, flying short distances from perch to perch or making actual large-scale shifts of several hundred meters in moving from one site or habitat to another (Bingman 1980; Rappole 1995). Such large groups seldom remain at a given site for more than a few minutes and usually depart the area entirely to continue migration when night comes (for nocturnal migrants) unless grounded by weather (Rappole and Warner 1976). Birds in these flocks may actively forage and consume food as they are moving.	It is unknown whether or not birds that forage while in this state (1) meet their current energy needs, (2) add to their energy stores, or (3) depend largely on already-stored reserves. However, recaptures of birds remaining only 12 hours prior to continuation of migration demonstrate that such birds do not show the 5–10 percent daily gains in mass found in birds that are in *Zugdisposition* (Rappole and Warner 1976).

STATE	BEHAVIORAL CHARACTERISTICS	PHYSIOLOGIC CHARACTERISTICS
"Feeding" state (*Zugdisposition*)	Birds forage and eat intensively (Cochran and Wikelski 2005), increasing food intake by as much as 40 percent above normal (maintenance) daily foraging. Individuals that normally forage alone (as opposed to birds that forage in flocks) restrict movement to a relatively small area (Rappole and Warner 1976; Cochran and Wikelski 2005), may be aggressive toward conspecifics, and defend short-term feeding territories (Rappole and Warner 1976). Habitat choice is characteristic of the foraging habits of the species (Rappole and Warner 1976). Individuals may show short-term (days) site fidelity to foraging sites even if displaced short distances (1 km) (Rappole and Warner 1976).	Rapid deposition of energy reserves in the form of subcutaneous fat and/or protein produces mass gains of up to 50 percent in 6–12 days (King 1972; Rappole and Warner 1976). Heart rate can be as high as is observed during migratory flight (Cochran and Wikelski 2005).
"Resting" state ("fasting" of Jenni and Jenni-Eiermann 1992)	In addition to sleeping during hours of darkness while at stopover sites, it has been noted that birds that have been in apparent *Zugdisposition*, alternating between feeding and resting bouts of 10–20 minutes for hours or days, stop moving and roost for several hours on the afternoon prior to nocturnal continuation of the migratory journey (Cochran and Wikelski 2005).	Heart rate drops considerably relative to foraging or flying states (Cochran and Wikelski 2005); key metabolites differ from those of birds in "feeding" or "flying" states (Jenni and Jenni-Eiermann 1992).

PROXIMATE FACTORS

• *Time of day.* As discussed in chapter 4, migratory flight occurs during specific portions of the 24-hour daily cycle, and most birds are either diurnal or nocturnal travelers. For nocturnal migrants, stopover begins at some point during the night, often shortly after midnight, but no later than daybreak unless the birds find themselves over entirely inappropriate habitat (e.g., water or desert) (Kerlinger and Moore 1989). For diurnal migrants, stopover usually begins at some point during the day (Dolnik and Blyumental 1967; King 1972) or at nightfall and lasts at least until daybreak of the following day. Persistence at a site beyond the normal time period for daily stops for either nocturnal or diurnal migrants depends on a number of factors as discussed in the following.

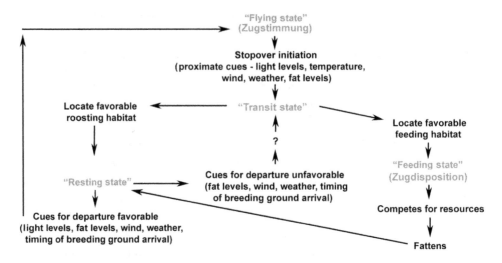

FIGURE 6.6   Hypothetical alteration of physiologic/behavioral states in response to proximal and endogenous cues for spring migration of a nocturnal migrant.

• *Habitat.* Transient migrants in stopover often occur in habitats that appear completely inappropriate for their species (U.S. National Park Service 2011). Nevertheless, extensive studies by a number of authors document various levels of apparent discrimination in habitat occupancy by transients (Parnell 1969; Power 1971; Moore et al. 1995:133; Rappole 1995:90; Chernetsov 2006). These conflicting results raise questions concerning what migrants actually require in terms of stopover habitat. Reference to table 6.1 is helpful in this regard because it shows how the same bird in different physiologic states might have completely different habitat requirements.

For instance, consider a prototypical, forest-related songbird—for example, the Blackburnian Warbler (*Setophaga fusca*) migrating northward in spring along the western coast of the Gulf of Mexico. As daylight comes, it is over water, so it turns and heads west toward the nearest landfall, a barrier island called Mustang Island, which is covered mostly with grassland habitat. It lands briefly in grassland (shifting from *Zugstimmung* ["flying" state] to a "transit" state) and then flies westward looking for woodland habitat, the first encountered of which would be oak savanna on the inland shore of Copano Bay. If fat reserves are sufficient, it continues in a transit state, moving northward along the coast, more or less in foraging mode, hopping from branch to branch with occasional short flights to the next grove of oaks, often as a member of large flocks of other migrants consisting of both conspecifics and other species. If fat reserves are below some critical level, however, it searches farther inland for more suitable habitat for building of fat reserves, perhaps riparian forest or mesquite thorn forest. If it finds appropriate habitat, it switches from a "transit" state to a "feeding" state (*Zugdisposition*) and ceases its gregarious, northward travel and begins intensive, solitary foraging within a relatively restricted piece of ground, perhaps even chasing away other conspecifics.

Depending on the status of its fat reserves, it may remain at this site for several days, gaining mass at 5 to 10 percent per day. Each night during this stay, the bird will switch from a "feeding" state to a "resting" or "roosting" state. Habitat required for roosting may be the same as that for foraging, and probably is for a Blackburnian Warbler, although the microhabitat is probably different. For other migrants (e.g., Northern Pintail [*Anas acuta*]), habitat use during the "feeding" state (grain fields) is completely different from habitat used in a "resting" state (open water) and separated by 4 to 5 km (Bergman and Rappole, unpublished data). On the afternoon of the day when our Blackburnian Warbler has rebuilt fat reserves sufficient for continuation of migration, it will seek out appropriate habitat for a "resting state," that is, one that provides shelter from inclement weather or predators. As darkness approaches, the bird switches to a "flying" state, joining loose flocks of other Blackburnian Warblers as they continue their migratory journey.

The behaviors described in this scenario are typical for forest-related songbirds in spring migration along the western Gulf coast. Although the association intimated with the various physiologic states is hypothetical, all of the habitat occupancies described have been documented for many migrant species, including the Blackburnian Warbler, by many observers (Packard 1951; Hagar and Packard 1952; Rappole 1978; Rappole and Blacklock 1983; Blacklock 1984; U.S. National Park Service 2011). The point of this exercise is to illustrate how observations of transients can show that they can appear to be selective with regard to habitat, as in when they are in *Zugdisposition*, or nonselective, as in when they are in a "transit" state or just coming out of a "flying" state. In addition, they can appear to be selective for different habitats or microhabitats at different times, depending on whether they are in a "feeding" state as opposed to a "resting" state. Without knowing what physiologic/behavioral state the bird is in, a simple observation of a transient in a given habitat may have little meaning in terms of trying to understand habitat preference or needs.

• *Food use.* As noted, our basic assumption for migration is that stopover areas containing superabundant food resources (i.e., those that cannot be harvested completely by residents) must be available for the migrant along the route during the period of passage. This assumption does not mean that the foods used are necessarily the same as those used primarily on the breeding or wintering areas, and, in fact, often they are not. As an example, many migrants that appear to depend largely on arboreal arthropods on the breeding ground (e.g., Eastern Kingbird [*Tyrannus tyrannus*]) may use fruits extensively at stopover or wintering areas (Morton 1971). On the coast of Louisiana, 44 percent of spring migrant species ($n = 61$) stopping over consume fruit. This high percentage is, in part, the result of migrants feeding on persistent fruits remaining from the winter crop. About 70 percent of all spring migrants were observed feeding at least once on flowers (nectar), and 25 to 70 percent of total foraging observations included nectar for some species (e.g., Ruby-throated Hummingbird [*Archilochus colubris*], Tennessee Warbler [*Oreothlypis peregrina*], Baltimore Oriole [*Icterus galbula*], and Orchard Oriole [*Icterus spurius*]) (Barrow et al. 2000b).

Researchers have questioned the effect of migrants on their resource base at stopover areas, and studies of birds that occur at high concentrations at relatively restricted refueling sites (e.g., Red Knot [*Calidris canutus*]) have shown that their impact can be significant, causing declines greater than 70 percent in favored prey (Piersma et al. 2005b).

Observational data on feeding behavior of free-flying migrants suffer from the same kinds of problems as those related to determination of transient habitat use; that is, a bird in a "flying," "transit," or "resting" state may have quite different food requirements from that of a bird in a "feeding" state. Despite this difficulty, several studies have documented a role for intraspecific competition in some species of birds in apparent *Zugdisposition* at stopover sites (Rappole 1995:35–37). These observations raise the question of just exactly what birds are competing for if resources are superabundant at a stopover site that is used for refueling? We suggest that they are competing for resource access within a particular time period (i.e., a specific refueling rate that will allow return to the breeding area according to an optimal timetable for an individual within a given sex or age group) (Alerstam 1991). As we have noted elsewhere, any potential migrant probably can completely avoid intraspecific competition simply by remaining on the wintering ground through the coming breeding season or by delaying migration until later in the season. The cost for this behavior is decreased probability of successful breeding, which the individual must balance against survivorship probabilities.

• *Refueling.* Several studies, both in the laboratory and field, have shown that fat storage levels can affect timing of departure from stopover sites (King 1972). Cochran and Wikelski (2005), for instance, have found that *Catharus* thrushes remained at stopover areas until fat scores exceeded a certain characteristic level. However, Alerstam (1991) has demonstrated that the relationship between fat storage levels and stopover duration can be complex. He found that birds often depart stopover sites at fat storage rates well below capacity, requiring more stopovers than would be necessary if they stayed longer at any given stopover site. He hypothesized that variation in departure energy-storage levels results from balancing migration speed (increased stopovers require more migration time; low-quality stopover sites require more time to achieve appropriate storage levels) against costs (increased fat storage mass decreases efficiency in that more energy is devoted to carrying the extra weight relative to actually powering flight). Thus, given a particular time schedule for movement, it may be more advantageous for a bird to leave a lower-quality site with lower fat levels earlier rather than delaying flight until higher levels are reached.

• *Weather.* Weather is often involved in stopover initiation decisions where rain and wind have been observed to cause migrants literally to "drop from the sky" (see chapter 4). Its role in departure decisions in tropical latitudes is less well understood (Piersma et al. 1990). However, it plays an increasingly important role with increasing latitude. Indeed, for most short-distance, Temperate Zone birds that migrate early in the season, weather appears to be the most important proximate factor (Belrose 1976; Tryjanowski et al. 2005). Even for putative calendar migrants, weather plays a large role in stopover decisions (Richardson 1990). Cochran and Wikelski (2005:274) found that *Catharus* thrushes would not depart from spring stopover in central Illinois until maximum daily air temperature exceeded 21°C and wind at takeoff was less than 10 km/h. In addition, if birds in migratory flight confronted cold fronts (temperature drop ≥2°C), they stopped migrating and began stopover, regardless of the time of night.

• *Predation.* There are few data on proximate effects of predation on stopover decisions, but those that are available indicate that predation can serve as a factor affecting stopover location and duration (Rudebeck 1950; Walter 1979; Kerlinger 1989; Lindström 1989; Alerstam and Lindström 1990; Moore et al. 1990). On the basis of what data are available, it seems likely that predation can play a role in both habitat selection for the various physiologic/behavioral states experienced during stopover (e.g., Bergman and Rappole, unpublished data) and allocation of searching, foraging, resting, and roosting time and behavior. Optimality theory provides various predictions regarding balancing of foraging movement based on various risk/reward strategies (Bednekoff and Houston 1994; Houston 1998), but given our crude level of knowledge regarding factors potentially affecting stopover decisions, modeling may be problematic. Even if we restrict model conclusions as pertaining solely to birds that are actively foraging, attempting to use them to understand real-world behavior could potentially be misleading. For instance, as described earlier (table 6.1), birds in a "transit" state and birds in a "feeding" state both forage, but birds in a "feeding" state forage at rates up to 40 percent greater than birds in a "transit" state. It seems certain that predation plays a part in shaping foraging behavior in these two different states, yet its role relative to habitat type, time of the season, sex, age, and energy needs is unknown. Models will have to incorporate more of these factors before they provide significant illumination. The fact that transients often occur in mixed-species foraging flocks (Chen et al. 2011) may be indicative of the role predation may play in shaping transient behavior.

• *Within-season and between-season differences in timing of movement relative to position along the route.* Spring migration is more rapid than fall migration for many migrant species, and rate of daily migration movement increases as time grows later in the spring migration season and the bird gets closer to its destination, at least on a population basis and especially for some calendar migrants (Cooke 1915:43–47; Lincoln 1952; Dorst 1962; King 1972:211; Piersma et al. 2005b). No such change in movement rate is known for fall migration. This rate change could be simply a reflection of the rate at which spring advances in temperate regions (i.e., greater prey availability at stopover sites allows more rapid refueling resulting in more rapid migration as the season progresses) (Alerstam 1991). However, this hypothesis does not explain early departure from the wintering ground or differences in movement rates by sex and age. The fact that late-departing birds could arrive at breeding sites at the same time as early-departing ones simply by taking advantage of greater prey availability at stopover sites indicates that something more is involved than prey availability. We suggest that differences in rate result from differing selection pressures. Adult males must arrive early enough to secure good breeding sites, but not so early that they will be unable to survive. They leave early from the wintering ground, making a gradual approach to breeding areas commensurate with favorable temperatures for prey availability at stopover sites. The closer they get to breeding areas, the more likely that ambient temperatures and related prey availability reflect that of the breeding site, favoring faster movement. Thus, movement rate changes result from balancing selection for reproduction and survival. This may also show why birds can change movement rates quickly in response to rapid climate change; that is, movement rate is actually governed by temperature.

CLIMATE CHANGE

A central thesis of our work is that migration is a response to the availability of new environments whose seasonal occupation for breeding purposes can result in higher fitness than is possible for resident populations of the same species. In general, we tend to think of the environmental changes promoting this response as occurring on geologic timescales, covering hundreds or thousands of years (Mengel 1964) (although see Turney and Brown [2007] on the potentially rapid climate changes associated with sudden draining of glacial Lake Agassiz, roughly 8,500 ybp). Extraordinarily rapid environmental change has occurred over the past few decades, most evident in sharp increase in worldwide mean annual temperatures (global warming) (Parmesan 2006; Gordo 2007; Norwine and John 2007). On the basis of our hypothesis, one would predict that rapid changes in seasonal environments would produce rapid response from migrant populations, and this has, in fact, been the case. Two major kinds of changes have occurred in migratory birds: (1) changes in migration timing and (2) changes in breeding range (Boucher and Diamond 2001; Matthews et al. 2004; Robinson et al. 2005; Gordo 2007; Rappole et al. 2011c).

MIGRATION TIMING

Earlier spring passage and arrival dates have been documented for many species of migratory birds in a number of studies for which advance in mean spring passage (msp) timing generally appears correlated with higher mean temperatures at temperate stopover or breeding occurring earlier in the year (Przybylo et al. 2000; Cotton 2003; Hüppop and Hüppop 2003; Gordo 2007). Rapid change in msp or arrival for short-distance migrants associated with rapid temperature change is not surprising. Although the basic timing of spring migration for short-distance migrants likely is under a broad degree of endogenous control, even for facultative migrants, specific departure timing is known to be triggered by local weather conditions. Such rapid change in timing for temperate msp in long-distance migrants from the tropics, however, would be surprising because experimental evidence has demonstrated that departure from tropical wintering grounds is under tight, endogenous control (Gwinner 1972; Berthold 2001; Pulido 2007). Various studies confirm this dichotomy in response (Sokolov et al. 1998; Hüppop and Hüppop 2003), and several also demonstrate that timing of arrival for some long-distance migrants has not tracked earlier spring arrival and accompanying peaks in resources accurately, resulting in timing of breeding in these species different from what would be optimal in terms of available resources (Visser et al. 1998; Both and Visser 2001; Coppack and Both 2002; Tryjanowski et al. 2002). A long-term study of transients on the island of Helgoland in the North Sea is instructive regarding this dichotomy. The msp times for migrants in stopover have been recorded on the island since 1909. Hüppop and Hüppop (2003) found that msp had advanced by 2 to 12 days over the past 40 years for 24 species, 12 short-distance migrants and 12 long-distance migrants, at this site. Earlier arrival timing for short-distance migrants appeared correlated with increase in mean local temperatures, whereas earlier arrival by long-distance migrants showed better correlation with timing of the North Atlantic Oscillation (NAO), a large-scale climatic phenomenon governing favorability of spring weather conditions in Europe. Other investigations have reported changes in spring passage and arrival times for tropical-wintering, long-distance migrants as well (Cotton 2003).

We see at least five possible explanations for earlier msp at temperate stopover sites or earlier arrival on breeding areas by long-distance migrants that winter in the tropics.

• *Tropical departure timing is under tight endogenous control and has not changed, but migrants are able to move faster because earlier occurrence of warmer temperatures associated with NAO provides greater prey density at stopover, allowing more rapid refueling* (van Noordwijk 2003). Cotton (2003) found no effects of NAO on arrival date for migrants in Oxfordshire in the United Kingdom, although this result does not rule out the possibility of changes in potential refueling rates associated with increased prey availability at stopover sites associated with long-term global warming.

• *Tropical departure timing is not under tight endogenous control, and has changed in response to local African weather changes occurring in response to the NAO* (van Noordwijk 2003). Cotton (2003) found no evidence of a relationship between NAO and African weather.

• *Tropical departure timing is under tight endogenous control, but natural selection has altered the timing of the genetic response* (van Noordwijk 2003). This prediction is based on laboratory experiments that show both the genetic nature of migration timing and rapid change in timing resulting from artificial selection (Pulido et al. 2001; Pulido and Berthold 2003; Pulido 2007). On the basis of these findings, one would predict that any change in migration timing in long-distance migrants must reflect genetic change. Pulido and Berthold (1998) have argued that the "migration timing gene" appears to control a suite of characters. If a change were to occur to this hypothetical gene in response to natural selection (i.e., differential fitness), some members of the population would express the new form of the gene, whereas others would possess the old form. Unlike in laboratory experiments, it is highly unlikely that fitness for individuals possessing the old form would be zero; presumably, it would be some percentage of difference from the new form, and this percentage likely would be very different for different species of long-distance migrants, resulting in very different rates of change. This scenario is not what has been recorded. Rather, whole groups of long-distance migrants representing a wide range of taxa and ecological groupings ranging from shorebirds to songbirds have shown similar changes in migration timing (Cotton 2003; Hüppop and Hüppop 2003) providing strong evidence that the response is phenotypic (Cotton 2003). Nevertheless, the concept of reaction norms may apply in some way not as yet understood.

• *Tropical departure timing is not under tight endogenous control: migrants depart earlier because of advance in seasonal change on tropical wintering grounds.* Cotton (2003) found strong evidence that climatic change in Africa was correlated with the earlier arrival of long-distance migrants. However, he presented no actual data on mean changes in African departure date, and his arrival data are also correlated with average temperature increases throughout the entire range for migrants: breeding, wintering, and stopover (Gordo 2007). Therefore, attributing advanced arrival to advanced seasonal change in sub-Saharan Africa is equivocal at best.

• *Timing of departure from tropical wintering quarters is under tight, endogenous control, but length of migration period is subject to modification according to specific weather conditions encountered along the route.* This hypothesis is similar to the first hypothesis, except that there is no implied relationship between NAO (which is largely a European phenomenon in any case) and rate of movement. According to this hypothesis, genetically based timing of the migration period could have sufficient flexibility to allow for the individual to respond to very specific aspects of the environment. The fact that shorebirds and many other species of long-distance migrants are able to increase their rate of movement as they approach their breeding area is at least suggestive that such flexibility exists (Cooke 1915:43–47; Lincoln 1952; Dorst 1962; King 1972:211; Piersma et al. 2005b).

Of course, each of these explanations could apply to a greater or lesser degree to different species of long-distance migrants in different migration systems. As van Noordwijk (2003:29) said, "I eagerly await further results on changes in spring migration."

RANGE EXPANSION

Thomas and Lennon (1999) were among the first to document breeding range expansion by migratory birds in apparent response to climate change. Since then, many studies have documented such shifts (Robinson et al. 2005). Migratory bird range changes have been modeled based on climate change data, and predictions are that range changes will accompany shifts in ecological communities (Matthews et al. 2004; Jetz et al. 2007; Sekercioglu et al. 2008). However, intensive studies of communities in which rapid shifts in breeding range are occurring have not revealed this type of change. For instance, northward range shifts have been documented in several migrants that formerly reached the northern end of their breeding range in the Texas subtropics (Lockwood and Freeman 2004; Rappole, Blacklock, et al. 2007, 2011c). However, no corresponding shrinkage has been recorded in the range of Temperate Zone species. In terms of birds, the subtropical and Temperate Zone communities now overlap with species from both communities present during the breeding period (Rappole et al. 2011c). Similarly, migratory birds that breed in western New York State in the Carolinean Zone (<200 m) have expanded their breeding range in this region upslope into the Transitional Zone (200 to 600 m) and even the Boreal Zone (>600 m) based on data from the New York State breeding bird atlas surveys taken 20 years apart (Andrle and Carroll 1988; McGowan and Corwin 2008). These atlases record no shrinkage in the Transitional Zone or Boreal Zone bird communities in response to expansion of the Carolinean Zone avian community, contrary to predictions. We see two possible explanations. The first is that it will take time for the new community to displace the old one. The second is that no simple displacement will occur. Rather, a new community will develop that is different from both of the old communities. We predict the second outcome is the more likely, and it is likely to have a profound effect on migratory bird populations worldwide over time as these new communities evolve. The first indication of such profound changes is likely to occur in the genetics of closely related species, whose breeding ranges were previously separated by latitude or altitude. We see three possible outcomes resulting from newly overlapping populations of such superspecies groups:

1. The southern or lower-elevation species will displace the northern or higher-elevation species.
2. The two species will live together in a new composite community derived from a combination of the old communities.
3. The southern or lower-elevation species will genetically swamp the northern or higher-elevation species (Rappole et al. 2011c).

It is possible that each of these scenarios could occur in different species pairs. However, we believe that the third hypothesis is the most likely, for which the current status of the Blue-winged Warbler (*Vermivora cyanoptera*) and the Golden-winged Warbler (*Vermivora chrysoptera*) may represent a paradigm.

CHAPTER 7

POPULATION ECOLOGY

T HE HISTORICAL, and perhaps still most prevalent, understanding of migratory bird population ecology assumes that density-dependent competition for limited resources occurs mainly or solely during the breeding period and serves principally to control *breeding population size*, or the number of individuals that actually participate in production of offspring. *Total population size*—all the individuals that compose a population whether or not they reproduce—is assumed to be controlled largely by density-independent factors (e.g., predation, disease, or accidents acting mostly during the nonbreeding period) (von Haartman 1971; Sherry and Holmes 1995). Even in situations in which competition during the nonbreeding period is known to occur, it is assumed that the main, density-dependent effects are exerted during the subsequent breeding period (Runge and Marra 2005).

We agree that breeding period competition is important, although its relationship to a complete understanding of migratory bird ecology is complex. For instance, as we have noted in other chapters, there is extensive evidence of the potential for density-dependent population limitation during parts of the nonbreeding period:

• *Postbreeding period.* If one considers the postbreeding period to be essentially separate and distinct from the breeding period (see chapter 3), then, like all other portions of the migrant annual cycle, it should have its own set of population parameters; that is, the postbreeding habitat should be examined from the perspective of having a birth rate (immigration in this case), a death rate, and critical habitats possessing their own species-specific carrying capacity. This portion of the life cycle, in general, is poorly known, but what is known indicates that resources typically are not limiting for most migrants; that is, there is little evidence of intraspecific competition, in which case one must assume that most migrant species are below carrying capacity in terms of critical habitat during this period. It should be noted that this conclusion is not the same as concluding that postbreeding-season habitat might not be limiting for certain migrants under certain circumstances. For instance, clearly many waterfowl and shorebirds (Ankney 1984; Jehl 1990), and perhaps some members of other groups of migrants (Butler et al. 2002), require very specific habitat types for molting during the postbreeding period, which, if limited, could limit population size.

• *Migration season.* As for the breeding and postbreeding periods, one must assume that resources are likely to have been superabundant along the migration route or migration would not have developed in the first place. Thus, in most cases, the series of different habitats occupied during the course of the journey are unlikely to exert control over population size. Nevertheless, there are examples to indicate that transient habitat can be limiting for some species. Waterbirds are especially vulnerable to loss of wetland habitat along the migration route (van der Graaf et al. 2006), a principle clearly understood since at least the early 1900s in the United States, when broad-based efforts were put in place to establish systems of refuges throughout the country to provide critical stopover habitat for migratory waterfowl (Bellrose 1976). Shorebirds, too, have been recognized as having very particular requirements in terms of stopover habitats, reduction in which can threaten entire populations (Harrington 1996; Piersma et al. 2005b). Another indication of the importance of stopover areas is observation of site fidelity to such sites and territorial defense (e.g., Recher and Recher 1969; Rappole and Warner 1976; Kodric-Brown and Brown 1977; for a summary, see Rappole 1995:35–37). Clearly, the transient period must be examined with care for each migrant species to determine those in which stopover habitat is potentially limiting. Presence of transient territoriality is certainly indicative of the potential importance of stopover habitat but does not necessarily indicate that the habitat is limiting. There are at least two factors that might trigger defense of resources by an individual: (1) limited resources or (2) limited time. It could be that time spent on migration is an important factor and that migrants might compete to shorten that time to reach breeding or winter sites at the proper moment. In other words, in-transit territories might be related to competition for quality breeding or wintering sites.

• *Wintering period.* Extensive evidence of the potentially critical nature of wintering season habitat for migrants has been provided in chapter 5. Many species show site fidelity to wintering sites, and many demonstrate intraspecific territoriality during this period. In addition, there are indications that winter habitat may be limiting for some migrants based on inferences regarding relative amounts of breeding versus wintering habitat and population trends (Rappole and McDonald 1994, 1998; Goss-Custard et al. 1995a, 1995b; Newton 1998:68–69; Rappole et al. 2003b).

Theoretical and field evidence of intraspecific competition during the non-breeding period in many migrant species (e.g., Fretwell 1972; Williamson 1972; Goss-Custard et al. 1995a, 1995b; Sutherland 1996, 1998) forces a reconsideration of the view that the sole density-dependent control over migrant population ecology occurs during the breeding season. However, existing models for density-dependent population limitation require extensive modification to accommodate the possibility for multiple carrying capacities in multiple critical, geographically separate habitats occupied sequentially over the course of the annual cycle. We present such a model.

THEORY

The basic equation for population growth states that change in the population (dN) over time (dt) equals the birth or natality rate (bN) minus the death or mortality rate (dN) (Emlen 1973:234):

$$dN/dt = bN - dN \tag{7.1}$$

Birth rate and death rate from equation 7.1 can be combined into a single term, r, as shown in equation 7.2:

$$dN/dt = rN \tag{7.2}$$

Theoretically, if for any given population there is little apparent interaction between the birth rate (b), death rate (d), and population size (N) variables, the population continues to grow regardless of the number of individuals in the population until the resources on which it depends are entirely exhausted, at which point all individuals in the population die. Bacteria on a food medium in a Petri dish approximate this situation; their population grows until the food medium is exhausted, and the colony dies. This type of population growth is referred to as "density independent." Although no population is truly independent of its size, as population size is what birth and death rates act upon, in density-independent populations there is little interindividual effect on members based on their numbers alone. All individuals continue to have equal access to critical resources, regardless of their numbers, until the resources are gone. These kinds of density-independent populations are controlled mostly by stochastic factors affecting the birth rate and death rate—that is, the r term in equation 7.2, which is also referred to as the "intrinsic rate of natural increase" or "intrinsic growth rate" (Emlen 1973:235). Species whose populations are controlled principally by *r* are known in population biology terminology as "*r*-selected." This term means that the focus of their reproductive life history is on production of as many offspring as possible in the shortest amount of time (Emlen 1973:327–328).

For many kinds of organisms, the individuals in their populations differ in their ability to compete for resources, which means that some individuals can sequester resources better than others. Populations of these organisms are referred to as "density-dependent," which means that individuals compete for resources when they are limited by some aspect of the environment. Verhulst (1845) was the first to incorporate a limiting factor into population growth equations. His equation, often referred to as the logistic growth equation, includes a term, *K*, to represent the limiting factor imposed on the population by the environment:

$$dN/dt = rN(K - N)/K \qquad (7.3)$$

This equation states that a population will grow exponentially as long as birth rate exceeds death rate, until it reaches a saturation point. This saturation point will be determined by (1) competition among individuals in a given environment or habitat and (2) the total amount of a critical resource in that environment. Populations under density-dependent control do not die out as resources become limiting. Rather, a portion of the population is denied access to the critical resource (e.g., food, nest sites, or mates) through competition. This competition allows the remainder of the population to function normally, and the population achieves a steady state where birth plus immigration rate equals death rate plus emigration. The combination of total amount of a critical resource and density-dependent, intraspecific competition for that resource produces a population limit called "carrying capacity" that is characteristic for any given population in a given environment. Theoretical population growth in populations subject to density-dependent population limitation by carrying capacity of the environment produces a sigmoid curve, as shown in figure 7.1.

In the real world, there is a broad continuum both among and within taxonomic groups of species with regard to the degree to which life history is focused on maximizing reproductive rate (*r*-selected) versus those whose life history is focused on individual survival (*K*-selected). Normally, *r*-selected species occupy unstable environments so the life history emphasis is on producing the largest number of individuals in the shortest period of time, whereas *K*-selected species occupy more stable environments where intraspecific competition (density dependence) is important (Pianka 1970). However, whereas bacteria (extremely *r*-selected) and elephants (extremely *K*-selected) might be considered to be at opposite ends of the spectrum when all species are considered, there is a very broad continuum both within and between major groups of organisms. For instance, one could find examples among migratory bird species in which some appear to be more

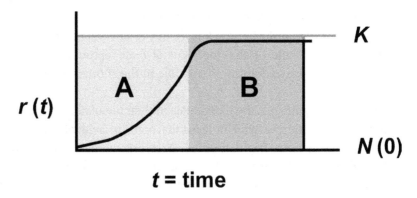

FIGURE 7.1   Population growth (*r*) subject to a limiting factor (*K*) over time (*t*) beginning at a population size (*N*) of zero. Shaded area *A* shows a period of density independence for the population, whereas shaded area *B* shows a period of density dependence when population growth is controlled by competition and carrying capacity (*K*).

r-selected (e.g., short-lived songbirds that produce two or more clutches of three or four young per season) as opposed to long-lived species (e.g., Whooping Crane [*Grus americanus*], which usually produces one clutch of two eggs per season [Lewis 1995]).

Population limitation for all organisms falls into one of these two categories: *density independent* or *density dependent*. For extremely *r*-selected species, control appears to be mostly or entirely density independent (Andrewartha and Birch 1954; Andrewartha 1971; Emlen 1973:267–269). However, for *K*-selected species, control can be either density independent or density dependent, depending on the status of their population size and amount of available, critical habitat at any given moment in time. If more critical habitat is available than can be used by the existing population members, then the population is governed largely by density-independent factors affecting birth and death rate. Only when population size exceeds amount of available critical habitat does carrying capacity have an effect on *r* (figure 7.1) through the medium of intensified intraspecific competition.

A major theme of this book is that migration is a strategy used by dispersing individuals to take advantage of seasonally superabundant resources distant from their birth place in order to minimize competition and maximize reproductive potential. According to this hypothesis, one would expect that invasion of a new, seasonal breeding habitat would be followed by exponential growth in the migratory population until carrying capacity of the breeding habitat is reached (figure 7.1). Thus, the logistic growth curve would seem to provide an adequate model for the behavior of migratory populations. However, for this model to be valid, a critical assumption must be met; namely, that population control outside the breeding season (i.e., during the nonbreeding period) must be density independent. As long as this assumption is true, the only density-dependent factor necessary to be considered in terms of population control is breeding habitat carrying capacity. For all migrants that behave like "fugitive species" during the nonbreeding period, as proposed by MacArthur (1972), shifting wintering sites in attempting to track temporary resource flushes, this model would seem to fit. However, if a migrant species were to occupy one or more specific habitats potentially limiting to populations through density-dependent effects (i.e., competition resulting from limited carrying capacity), then the simple logistic no longer provides a valid description of migrant population control. Thus, whereas the carrying capacity concept has proved to be extremely useful—at least from a heuristic perspective—in understanding how populations of apparently density-dependent organisms, like most birds, expand, contract, and disappear in response to environmental change, its validity for migrants is limited to very specific situations. Consider, for instance, that it was devised for a single species with a fixed r in an isolated habitat with constant K. Unfortunately, this equation is seriously flawed for migrants where the birth and death rates and the habitats, with their corresponding carrying capacities, vary seasonally. This problem may easily be seen in the nonbreeding season situation when the birth rate (b) is zero and the intrinsic growth rate (r) is equal to a negative of the death rate ($-d$), and where the $(K - N)/K$ factor in equation 7.3 implies that the effective death rate decreases with decreasing population size. Moreover, during the nonbreeding period, if the number of migrants from the breeding grounds exceeds the carrying capacity of a nonbreeding habitat, then $(K - N)/K < 0$, which, coupled with the negative r, would result in population *growth*. These unrealistic properties require a new approach to the logistic model for application to migrants.

DEVELOPMENT OF AN ALTERNATIVE MODEL FOR DENSITY-DEPENDENT POPULATION LIMITATION IN MULTIPLE HABITATS

In appendix A, Alan Pine describes a number of essential modifications of equation 7.3 to circumvent the anomalies of the logistic equation for periodic breeders occupying multiple, critical habitats over the course of the annual cycle. First, he allows for separate density-dependent factors for the birth and death rates, decreasing with N for b and increasing for d, each with distinct carrying capacities, K_b and K_d, respectively. Some sample analytical functions are given for the birth and death density factors, which may correspond to different physical mechanisms, such as available nest sites or competition for food, leading to their respective K. If more than one birth or death mechanism occurs, then they may be included by multiplying their respective density-dependent factors. On the basis of this equation, the smallest carrying capacity encountered over the course of the annual cycle has the most influence on the ultimate population size. An analytical expression can be obtained for this ultimate or equilibrium population N_{eq} averaged over the yearly cycle as shown in equation 7.4:

$$(N_{eq})^p = \left[\sum_{m=1}^{M} [b_m - d_m] \tau_m \right] \Big/ \left[\sum_{m=1}^{M} [b_m/(K_{bm})^p + d_m/(K_{dm})^p] \tau_m \right] \quad (7.4)$$

Here, m labels the season, M is the total number of seasons in the annual cycle, τ_m is the season duration, and p is the "saturation" power. This relationship requires specific assumptions about the functional form of the density dependencies of the seasonal birth and death rates (see appendix A, equation A.6b), though a more general result may be obtained numerically. Pine also discusses in appendix A how to incorporate threshold effects, such as finding mates, flocking for protection, and cooperative hunting.

Another difficulty with the differential logistic equation is that the birth and death rates apply to the entire population. This implies, for example, that newborns have the same fecundity and mortality as adults, which is not true for most species. Pine shows in appendix A that natal fertility can be postponed until the next breeding season by substituting a discrete seasonal difference relation, $(N_{n,m+1} - N_{n,m})/\tau_m$, for the continuous derivative, dN/dt. Here, the subscripts n and m refer to the year and season, respectively, noting that $n \rightarrow n + 1$ and $m \rightarrow 1$ after the last season, $m = M$. This delayed response results in a slightly slower growth but a somewhat higher ultimate population than obtained for the corresponding differential equation. For extremely high birth rates, Pine demonstrates that these finite delays can lead to unstable or chaotic populations, much like the iterative intergenerational models discussed by May (1976) and others. Also, Pine notes that the discrete difference model can be applied to the various phases of metamorphic and spawning species.

Although the discrete difference models can represent a seasonal delay in breeding, they do not properly account for age-dependent birth and death rates seen in many avian species. In appendix B, Pine has extended these models for age-structured periodic breeders. This requires the added complication of keeping track of the demographics (i.e., the age distribution of the population). The models described in appendix B follow the earlier age-structured matrix methods for a single habitat given by Leslie (1945, 1948). Here, the single Leslie matrix is replaced by a chain of seasonal matrices. Once this chain is multiplied together, we find annual results similar to the single-habitat behavior. First, for an arbitrary initial age distribution, the total population exhibits waves on the timescale of the peak reproductive maturation age. Eventually, these waves are damped out, and a stable youth-oriented age distribution (characteristic of the dominant eigenvalue of the compound matrix) emerges. Once this stable distribution is reached, the total population closely follows the dynamics of the much simpler age-independent discrete difference models given in appendix A.

IMPLICATIONS OF DENSITY-DEPENDENT POPULATION CONTROL VIA MULTIPLE CARRYING CAPACITIES

The number of critical habitats occupied over the course of an annual cycle by members of any given migratory species is large, even for species whose total population size is tiny. For instance, the Whooping Crane has a total population size of less than 600 individuals, but has breeding and wintering ranges covering

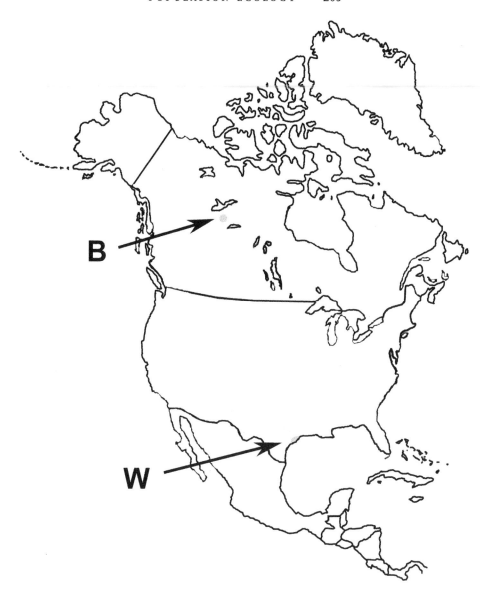

FIGURE 7.2   Principal breeding (*B*) and wintering (*W*) areas (*circles pointed out by arrows*) for the wild population of the Whooping Crane (*Grus americanus*).

several thousand square kilometers respectively and a 3,800-km migration route that includes an additional several thousand square kilometers of potentially critical stopover areas (figure 7.2).

Pine's equation predicts that if population size exceeds carrying capacity in one of these critical environments, then that specific environment will effectively limit total population size via some combination of density-dependent mechanisms. However, if population size is actually *below* carrying capacity for *all* critical environments, then population size is limited mainly by density-independent factors.

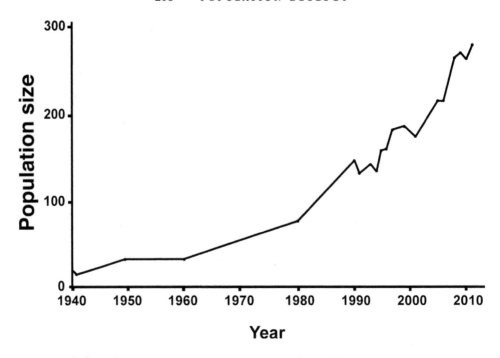

FIGURE 7.3   Whooping Crane (*Grus americanus*) population change since 1940.

Using the Whooping Crane as an example, this species formerly (pre–European colonization) had a breeding range that covered a large portion of prairie wetlands of north-central North America, extending from Illinois to the Canadian Northwest Territories, and a wintering range covering much of the coastal region of the Gulf of Mexico (Natural Resources Conservation Service 2011). A density-independent factor, hunting (Lewis 1995), caused population decline presumably of several orders of magnitude. Beginning in about 1940, the population has grown, undergoing exponential increase since about 1980 (figure 7.3).

Note, however, that, even during the period of exponential increase when populations obviously were still far below carrying capacity for all critical breeding, migration, and wintering habitats, declines were recorded in some years. Furthermore, analysis of these declines is indicative that some density-dependent factors associated with amount and quality of a critical habitat may be operative. For instance, consider the following statement from the *Los Angeles Times* of April 2, 2009:

> This past winter was the worst on record in terms of bird deaths [for Whooping Cranes], according to Tom Stehn, whooping crane coordinator with the U.S. Fish and Wildlife Service.
>
> "Total winter mortality is estimated at six adults and 15 chicks, a loss of 7.8% of the flock," Stehn stated. "When added to the 34 birds that left Texas in spring 2008 and failed to return in 2009, 20% of the flock was lost during the last 12 months."

Stehn attributes the winter losses to poor habitat conditions at the Aransas National Wildlife Refuge in Texas, which the birds migrate to each fall. Low rainfall totals resulted in saltier bays and also fewer blue crabs, the primary food source for the cranes. (Burgess 2009)

The fact that deaths among young birds in the population was more than double that of adults may indicate that competition for limited resources during the wintering period affected even this small population, even though total population size remains well below total carrying capacity of the winter habitat, which formerly included much of the shoreline of the northern Gulf of Mexico, from Florida to Texas.

This observation may have some relevance to understanding how Pine's equation might illuminate population limitation for other migratory bird species. Consider the following:

1. Most migrant bird species have breeding, postbreeding, wintering, and stopover areas that cover vastly greater areas than those of the Whooping Crane.
2. Every such area is likely to be somewhat different from every other area in terms of "quality" (i.e., value in terms of survivorship or reproductive output).
3. Survivorship or reproductive output for losers in competition for higher-quality habitats in any given area will not be zero; rather, these fitness values will be reduced by some factor that is characteristic for that area, as losers are pushed into lower quality habitats.
4. Five major categories of habitats are potentially critical for migrant species: breeding, postbreeding, fall transient stopover, wintering, and spring transient stopover.
5. To assess the probability that habitat availability is limiting for any given portion of the life cycle, one must sum the habitat limitations on population size for different areas within each of the five seasonal subsets listed previously.

Looking at migrant population limitation in this way provides considerable conceptual power in examining factors likely to limit a particular migrant species. First and foremost, this approach emphasizes the need for detailed information for all parts of the life cycle because *any* portion is potentially limiting. In addition, it makes clear that different metapopulations of the same migrant species could be limited during different portions of the life cycle depending on the degree to which they are geographically isolated throughout the entire annual cycle.

OTHER MODELS

As already noted, early models for migrant population limitation focused on breeding habitat availability (e.g., von Haartman 1971). Nonbreeding-period effects on population size were assumed to be largely density-independent, and only under

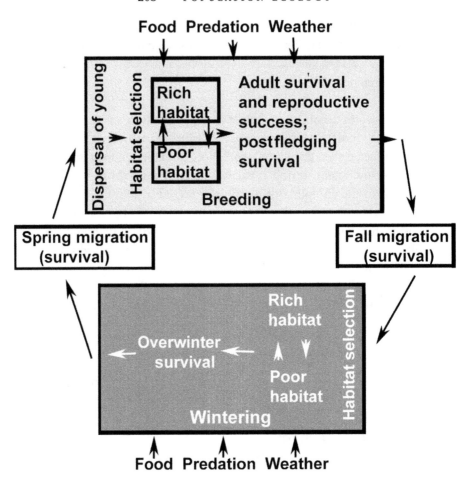

FIGURE 7.4   The Sherry and Holmes (1995) model for migratory bird population control. See text for discussion.

rather special circumstances (e.g., excessive hunting) were populations expected to be held below breeding habitat carrying capacity. Sherry and Holmes (1995) provide a formalized version of this hypothesis in which migrant breeding population size is governed by density-dependent factors whereas stochastic processes, reflecting a combination of influences on birth rate, immigration, emigration, and all mortality factors encountered throughout the annual cycle, govern total population size (figure 7.4). Viewed in the context of Pine's equations, this finding would be true only in the situation in which the population never reaches carrying capacity in *any of the habitats occupied* over the course of the annual cycle (except the breeding area, where density dependence affects birth rate but not mortality).

Runge and Marra (2005) propose a similar, quasi-density-dependent (breeding period) model (figure 7.5) based largely on work with the American Redstart (*Setophaga ruticilla*) on breeding grounds in New Hampshire and wintering grounds in Jamaica. This model is important not only because of what it says about

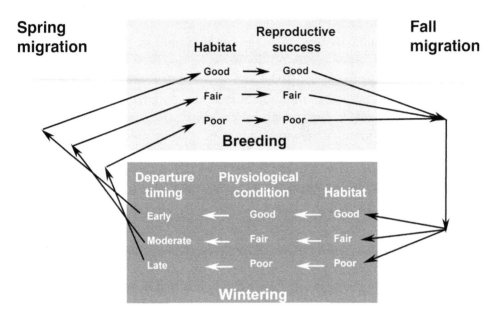

FIGURE 7.5   The Runge and Marra (2005) model for migratory bird population limitation. See text for discussion.

how migrant populations are controlled but also because it goes to the heart of several critical issues regarding larger concepts of migrant population biology. For these reasons, we next examine their model and its assumptions in detail.

• *Winter habitat occurs in two types: "good" or "poor," which do not differ in terms of survivorship for individuals occupying them. "Good" habitat is limited but "poor" habitat is unlimited.* This assumption is based on wintering redstarts in Jamaica in which most males were found in mangroves and most females were found in scrub. However, banding–recapture data found no significant difference in survivorship between "good" and "poor" habitats (Marra and Holmes 2001). In addition, data from a single study site do not provide sufficient information to support this assumption for the redstart, which has a breeding range that covers a large portion of temperate North America and a winter range that extends throughout the Caribbean, Middle America (Central America plus Mexico), and northern South America, let alone for any other migrant species.

• *The difference between "good" and "poor" winter habitat lies in the quality of physiologic preparation for migration and breeding; birds that winter in "good" habitat are more likely to arrive early and obtain good breeding territories than those wintering in "poor" habitat.* This assumption is based on papers that purport to establish a link between certain physiologic differences, mainly corticosterone levels, measured in individual redstarts in the two different habitat types (Marra and Holberton 1998; Marra et al. 1998). The authors have no data on differences in survivorship among individuals that differ in these physiologic measures. They base their conclusion (i.e., that winter habitat occupancy affects time of arrival on the breeding ground, which affects reproductive output) on the fact that adult males tend to winter in "good" habitat and arrive earlier on the breeding grounds; females and young males tend to winter in "poor" habitat and arrive later. However, there is no evidence that early breeding ground arrival is related to quality of winter habitat occupied for redstarts. In fact, protandry (early male breeding ground arrival) is found in many migrant species, probably is genetically programmed (Coppack and Pulido 2009), and is likely related to differences in male and female breeding season roles (see chapter 2). Within-sex differences in terms of habitat occupancy and arrival may mean something, but until valid empirical data are available, the possible contribution of such differences remains unknown and cannot be assumed to be acting, either for, against, or as a neutral factor. In addition, if investigated, findings likely would show a strong age effect, confounding any fitness inferences that might be made.

• *Birds wintering in "good" habitat have higher survivorship during migration than birds wintering in "poor" habitat.* As pointed out by reviewer Darren Fa, this statement is true by definition and is, in fact, circular. What is assumed for redstarts is that the term "good" can be applied to woodland wintering habitat in general, whereas the term "poor" can be applied to scrub or grassland wintering habitat. Although the statement regarding the effect of good and poor wintering habitat on transient survival may be true, there are no data on redstarts or any other migrant of which we are aware to validate the assumption that one specific wintering habitat is superior to another in terms of transient survivorship.

• *Breeding habitat also occurs as "good" or "poor." There is no difference in survivorship in terms of male birds occupying the two habitats, but male birds in the "good" habitat have higher reproductive rate.* The odd aspect of this assumption is that an entire sex class (females) has been presented as likely to have wintered in "poor" habitat (although the model matrix parameters are the same for males and females: Wmg for males wintering in "good" habitat; Wmp for males wintering in "poor" habitat; Wfg for females wintering in "good" habitat; Wfp for females wintering in "poor" habitat) (Runge and Marra 2005:378–381).

• *Total amount of breeding habitat (= "good" + "poor") is assumed to be limited. Females unable to settle in one or the other are assumed to die. Males unable to obtain territories do not die, but do not breed.* Presumably, this assumption is based on the many studies that show males unable to obtain breeding territories occur as "floaters" whereas females do not (see chapter 2). However, there are no data to support differential survivorship for individuals occupying different habitats during the breeding period for any migrant of which we are aware.

In essence, the Runge–Marra model predicts that winter habitat exerts some influence over population size through effects on migration and breeding ground survivorship and reproductive rates (although the principal influence is breeding habitat availability). Lack of data on actual survivorship or reproductive rates for the species in question (i.e., American Redstart [*Setophaga ruticialla*]) raise questions regarding the underlying assumptions.

Despite the problems discussed earlier, the model has conceptual value in indicating the possibility of carryover, density-dependent effects from one season to the next. That carryover effects are possible is well known and documented, at least in some species of waterfowl, in which breeding success appears dependent upon amounts of fat reserves with which the females arrive on the breeding grounds (Raveling 1979; van der Graaf et al. 2006), and in shorebirds, in which timely arrival on breeding areas appears to be governed by success in obtaining food resources along the migration route (Drent and Piersma 1990; Drent et al. 2003), although it should be noted that the importance of competition in these examples (i.e., density-dependent effects) is not known.

The principal focus of the Runge–Marra model is on possible effects of density-dependent competition for high-quality winter habitat and how this competition might affect breeding habitat occupancy (and subsequent reproductive success). However, there is no reason to believe that such carryover effects would be limited to these seasons. Indeed, density-dependent, carryover effects could occur between any season and its successor, at least in some migrant species. For instance, quality of the breeding territory in terms of providing adequate provisioning for offspring (prefledging and postfledging buildup of fat reserves) is likely to affect offspring survival during the postbreeding period, thereby affecting parental fitness; quality of postbreeding, molting, and premigratory preparation habitat or timing of molt (Norris et al. 2004) is likely to affect migration success; and quality of stopover sites is likely to affect arrival time on wintering sites (and access to quality resources), and so forth. In addition, as noted by a reviewer (Darren Fa), the system is iterative in that population data are fed sequentially into the next stage of the life cycle to provide a temporal data stream. Such data (e.g., population numbers, age and sex distributions, etc.) will invariably determine the "quality" of the various environments encountered, which themselves vary over time, making the establishment of whether density-dependent or density-independent factors operate a question the answer to which will vary on a temporal basis.

POPULATION CHANGE

Normal populations of migratory birds (i.e., those not in stratospheric increase or catastrophic decline) can be expected to increase during the breeding season and decline for the rest of the year, with total breeding population size fluctuating on a larger scale of years or decades. This annual variation in population size produces a sawtooth curve whose lowest point is always reached at the beginning of each breeding season, which means that even in populations that are healthy (i.e., experiencing no serious, long-term decline), a major portion, often equal to half or more of the total postbreeding population size, will die between the end

TABLE 7.1   Human- and Cat-Related Mortality Sources for Birds in the United States

MORTALITY SOURCE	ANNUAL MORTALITY ESTIMATE (N)	ESTIMATED PERCENTAGE OF TOTAL HUMAN- AND CAT-RELATED MORTALITY SOURCES (%)
Buildings	550,000,000	58.2
Power lines	130,000,000	13.7
Cats	100,000,000	10.6
Automobiles	80,000,000	8.5
Pesticides	67,000,000	7.1
Communications towers	4,500,000	0.5
Wind turbines	28,500	<0.01
Airplanes	25,000	<0.01
Other (e.g., oil spills, fishing by-catch, etc.)	Not calculated	Not calculated

Source: Erickson et al. (2005).

of one breeding period and the beginning of the next. As a result, observations of individual mortality factors known to cause deaths of large numbers of individual birds, even if repeated year in and year out, do not provide sufficient information to allow us to draw conclusions regarding their overall effect on any given population (Arnold and Zink 2011).

As an example of the difficulty in understanding population effects of these characteristics of migrant population variability and control, consider data from a paper detailing "anthropogenic" factors causing songbird mortality (Erickson et al. 2005). A table presented in this paper (table 7.1) summarizes major known mortality factors for U.S. birds.

There are several problems with the data in table 7.1 from the perspective of trying to understand their meaning:

1. They are based on crude sampling procedures the validity of which is unverifiable.
2. They bear no relevance to any particular population of any species.
3. They are not considered in the framework of total mortality resulting from all factors.
4. Natality is not taken into account.

5. The data are not placed into any meaningful context of population change (i.e., everything dies, so the pertinent question is whether any one mortality factor poses a threat to the population).

Nevertheless, these and similar data (e.g., Coleman et al. 1997) have large effects on media reports (Clark 2011) as well as on summaries of threats to migrants in the popular literature (e.g., Hughes 2009:168, 182), which, in turn, can place considerable pressure on both management and conservation policy.

As an example of how such data can distort understanding of actual migrant population limitation, reports on avian mortality due to cats has resulted in various municipal efforts at cat control across the United States (e.g., Myers 2011) as well as a nationwide program to control cat predation in the interests of bird conservation (American Bird Conservancy 2010). As long as these cat-control efforts are promoted as potentially reducing individual bird deaths from this particular mortality factor in a small, defined area, there can be little scientific quibble with the program. If a person values the lives of individual birds in their neighborhood over the freedom of cat-lovers to allow their pets to roam at will, then that is a personal value judgment. However, if the cat-control efforts are promoted as likely to enhance migratory bird populations range-wide, that is a different matter. Then the question becomes whether or not data support the contention, which brings scientific analysis into the issue (Arnold and Zink 2011).

A recent exchange that took place in the pages of the newsletter of the Audubon Naturalist Society, a Washington, D.C., area conservation group, summarizes these different viewpoints. An opinion piece for the newsletter stated that although cats, wind turbines, glass buildings, and the like may cause large numbers of migratory bird deaths, there were few or no data to confirm that these factors affected overall population size for any migrant species; that, in fact, for most declining migrant populations, habitat loss or change was the likely culprit (Rappole 2005b). In a letter to the editor, an irate reader responded, "[T]o suggest that habitat is the one factor and that other factors likely will have no effect is both factually wrong and very irresponsible. Bird populations are declining for many reasons, and each species must be studied on a case-by-case basis. In addition to studying habitat-related factors, one must examine all other factors that affect mortality and how these factors interact with habitat losses" (Young 2005). Young was correct, of course, in stating that each declining species must be studied on a case-by-case basis to determine the cause but was incorrect, at least theoretically, in stating that if the ultimate cause is insufficient critical habitat at any point during the life cycle, reduction or elimination of a set of proximate mortality causes will ameliorate the problem.

We demonstrated earlier in this chapter that lack of a critical habitat encountered during any portion of the annual cycle can, in theory, limit population size of a migratory species. However, determination of how any given population is limited requires specific, long-term knowledge of the ecology, life history, and annual changes in breeding population size.

MEASUREMENT

The starting place for study of any population is measurement. Under certain circumstances, obtaining a good idea of trends may not be difficult, as in the case when entire populations of a species (e.g., Snowy Egret [*Egretta thula*] in late-nineteenth-century North America) disappeared from entire regions (Parsons and Master 2000). However, often the population status of migrants is not so clear. Current efforts for migrants focus largely on determination of breeding population size and the use of these data to establish long-term population trends. Such procedures have been in place for migratory game birds since the early twentieth century in the United States (Bellrose 1976) and for all breeding birds in the United States since 1966 through the auspices of the North American Breeding Bird Survey (BBS) (Sauer et al. 2011). No long-term, continent-wide survey of migratory bird populations exists for Europe. Nevertheless, some European countries have (or had) country-wide surveys including the United Kingdom, Sweden, Denmark, the former Czechoslovakia (Czech Republic), Finland, Estonia, and The Netherlands (Hustings 1988; Janda et al. 1990; Marchant et al. 1990, Jacobsen 1991; Marchant 1992; Gregory et al. 2002; Thaxter et al. 2010).

Breeding bird surveys depend, for the most part, on counts of various design of the number of singing birds in a specific area (Gregory et al. 2004). For the North American Breeding Bird Survey, counts are made by volunteers along a specified route, essentially, a form of line-transect, in which the number of singing males is estimated at each stop along the route. Each survey route is 39 km in length; the observer makes 50 stops of 3 minutes along the route, counting each bird seen or heard (Ziolkowski et al. 2010).

Singing bird counts suffer from a number of difficulties regarding the assumptions on which they are based—for example, that each singing bird represents a breeding pair, that birds sing at the same rate regardless of age, population density, or proximity to a road, and that all of the observers involved in conducting the counts have equal skill in detecting and identifying species heard. All of these assumptions have been shown to be more or less invalid (Nolan 1978; Rappole and Waggerman 1986; Swanson 1989; Rappole et al. 1993; Sauer et al. 1994:50; Rappole 1995:139; McShea and Rappole 1997; Kociolek et al. 2010), and certainly short-term trends over small areas should be interpreted with caution. Nevertheless, breeding bird surveys can serve to detect major, long-term trends, at least for the period during which they were performed.

TRENDS

For most migratory species, it is impossible to count all individuals every year (the Whooping Crane is an exception). Thus, assessment of populations requires long-term data sets that can be analyzed to reveal trends. Two regions that have such data sets are Europe and North America, as discussed in the following.

• *European.* Gregory et al. (2002) provided summaries of major population trends for British birds relevant to the International Union for Conservation of Nature's "Red List" and "Amber List" (roughly equivalent to "Endangered" and "Threatened," respectively, in the U.S. Fish and Wildlife Service classification system [U.S. Congress 1973]), which showed significant declines in 32 species, most of which were migrants whereas 15 species showed significant increases. Sanderson et al. (2006) published the first continent-wide analysis of European breeding birds using data from several sources covering the period from 1970 to 2000. They found significantly higher, negative long-term trends in trans-Saharan migrants than in either residents or short-distance migrants. Heldbjerg and Fox (2008) analyzed abundance data on 62 species of breeding birds in Denmark from 1976 to 2005. They found that trans-Saharan migrants as a group declined by an average of 1.3 percent per year, whereas short-distance migrants and residents increased an average of 1.4 and 1.0 percent, respectively. Similarly, Both et al. (2010) report long-term declines based on breeding bird surveys in The Netherlands for forest-related, trans-Saharan migrants as did Thaxter et al. (2010) for trans-Saharan migrants in England.

• *North American.* Sauer et al. (2011) provide survey-wide summaries for more than 400 species of North American breeding birds, most of which are migrants. (Regional and survey-wide trends can be obtained for each of the 400-plus species by reference to Sauer et al. online). General trends for major groups have been summarized by the North American Bird Conservation Initiative (2009) (figure 7.6).

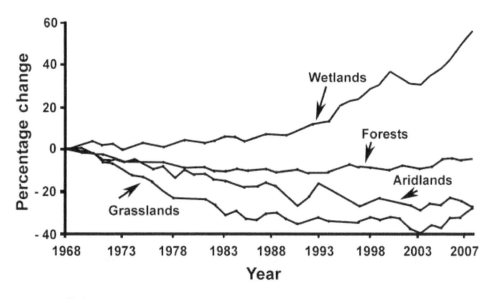

FIGURE 7.6   Population trends for North American breeding birds grouped by breeding habitat, 1968–2007 (based on North American Bird Conservation Initiative 2009).

TREND ANALYSIS

Analysis of nationwide trends by taxonomic or ecological groupings can be revealing, but mostly they signal the need to take a closer look at the data on a species-by-species basis. As an example, the North American Bird Conservation Initiative (NABCI) trend analysis (figure 7.6) shows wetland-breeding birds in North America enjoying a distinct upward trend, driven largely by population increases in geese, hole-nesting ducks, and some other wetland-generalist species that have benefited from prairie-region conservation programs (Reynolds et al. 2001). However, 24 percent of U.S. wetland birds, mostly migrants, are species of conservation concern, and 10 are on U.S. Fish and Wildlife Service "Threatened" or "Endangered" species lists (North American Bird Conservation Initiative 2009:18). For the mid-Atlantic region of the United States, wetlands constitute the single most threatened major habitat type (Rappole 2007:14–15).

Accepting that there are potential problems involved with field procedures, statistical methodologies, and the lumping of population trends for groups of migrants, there are, nevertheless, patterns of decline that appear in the breeding bird survey data for both European and North American migrants that warrant closer examination (Thaxter et al. 2010; Sauer et al. 2011). For instance, downward trends in North American grassland species (including other early successional-stage species, e.g., shrubland-breeding birds) generally appear to be accepted as posing significant conservation problems, although, as discussed later, exactly where during the annual cycle declines are caused, and how to rectify them, are not necessarily obvious (Askins et al. 1990; Askins 1993; Askins et al. 2007). Declines in forest-related, long-distance migrant species have been somewhat more controversial with regard to their causes and meaning, at least in North America (Hutto 1988; James et al. 1992; DeGraaf and Rappole 1995; Latta and Baltz 1997; Faaborg et al. 2010a:17).

One problem characteristic of all major country-wide or continent-wide breeding bird surveys is that none dates back much more than one-half century. As an example, the North American Breeding Bird Survey is now 56 years old (1966–2012). Yet forest cover was declining in Middle America (Central America plus Mexico) long before this date (figure 7.7). This fact has potential significance for songbird populations, 60 to 80 percent of which are forest-related in eastern and western North America (Powell and Rappole 1986) in that major changes in migrant populations could have predated the BBS survey. King et al. (2006b) addressed this issue using North American breeding bird census (BBC) data from 46 sites dating back to 1940. These data show that for several species of long-distance, forest-related migrants that winter in the neotropics, BBC data (1940–1995) show significant declines whereas BBS trend data (1966–1995) are essentially flat, showing no change. This finding indicates that declines in some migrant species predated the origin of the BBS (table 7.2).

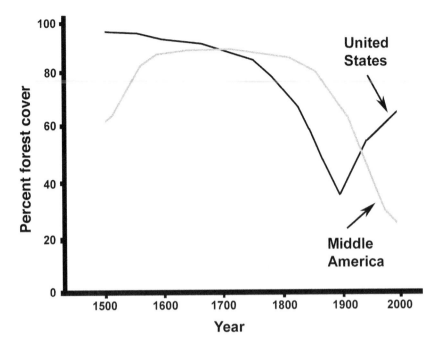

FIGURE 7.7   Percent forest cover in the United States (*black line*) and Middle America (*gray line*), 1500–2000 (Powell and Rappole 1986).

Despite all of the problems and issues related to measurement and analysis, trends remain the best means available at present to assess population health for large numbers of migratory species.

CAUSES OF MIGRANT POPULATION CHANGE

As discussed, there are two major categories of factors governing population change in migrants: (1) density independent, and (2) density dependent. The same factor can, of course, serve as a density-independent factor under some circumstances and a density-dependent factor under other circumstances. For example, Thaxter et al. (2010) found that breeding populations of ground nesters in England declined more than those of cavity nesters over the period from 1966 to 2007, which they attribute to the possibility of increases in predator populations over that period. If populations in all breeding habitats occupied by ground nesters suffer equally from predation, then it would seem that predation is largely density-independent in terms of its effects (i.e., competition plays little or no role). However, if predation rates are higher in poorer breeding habitats, then predation may be serving as a density-dependent factor. Indeed, as reviewer Darren Fa has noted, density dependence or independence will vary as a function not only of the dynamics of prey populations but also of those of their predators (and other associated species).

TABLE 7.2 Trend Estimates (SE) and 95 Percent Confidence Intervals of Selected North American Migrant Species

Species	BBC (1940–1995)	BBC (1966–1995)	BBS (1966–1995)
Eastern Wood-Pewee	0.99 (0.01)*	1.00 (0.01)	0.983 (0.0003)*
(Contopus virens)	(0.98–0.99)	(0.98–1.03)	
Blue-gray Gnatcatcher	0.93 (0.05)	1.00 (0.01)	1.013 (0.0001)*
(Polioptila caerulea)	(0.87–1.01)	(0.98–1.04)	
Wood Thrush	0.77 (0.11)*	0.97 (0.02)	0.982 (0.0001)*
(Hylocichla mustelina)	(0.73–0.99)	(0.94–1.00)	
Yellow-throated Vireo	0.94 (0.02)*	0.99 (0.03)	1.007 (0.0001)
(Vireo flavifrons)	(0.92–0.97)	(0.94–1.06)	
Blue-headed Vireo	0.77 (0.11)	0.96 (0.08)	1.052 (0.0001)*
(Vireo solitarius)	(0.75–1.02)	(0.75–1.05)	
Red-eyed Vireo	0.99 (0.01)*	0.98 (0.03)	1.012 (0.0001)*
(Vireo olivaceus)	0.97–0.99	(0.99–1.01)	
Black-and-white Warbler	1.01 (0.01)	1.01 (0.00)*	1.005 (0.0002)
(Mniotilta varia)	(0.99–1.02)	(1.01–1.02)	
Ovenbird	0.96 (0.02)*	0.99 (0.02)	1.008 (0.0006)*
(Seiurus aurocapilla)	(0.93–0.99)	(0.96–1.03)	
Hooded Warbler	0.96 (0.04)*	0.98 (0.01)	1.004 (0.0000)
(Setophaga citrina)	(0.88–0.99)	(0.93–1.03)	
Scarlet Tanager	0.98 (0.02)	1.00 (0.01)	1.000 (0.0000)
(Piranga olivacea)	(0.95–1.01)	(0.98–1.02)	

Source: King et al. (2006b).

Note: Table data derived from route-regression analyses of Breeding Bird Census (BBC) data from 46 Sites in North America, as well as trend estimates (SE) from the Breeding Bird Survey (BBS), Trend values of >1 indicate that the population is increasing, whereas values of <1 indicate that the population is declining (see text). Significant trends ($\alpha = 0.05$) are indicated by an asterisk (*).

Suggested Density-Independent Causes
of Migrant Population Change

There are many examples of migrant species whose population size appears governed largely or entirely by density-independent factors (i.e., those governing natality and mortality). Among the most obvious are those whose populations are held below the carrying capacity of any environment occupied during the year. The exemplary list given in the following is of factors that have been suggested as being responsible for the decline or extirpation of one or more migrant populations. We categorize them as "density-independent" because competition does not appear to play a significant role in terms of their effects on populations.

• *Hunting.* Perhaps the best evidence of populations controlled by density-independent factors comes from historical data and current research on species that were, or are, harvested commercially. Market hunting during the late 1800s and early 1900s caused extirpation of waterfowl, shorebirds, herons, and egrets from large portions of the United States (Eaton 1910; Bull 1974; Nichols et al. 1995). Protective legislation has allowed recovery of many of the species affected. Biologists in many countries monitor migratory gamebird populations carefully and set "take" or "bag" limits according to statistical analyses that calculate "compensatory" versus "additive" mortality (Nichols et al. 1995). "Compensatory" mortality represents natural mortality that would have taken place anyway at some point during the annual cycle, even if the bird had not been shot. For example, if 100,000 adults and young of a species head south in fall, but the wintering habitat will only support 50,000, then 50,000 birds will die from various mortality factors. Whether these birds are shot, eaten by predators, or die of starvation is irrelevant because 50,000 were doomed in any event. Natural mortality is reduced in compensation for the imposed shooting mortality, so that total annual mortality remains the same. "Additive" mortality means mortality that is not compensated for by other mortality factors. For instance, if 100,000 birds head south and the winter habitat can support 150,000 birds, then every mortality factor will contribute to the size of the spring population that heads north. In this example, hunting serves as an added mortality factor in addition to natural mortality factors and can have a significant effect on population size. Populations in which compensatory mortality applies are assumed to be limited by density-dependent causes (e.g., habitat availability). In additive mortality, populations are controlled by density-independent causes.

• *Pesticides.* The role of pesticides in the decline of certain migrant species is well known and documented (Newton 2008:175). Populations of raptors, waterbirds, and some other migrants declined precipitously beginning shortly after the introduction of DDT into widespread use as a pesticide in the mid-1940s (Cottam and Higgins 1946). The main effect of this and other chemically related toxins was on natality rate, in that high blood levels of the agent, resulting from consumption of prey that had eaten poisoned foods or water, caused severe reductions in nesting success (Blus et al. 1971). Pressure on many migrant populations was extreme at least up until the mid-1970s (Bildstein 2006) when use of many pesticides known to have negative effects on bird populations was banned. However, problems persisted among some migrant populations that traveled to countries where widespread use of pesticides banned elsewhere continued at least up until the late 1980s (Henny and Herron 1989).

• *Diseases or parasites.* Observation of large numbers of bird deaths associated with outbreaks of disease has led some to conclude that some populations of some migrant species could be threatened (LaDeau et al. 2008). In particular, when West Nile virus first appeared in North America, tens of thousands of individuals of several migrant species died, and some serious declines in some metapopulations were reported (e.g., American Crows [*Corvus brachyrhynchos*] in central Illinois) (Yaremych et al. 2004). However, breeding bird survey data have shown no persistent decline that appears disease-related for migrants (including crows) with continental distributions (Sauer et al. 2011). The best data documenting effects of disease on birds (not necessarily migrants) come from studies of small, isolated island populations subjected to introduced pathogens (e.g., avian malaria on Hawaiian honeycreepers). For some of these species, the effects have indeed been severe (Beadell et al. 2006).

• *Migration mortality.* Density-independent mortality factors associated with migration are often thought to affect migrant population size. Newton (2008:777) expresses the prevailing sense of the "perils of migration" in his statement, "It would indeed be surprising if bird breeding numbers were unaffected by conditions on migration," a sentiment that has been expressed by many authors and students since humanity first began to develop an understanding of the phenomenon (Cooke 1915; Wetmore 1926:121; Moore 2000:1; Hughes 2009). Sillet and Holmes (2002) present data in support of this idea for the Black-throated Blue Warbler (*Setophaga caerulescens*), reporting migration-related mortality rates to be 15 times higher than either breeding or wintering season mortality rates in this species based on banding–recapture/resighting information obtained at breeding sites in New Hampshire and wintering sites in Jamaica. Such numbers, of course, are meaningless in terms of long-term, range-wide population control without some sense of their relationship to birth rate, overall population size, and relative carrying capacity of critical environments occupied over the course of the annual cycle.

The catalog of potential threats encountered by transients is impressive: storms, adverse winds, predators, disease, orientation errors, competitors, and stopover-site availability cause the deaths of myriad migrants annually (Newton 2008: chaps. 27 and 28). Nevertheless, the number of species whose long-term population trends are known (or even suspected of being or having been) driven by migration season factors is small (Hutto 2000). At the top of the list are species whose breeding and wintering populations lie largely outside human influence (so far as is known) that have nevertheless suffered obvious long-term, range-wide declines (e.g., Eskimo Curlew [*Numenius borealis*]). This bird is an example of how a direct cause of mortality (market hunting) encountered during the migration season apparently caused the near-extinction of a species (Gill et al. 1998). Many other migrant gamebirds likely were decimated by market hunting during the nineteenth century in North America, some of whose populations are only now beginning to recover (e.g., Trumpeter Swan [*Cygnus buccinator*], Tundra Swan [*Cygnus columbianus*], and Hudsonian Godwit [*Limosa haemastica*]). However, aside from hunting, we know of no evidence to indicate that long-term change in any migrant population has been or is being caused by mortality factors encountered during migration.

   • *Climate change.* Climate change, whether as a result of regional cycles occurring over a period of years or decades as opposed to long-term shifts in global patterns (see chapter 8), can have profound effects on migratory bird populations. Blake et al. (1992), for instance, reported that breeding populations in Wisconsin and Michigan (north-central United States) of 8 of 11 species of neotropical migrants showed significant declines during a period of regional drought (1985–1989). Similarly, several authors have reported declines in European breeding populations of trans-Saharan migrants, presumed to have resulted from drought in their African wintering grounds (Heldbjerg and Fox 2008; Both et al. 2010; Thaxter et al. 2010). Changes in climate that force change in quality of a critical habitat seems likely to be density independent in its effects on populations, but determination of whether or not this is so would require investigation on a species-by-species basis.

SUGGESTED DENSITY-DEPENDENT CAUSES OF MIGRANT POPULATION CHANGE

We have argued that the standard logistic-carrying capacity model must be modified to allow for density-dependent factors to act in different environments at different times over the course of the annual cycle. However, we have not discussed how density dependence might influence populations in these different environments. The ways that density-dependent factors can influence populations differ radically during the breeding season compared with other times of the year. For all of the nonbreeding-season environments occupied, density-dependent factors act mainly on mortality (although carryover effects between seasons are possible). However, during the breeding season, density dependence can act through *both* natality and mortality, and in most populations, there is a heavy emphasis on natality; for example, young individuals hatched the previous year are denied access to reproduction (as shown in the model by Runge and Marra 2005). Alan Pine's equation (equation 7.4) allows for inclusion of factors affecting natality and mortality in different habitats and times of the year, but it is important to remember that effects on reproductive rate can be quite subtle and affect different subsets within the different age and sex groups differently. We are not aware of efforts to quantify these effects for any migrant population, but they will be necessary in order to develop a complete understanding of how breeding depression among certain classes of individuals might affect overall population size.

With these ideas and caveats in mind, we present in the following a list of the density-dependent factors that have been suggested in the literature as potentially responsible for significant long-term decline in one or more populations of migrants. The list of factors and examples presented should be considered as exemplary rather than exhaustive.

• *Breeding habitat loss.* Loss of agricultural habitat in England has been suggested as responsible for decline in some migrant populations (Thaxter et al. 2010). Similarly, migratory waterfowl that breed in small ponds ("potholes") in the prairie regions of North America underwent sharp declines, apparently associated with drainage of millions of hectares during the twentieth century (Bellrose 1976:46). Recovery of some of these populations has been attributed to government programs that provide financial incentives to farmers to maintain this habitat type (Reynolds et al. 2001). Additional examples of the potential importance of breeding habitat for migrants are given in chapter 8 in which historical distribution changes of the Bachman's Sparrow and Bewick's Wren seemingly provide clear evidence of the effects of breeding habitat loss on populations of migratory species. Both species were resident in the southeastern United States at the time of European colonization, and both developed migratory populations to more northern regions of the eastern United States that took advantage of the conversion of forest to crops and second growth. However, as forest returned to these areas, these migratory populations have disappeared. Similar effects on other migrants that breed in early successional stage habitats in North America have been documented: 17 of 28 grassland species have undergone significant declines (Knopf 1994; Sauer et al. 2003; Askins et al. 2007) as well as many shrubland-breeding species (Askins 1993). However, as with other types of migrants, the cause and conservation meaning of declines are not necessarily clear. For birds like the Bachman's Sparrow, Bewick's Wren, and Loggerhead Shrike, it seems unlikely that they were part of the native breeding fauna of many of the regions from which they have disappeared, apparently as the result of natural processes of succession. Also, the winter habitats of these and several other declining grassland-breeders have suffered as well, leaving unsettled the issue of which portion of the annual cycle was most critical in terms of causing declines (England 1998; Di Giacomo et al. 2005; Askins et al. 2007; Renfrew and Saavedra 2007).

• *Breeding habitat fragmentation.* Whitcomb (1977) and Lynch and Whitcomb (1978) proposed that observed declines in numbers of forest-related, Nearctic–neotropical migrants breeding in small, isolated forest patches in Maryland and Washington, D.C., accorded with island biogeography theory in which the number of species in a community is dependent on a balance between the relative rates of colonization and extinction. These rates are affected by the size (area) occupied by the community and its degree of isolation. The model predicts that in large sites with a low degree of isolation, an equilibrium between colonization and extinction will be established that is at the high end in terms of species richness, whereas small, isolated sites will show an equilibrium at the low end (Arrhenius 1921; Gleason 1922; Cain and Castro 1959; MacArthur and Wilson 1967; Andrén 1994). Hundreds of papers have since been published investigating this phenomenon, mostly focused on forest-breeding migrant birds of eastern and central North America (Newton 1998:123–142; Faaborg et al. 2010a). The majority of studies reported a so-called "area" effect in which size of the habitat patch was indeed related to apparent richness and density of breeding pairs of migrants, as predicted by the model (Galli et al. 1976; Ambuel and Temple 1983; Askins et al. 1990; Freemark and Collins 1992). In fact, some studies were able to provide minimal area requirements for patch size able to support particular species (Robbins et al. 1989a). According to the theory proposed by MacArthur and Wilson (1967), variation in species richness and density are stochastic, resulting from differing probabilities of immigration, emigration, and population size as these might vary according to size of the site and degree of isolation. Nevertheless, most researchers looked hard for site characteristics beyond size and isolation that would explain the effect, mostly settling on poorer quality of smaller patches in terms of reproductive success resulting from increased rates of social parasitism (by the Brown-headed Cowbird [*Molothrus ater*]) or nest predation (Faaborg et al. 2010a). On the basis of this work, several investigators suggested that observed declines in forest-related migrants resulted from breeding-ground habitat fragmentation (Robbins 1979; Butcher et al. 1981; Lynch and Whigham 1984; Sherry and Holmes 1992:432). However, to date, no range-wide, long-term declines in migrant species have been linked convincingly to fragmentation effects (Faaborg et al. 2010a:16). Notably, this focus on forest fragmentation as a potential cause of migrant declines has been largely a North American concept (Newton 2008:736).

Rappole and McDonald (1994, 1998) suggested that the area effect in migrants might be explained using a different logic. They reasoned that for migrants that moved hundreds or thousands of kilometers between breeding and wintering sites, island biogeography principles affecting habitat pieces separated by a few kilometers were unlikely to produce measurable differences in species richness or density. Therefore, they concluded that the area effect likely was due to the factors reported by many researchers, namely differences in habitat quality. However, if this assumption were true, and breeding habitat were limited, then one should expect differences in the *kind* of individuals occupying the less-suitable habitat, not the *number*; that is, individuals of lower competitive status, presumably younger, would be pushed into the lower-quality habitat as predicted by Fretwell's (1972:98) "ideal despotic distribution." If breeding habitat were limited for a given migrant species, then the lower-quality habitat could have numbers of breeders in it equivalent to those in higher-quality habitat (figure 7.8) (Van Horne 1983; Winker et al. 1995; Newton 2008:417). Only if nonbreeding factors were limiting the population *below* breeding habitat capacity would one expect suitable breeding habitat to be

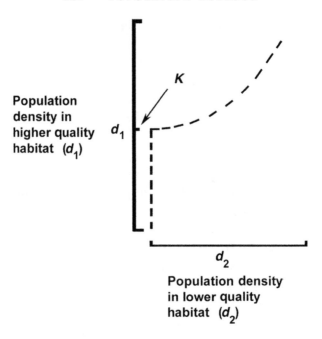

FIGURE 7.8   Predicted changes in density in two habitats of differing quality with increasing population resulting from immigration using Fretwell's "ideal despotic distribution" model (based on Fretwell 1972:107): d_1 = density in the higher-quality habitat, Habitat 1; d_2 = density in the lower-quality habitat, Habitat 2; K = carrying capacity of Habitat 1; *dashed line* = density change over time.

empty, even if it were of lower quality. A corollary of this prediction is that resident species breeding in the same habitats as migrants would not show the area effect, as they would not be subject to the same limiting, nonbreeding habitat as long-distance migrants (DeGraaf and Rappole 1995:28), a prediction supported by data from several studies (Whitcomb 1977; Lynch and Whitcomb 1978; Butcher et al. 1981; Askins and Philbrick 1987).

• *Postbreeding (molt) habitat loss.* In chapter 3, we present extensive evidence documenting the existence of distinct habitat needs during the postbreeding period for several migrant species. Whether or not these needs constitute carrying capacity limitation in any species is unknown.

• *Stopover habitat loss.* As in the case for postbreeding habitat loss, the logic behind the potential for population limitation is sound, but data are lacking for most species. However, some populations of the Red Knot (*Calidris canutus*) are evidently dependent upon very particular stopover sites that are widely separated geographically and whose resources are potentially threatened (Davidson and Piersma 1992; Harrington 1996; Piersma et al. 2005b).

• *Winter habitat loss.* Nearly 40 years ago, Fretwell (1972) and Terborgh (1974) suggested that migratory birds could be limited by winter habitat availability, a possibility considered by Lack (1954) as well. To date, however, attribution of long-term population declines to winter habitat loss or change for any migrant species is based almost entirely on inference. Nevertheless, the inference is strong in some cases. Newton (2008:716), for instance, summarizes data on population trends for the Greater Whitethroat (*Sylvia communis*) in Britain in relation to rainfall patterns in the Sahel region of Africa showing a close, positive correlation. The same problem has been suggested as causing declines in other long-distance migrants in Europe (Berthold 1999). However, Marchant (1992:119), using long-term survey data on 10 trans-Saharan migrant species from several European countries, found, "In general, . . . the correlations between rainfall and year-to-year population changes were weak and explained little of the total variation. Errors in measuring population change may be important sources of variation in the census data and, along with environmental noise, may confound the relationships being examined. More detailed rainfall indices, and better knowledge of bird movements between Europe and Africa, would improve the power of the analysis."

In a similar situation for New World migrants, Rappole et al. (1983:73–74) and others (e.g., Robbins et al. 1989b; Terborgh 1989) suggested that what the many species of declining forest-related migrants in North America shared in common was decimation of winter habitat. However, some of the same kinds of methodological problems mentioned by Marchant (1992) confront testing of this hypothesis. In addition, many researchers and conservation biologists, at least in North America, remain skeptical that winter habitat loss has a significant effect on migrant population size, presumably for reasons discussed in chapter 5 (Latta and Baltz 1997; Faaborg et al. 2010a:17). In truth, testing for where in the annual cycle declines are caused probably is best done on a species-by-species basis and is not simple even when so restricted (Haney et al. 1998; Rappole et al. 2003b). Because of this fact, Rappole and McDonald (1994, 1998) suggested that populations controlled during different periods of the annual cycle would have different characteristics that, unlike total amounts of breeding, wintering, and stopover habitat, could be measured with relative simplicity and, indeed, already had been measured for many migrants. They provided a list of 14 characteristics of breeding and wintering populations that could be readily measured to provide predictive information on likely source of declines. The most commonly measured and generally most readily available piece of information concerning migrant populations is breeding habitat occupancy. In general, following Fretwell (1972) as discussed earlier under "Breeding Habitat Fragmentation," they reasoned that if breeding habitat were limiting for a population, all available habitat would be filled, including that of poor quality, whereas if nonbreeding-season factors were limiting, apparently suitable breeding habitat would be vacant. Many investigators have documented the phenomenon predicted by Fretwell's (1972) ideal despotic distribution, in which highest-quality breeding habitats are occupied by the first-arriving males, followed by occupation of lower-quality habitats (Lundberg et al. 1981; Lanyon and Thompson 1986; Wiggins et al. 1994; Aebischer et al. 1996; Newton 1998; Currie et al. 2000). It should be recognized that the theoretical predictions deriving from

Fretwell's model are likely to be modified by species-specific circumstances—for example, those described by Schmutz (1987) for Swainson's Hawks (*Buteo swainsoni*) in which the principal component of breeding territory quality appears to be related to prey abundance rather than some specific structural aspect of habitat.

A CASE HISTORY

Determining whether or not any given population is limited by density-dependent as opposed to density-independent factors and where and when during the annual cycles such control might be exerted can be difficult questions to answer. The White-winged Dove (*Zenaida asiatica*) presents a situation illustrative of the problems involved in identifying precisely what is the cause of a long-term decline in a migrant population. Eastern populations of this species breed in South Texas, United States, and Tamaulipas, Mexico, and winter along the Pacific slope of Central America (figure 7.9).

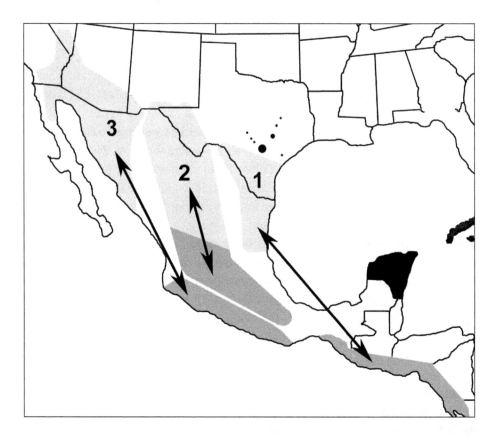

FIGURE 7.9   Breeding (*light gray*) and wintering (*dark gray*) range for three migratory populations of the White-winged Dove (*Zenaida asiatica*): 1 is referred to as the South Texas/Tamaulipas population in the text. Ranges of populations composed mostly or solely of resident birds are shown in black.

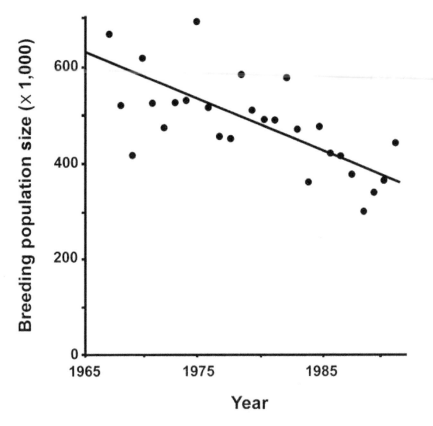

FIGURE 7.10   White-winged Dove (*Zenaida asiatica*) breeding population decline in South Texas, 1965–1993 (based on data in George et al. 1994).

Most of the migratory South Texas population occurs in the lower Rio Grande Valley (LRGV) along the border with Mexico. Since the 1960s, LRGV populations of the species have been under decline (George et al. 1994; Hayslette et al. 1996) (figure 7.10).

Amounts of breeding habitat (thorn forest and citrus groves) in the LRGV have fluctuated during this period, but there is no obvious relationship between these amounts and breeding population size (Hayslette et al. 1996, 2000). Populations are hunted in Texas, and the number taken annually varies from less than 30 percent to greater than 100 percent of the breeding population size (George et al. 2000); in other words, much or all of the annual productivity in the population is shot. Given this fact, one might conclude that populations are controlled by the principal, known mortality factor (i.e., hunting). However, there is a *positive* correlation between the number of birds killed in the fall and the size of the next year's breeding population (Rappole, Pine, et al. 2007). This finding seems counterintuitive. However, consider that the LRGV breeding population is only a small portion (5 to 10%) of the total size of the migratory eastern population of the species, most of which breeds in Tamaulipas (Nichols et al. 1986). We take these facts,

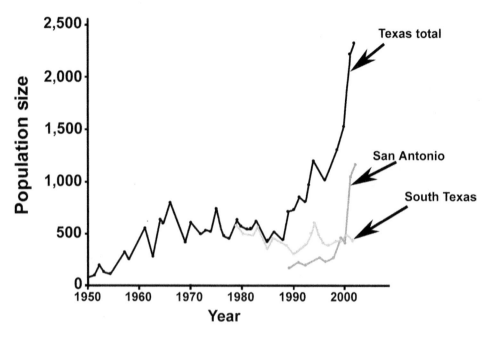

FIGURE 7.11   White-winged Dove (*Zenaida asiatica*) population change for migrants (*South Texas*), resident populations in the city of San Antonio (*San Antonio*), and total whitewing populations for the entire state of Texas (*Texas Total*, which includes all resident populations as well as the South Texas migrant population) (Rappole, Pine, et al. 2007).

when considered together, to mean that both the size of the fall LRGV kill and the next year's LRGV breeding population are controlled by the same factor—namely, breeding productivity for the entire eastern population (from which much of the next year's LRGV breeding population likely is recruited) during the season previous to the hunt. Whether or not this explanation is correct, the fact that there is a positive relationship between fall harvest and the next year's LRGV breeding population size does not support the argument that harvest rates have a negative effect on breeding population size.

An additional piece of information regarding Texas population change for the White-winged Dove is that prior to 1973, the entire population was migratory (Oberholser 1974). However, since that date, a resident population has appeared in several Texas cities (the San Antonio population now exceeds 1 million birds), which is not migratory. This population behaves as though it were a different species from the migratory population: During the same time period in which the South Texas migratory population (most of which lives in the LRGV) has declined, the Texas resident population has undergone exponential growth (figure 7.11). Clearly, food during the breeding period, mostly agricultural crops, which are available to both the resident and migrant populations, is not a limiting factor.

If neither hunting, breeding habitat loss, nor food availability during the breeding period is responsible for LRGV breeding population declines, then what is?

The answer to this question is not known, but Rappole, Pine, et al. (2007) have inferred the possibility of winter habitat loss based on the following reasoning:

> We propose that the migratory White-winged Doves may be controlled by wintering ground carrying capacity in southwestern Mexico and western Central America. In a migratory population, e.g., the rural LRGV White-winged Doves, at least two, separate carrying capacities potentially exert control over the population, i.e., breeding and wintering ground carrying capacity. In this situation, we propose that the habitat with the lower carrying capacity dictates population size (using an equation developed by Alan Pine [appendix A]). This possibility is illustrated in [figure 7.12] in which a hypothetical whitewing population is subject to control by: (1) a breeding ground carrying capacity (Habitat 1) during the summer; and (2) a wintering ground carrying capacity (Habitat 2) in winter. In this figure, a graph is shown for a model population of migratory White-winged Doves in South Texas (and Tamaulipas) for summer and winter habitats according to the equation:
>
> $$dx/dt = x \left\{ b_0 \exp[-(x/q_b)^{p_b}] - (d_0 - 1) \exp[-(x/q_d)^{p_d}] - 1 \right\}$$
>
> where x is the population N relative to the carrying capacity K over time t; b_0 and d_0 are the birth and death rates in a given habitat; q_b and q_d are the carrying capacities for the birth and death rates relative to K; p_b and p_d are the saturation powers for the birth and death rates representing the abruptness of the response at the respective carrying capacities; all parameters can vary with the habitat; initial carrying capacity in 1950 for both Habitat 1 and Habitat 2 are set as equivalent; carrying capacity for Habitat 1 [breeding ground] is considered to have remained the same from 1950 to the present while that for Habitat 2 [wintering ground] is considered to have declined linearly to roughly half its value in 1950 by 2005; the initial population size (x_0) in 1951 is set at 50% of the carrying capacity in Habitats 1 and 2 as of that date after the freeze of 1950; other parameters used are shown in [figure 7.12]. Annual birth and death rates for the population are based on a combination of the birth and death rates for Habitats 1 and 2 (note that birth rate in Habitat 2, the wintering ground, is assumed to be zero). The shape of the curve in [figure 7.12] will vary somewhat depending on the specific parameters used. However, so long as the average birth rate for the two habitats exceeds the death rate, the population will decline in accordance with whichever carrying capacity is smaller. The "carrying capacity" function in the equation essentially acts as a flexible death rate. The population goes to extinction regardless of carrying capacity if average death rate exceeds birth rate, unless offset by immigration, a factor not taken into account in this equation [although, as noted by reviewer Darren Fa, it would not be difficult to add immigration to the equation, which could be informed by empirical data].

Following the logic of this model, we propose: (1) that actual breeding habitat carrying capacity, although unknown, exceeds current population levels; (2) that

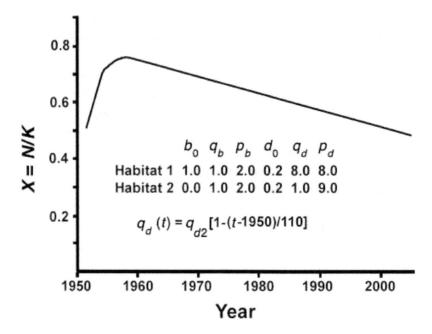

FIGURE 7.12   Hypothetical population growth curve for the migratory White-winged Dove population of South Texas as predicted by the equation $dx/dt = x\{b_0 \exp[-(x/q_b)^{pb}] - (d_0 - 1) \exp[-(x/q_d)^{pd}] - 1\}$, as described in the text for the time period 1950–2005. The seasonal variations (sharp annual "sawtooth" increases and declines associated with breeding-season reproduction followed by nonbreeding-season mortality) given by this sequential habitat model have been averaged for the purposes of graphic depiction (Rappole, Pine, et al. 2007:16).

actual winter habitat carrying capacity for migratory whitewings, although also unknown, is less than the number of birds migrating southward after years of good production; (3) that this winter habitat carrying capacity has been lower than breeding habitat carrying capacity and has been in decline since at least the early 1960s [Dinnerstein 1995]. These factors result in a curve showing an average long-term decline similar to that reported by Hayslette et al. (1996) (Rappole, Pine, et al. 2007:16–17).

CHAPTER 8

EVOLUTION AND BIOGEOGRAPHY

THE THEORY

The theory on which we base our understanding of the evolution of migration may be stated as follows: Migration is a form of dispersal in which resident, subadult birds entering their first breeding season are forced to leave by adult conspecifics. Dispersal is random in terms of both direction and distance, but some individuals move to areas distant from the original breeding area where resources (e.g., food, nest sites, or mates) allow successful reproduction. If the new area is aseasonal (i.e., in the sense that the individual can survive throughout the annual cycle and reproduce without having to go elsewhere), then dispersal is essentially colonization and range expansion; if the new area is seasonal (i.e., in the sense that the individual cannot survive throughout the entire annual cycle at the newly occupied site), then the birds return to their area of origin either when it is time to breed back on their natal site (if no mates were available at the newly occupied site) or after breeding when environmental conditions deteriorate, completing the first generation of migration. Intraspecific competition for breeding space and fitness benefits in terms of greater survival and/or reproductive output drive subsequent generations of migratory movement.

Theories on the origin of migration are generally assigned to one of two categories: "northern home," in which Holarctic migrants that breed in the Temperate or Boreal zones and winter in the tropics are thought to have originated from resident populations that were originally resident in the Holarctic but were pushed southward by deteriorating climates; and "southern home," in which resident tropical species invaded seasonal climates in the Holarctic (Gauthreaux 1982; Rappole 1995:102) (although see Cox 1985 for a "hybrid" theory). Our theory for the origin of migration through dispersal is neither northern or southern, and no change in climate at point of origin is required (contrary to Cohen 1967; Gauthreaux 1982; Newton 2008:370). All that is needed for it to occur is intraspecific competition for breeding space and a new, reachable environment where competition is less or absent and probabilities of survival and/or reproduction are greater than at the site of origin.

"Dispersive" migration (Newton 2008:513–518) may provide an example of an early stage in the development of migration through dispersal. In this type of migration, individuals of resident species, often disproportionately females and young, disperse in random directions and distances from breeding areas to wintering sites, returning in time for the next breeding season (e.g., Spruce Grouse [*Dendropagus canadensis*] [Herzog and Keppie 1980]). Presumably, survivorship is higher at the wintering sites for those birds that moved than would have been possible at the point of origin, but because the wintering sites are not suitable for breeding, whether from lack of mates or other circumstances, the birds return to point of origin in spring. This type of movement fits a "northern home" migration scenario, in which the movement is caused by competition at the point of origin, forcing birds away from the breeding area to sites where survival probabilities are greater. However, if the result of this movement were to result in occupation of sites with mates and suitable, seasonal breeding habitat, then one can see how the result could fit a "southern home" migration origin perspective. Clues as to ancestral point of origin for any given population of migrants are provided by their ecology, behavior, and phylogenetic relationships (Rappole 1995:111–112).

EXAPTATIONS VERSUS ADAPTATIONS FOR MIGRATION

A major theme of the treatment of avian migrants presented in this book is that the "dispersal" theory for the origin of migration requires no genetic change in the originating population of first migrants, which means that much of the morphologic, physiologic, and behavioral aspects deemed definitive of migrants are, in fact, exaptations evolved to enhance the fitness of ancestral resident populations. However, these exaptations, which make migration possible for the first generation of migrants, do not make migration an optimal strategy, only the better of two choices (in a fitness sense). Therefore, a corollary for this hypothesis is that natural selection may act quickly to modify these exaptations in ways that enhance fitness of migratory populations. Throughout the book, a number of these possible enhancements to exaptations have been suggested, many of which are summarized in table 8.1. Note that this list is exemplary and illustrative rather than exhaustive. It is likely that migration imposes a new selective regime on most of the phenotype, resulting in an increasing number of modifications with the passage of evolutionary time.

GENETICS OF MIGRATION

Nearly a century of field and laboratory work has documented hereditary endogenous control over major aspects of migrant life history, including onset of actual migratory flight (see chapters 4 and 6). Among the most important research programs devoted to investigation of this topic has been that conducted by researchers at the Max Planck Institute of Ornithology, led for many years by Eberhard Gwinner and Peter Berthold. Large parts of this work have involved artificial selection experiments with Blackcaps (*Sylvia atricapilla*), European Robins (*Erithacus rubecula*), and a few other species (Biebach 1983; for reviews, see Berthold 1988, 1999; Pulido and Berthold 2003; Pulido 2007). The focus of the selection in these trials is on the expression of *Zugunruhe* (migratory restlessness). Among various experiments, one involved the taking of birds into the laboratory from partially migratory wild populations, some of which were "migratory" (i.e., those that show *Zugunruhe*) whereas others were "nonmigratory." Migratory birds were mated with migratory birds, and nonmigratory birds were mated with nonmigratory birds. After three generations, all progeny produced were "migratory" in the migratory × migratory crosses, whereas nonmigratory × nonmigratory crosses took five generations for all progeny to be "nonmigratory." The results appear to confirm a genetic basis for migratory movement. However, they raise a number of issues that are not easily understood (see chapters 4 and 6). The central problem underlying interpretation of the results is the fact that *Zugunruhe* is not migration: It is a behavior caused by the frustration of migration that is used as a stand-in for actual migratory flight (*Zugstimmung*). Although the experimenter assumes that *Zugstimmung* (or its stand-in *Zugunruhe*) is a single trait, there is no way to know this, and we suggest that this assumption is highly unlikely to be correct. Onset of migration, and presumably initiation of migratory flight, are directly related to the timing of other life history parameters including completion of molt and preparation for migratory flight (*Zugdisposition*). Furthermore, extensive field studies have demonstrated that initiation of migratory flight, and therefore, presumably, *Zugunruhe*, is related not only to the timing of molt but also to the sex and age of the bird, as well as the individual's physiologic state (in terms of "readiness" to migrate) and environmental conditions, usually some aspect of weather, as a reflection of wind conditions aloft. In summary, we do not know what the mating procedures conducted in the experiment selected for: All we know is whether they produce progeny expressing

TABLE 8.1   Exaptations for Migration That Resident Species Possess, and the Likely Set of
Modifications of These Exaptations (Adaptations) That Result from Natural Selection
Resulting from Migration

Category	Exaptation	Adaptation
Flight muscle composition	Three major types of muscle fiber compose the flight muscles of birds, the proportion of which varies according to foraging and predator-avoidance needs.	The proportion of flight muscle composed of fast oxidative glycolytic (FOG) fibers will increase.
Flight	Wings	Wing shape has important effects on the efficiency of prolonged, horizontal flight; long-distance migrants tend to have highly convex, pointed wings.
Energy storage composition	Energy is stored by residents as triacylglycerol for overnight survival, fledging, dispersal, egg laying, cold weather, and other facotrs.	The location of energy deposits will change from what is quickest to what is more aerodynamic in terms of weight distribution; also, the proportion of energy stored as triacylglycerol as opposed to protein will change as a function of in-flight water needs.
Homing	Residents are capable of homing to point of origin using many of the same sensory cues as migrants.	Genetically based selection may shape the structure of migration routes that show specific inter-route directional changes (Helbig 1991) and differences between fall and spring.
Orientation	Residents are capable of orientation using many of the same sensory cues as migrants.	It is likely that modification of sensory-cue use occurs over evolutionary time, particulary in terms of response to proximate environmental cues triggering migratory movement (e.g., local weather). Another example may include Siberian-breeding populations of the Pectoral Sandpiper, which appear to make trans-polar migrations to winter in South America (Holmes and Pitelka 1998).
Endogenous control over timing of life history events	Residents possess endogenous cues to respond to local conditions to begin breeding and molt.	Migration likely imposes a completely different regime in terms of adaptive timing of breeding and molt. Comparison of endogenous timing of life history events between related populations of tropical resident versus temperate migrant is likely to provide information on the length of time for which the temperate-breeding population has been migratory.

CATEGORY	EXAPTATION	ADAPTATION
Molt	Residents generally have a single, protracted molt each year.	Migrants often have two or more molts per year. Comparison of molts between related populations of residents and migrants will likely provide information on how long the temperate-breeding population has been migratory.
Plumage coloration	Residents tend to show less plumage difference between the sexes than related migrants.	Degree of plumage difference (i.e., amount of polymorphism in male and female plumages) may be a reflection of the time period for which the population has been migratory, at least in some groups (e.g., Parulidae).
Sex roles	Most resident birds are monogamous and remain in pairs for large portions of the year, sharing in territorial defense.	Defense of the breeding territory is largely the responsibility of the male; he must arrive earlier than the female and remain on the territory longer after breeding is completed, often choosing not to migrate in partial migrant species.
Age roles	Young resident birds tend to remain with adults longer and may even assist in raising subsequent broods (i.e., serve as "helpers").	Young migrants generally reach independence less than 4 weeks after leaving the nest.

Note: Selection is predicted to vary by the length of the migratory journey and the evolutionary time span over which the population has made the journey. Note that the comparisons are general and meant principally to contrast "resident" exaptations with "migrant" adaptations for related species.

Zugunruhe or not. However, what if the selection were not acting on "the migratory syndrome of which *Zugunruhe* is a part" but on some aspect of the environment that triggers *Zugunruhe*? For instance, it may be that in some part of the Blackcap or robin population, *Zugunruhe* is triggered by lower temperatures than in another part. Because laboratory temperatures are kept constant (20°C), the selection experiments being conducted might be acting upon this trigger mechanism, not the "syndrome." We do not question the genetic nature of control over initiation of migratory flight; however, we do question the meaning of *Zugunruhe* experiments in terms of elucidating how that control is exercised or inherited.

Pulido and Berthold (2003:69–71), based largely on their work with Blackcap *Zugunruhe*, have argued that migration results from a single gene, which likely is common to all bird species. According to their theory, the "migratory" allele for this gene occurs at low levels in populations of nonmigratory species, but a change in the environment can exert selection pressure causing expression of the "migratory" allele to increase in frequency, resulting in conversion of the population from nonmigratory to partially migratory. Continued selection of increasing severity (e.g., a seasonally deteriorating environment caused, for instance, by an advancing ice age) results in greater and greater percentages of the population becoming migratory until a near 100 percent long-distance migratory population is produced.

These and similar findings and ideas regarding the genetics of migration have led to the formulation of a hypothesis in which all aspects of migratory movement are considered to represent an ancient, highly integrated bundle of adaptive traits termed the "migratory syndrome" (Dingle 1996; Pulido and Berthold 2003; Liedvogel et al. 2011; Zink 2011). The "migration as dispersal" hypothesis that we have developed differs fundamentally from the "migratory syndrome" hypothesis. As discussed throughout much of our book, the migratory lifestyle has profound effects on many different aspects of a bird's phenotype. Nevertheless, there are many bird species that show few or none of the adaptations for migratory flight that are listed in table 8.1 that, nonetheless, are migratory (e.g., Greater Prairie-Chicken [*Tympanuchus cupido*] [Cooke 1888]). According to our hypothesis, sedentary birds possess all of the genetic equipment necessary to allow dispersal to become migration: All that they require to complete the transformation is a reachable seasonal environment unoccupied by conspecific competitors where more offspring can be raised than in their ancestral environment. In our view, the fact that bird species that are apparently poorly adapted for migratory movement do migrate nonetheless argues strongly in favor of the view that migration is a behavior with potentially high fitness rewards if the right environment can be located, regardless of the quality of the bird's adaptations for movement. No migratory syndrome of co-adapted characters is required (Piersma et al. 2005a).

RAPID DEVELOPMENT OF MIGRATION

In addition to the fact that the theory for evolution of avian migration presented by Pulido and Berthold (2003) requires only a single gene, it is also a "northern home" hypothesis—that is, one that requires an environmental change at the ancestral point of origin for the migratory population in order for migration to begin (Gauthreaux 1982; Berthold 1999; Bell 2005). We have argued throughout our treatment that no such environmental change at point of origin is required in order for migration to appear in a population: All that is needed is a new seasonal environment within normal dispersal distance. Historical information on the rapid development of migration from sedentary, resident populations provides strong support for this hypothesis. In the following discussion, we examine rapid changes in distribution and seasonal movement patterns known to have occurred for selected bird species.

Establishment of permanent colonies by Europeans on the mainland of the New World began in the mid-sixteenth century, and by the mid-seventeenth century, several thousand colonists were arriving each year (Simmons 1981). In temperate North America, most colonists settled initially along the coastal plain, but after the American Revolutionary War (1775–1783), human populations rapidly expanded westward, clearing land for subsistence and market agriculture as they went. The "Longitudinal Median Center of Population" for the United States (i.e., the line of longitude selected so that half the human population of the country lives east of it and half lives west) was located near Baltimore, Maryland, about 150 km inland from the eastern coast in 1790; by 1890, the line was located along the Ohio–Indiana border, 900 km to the west (figure 8.1) (U.S. Bureau of the Census 1990). Forest clearing in the United States reached its historic maximum around 1900 when nearly 70 percent of forested regions had been cleared, mostly in the eastern half of the country (Powell and Rappole 1986). Thereafter, several factors combined to reverse this process, including large-scale abandonment of economically marginal farms in the east, and development of a conservation ethic (Noss et al. 1997). At present, forested area in the United States is roughly 60 percent (Powell and Rappole 1986), and in many eastern parts of the country, reforestation rates are much higher (Rappole and DeGraaf 1996; Askins 2000). Human effects on populations of a number of bird species from market hunting peaked around 1900 (Ogden 1978). At about this time, public concern became aroused regarding the disappearance of the seemingly boundless resources of the North American continent, resulting in passage of the Lacey Act and similar measures controlling wildlife and harvest levels (Noss et al. 1997). These and other legal and land protection changes allowed gradual recovery of a number of species that had been nearly or completely extirpated from large portions of their range, including many migratory birds (e.g., Snowy Egret [*Egretta thula*]) (Bent 1926; Ogden 1978). Hundreds of species of birds were affected by factors associated with European colonization of the New World. Remarkably, however, only a few were actually driven to extinction: Passenger Pigeon (*Ectopistes migratorius*), Carolina Parakeet (*Conuropsis carolinensis*), and, perhaps, Bachman's Warbler (*Vermivora bachmanii*), Ivory-billed Woodpecker (*Campephilus principalis*), and Eskimo Curlew (*Numenius borealis*). All other avian species native to the continent have made adjustments of one sort or another to the drastic environmental alterations associated with human effects. In some cases, these adjustments have been remarkable, involving rapid development of migration and radical changes in migration patterns. Five exemplary species in which these changes were especially marked are discussed in the following accounts. For each of these species, well-documented shifts in distribution and seasonal movement patterns took place on a continental scale in a matter of decades.

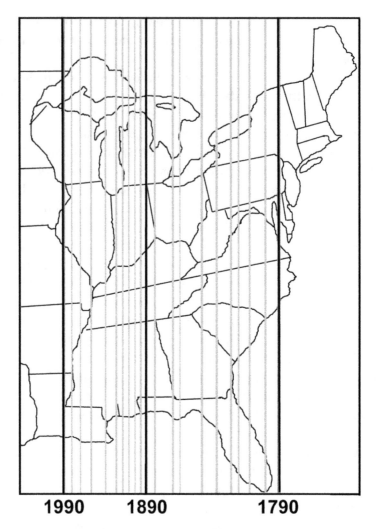

1990   1890   1790

FIGURE 8.1   The "Longitudinal Median Center of Population" for the United States, 1790–1990, as explained in the text.

• Bewick's Wren (*Thryomanes bewickii*). Bewick's Wren inhabits scrubby, low thorn forest, chaparral, thickets, and brushy second growth of the type often associated with small family farms (Ridgway 1889:92; Bent 1948). The bird was first described as "Bewick's Wren, *Troglodytes Bewickii*," by John James Audubon based on an individual collected near St. Francisville, Louisiana, in 1821 (Audubon 1839). Evidence from Audubon, and the lack of early records from eastern North America in the historical literature (Wilson and Bonaparte 1808–1814; Kennedy and White 1997), indicate that prior to expansion of European settlement across the continent, the Bewick's Wren was largely a sedentary, resident species whose distribution was restricted to the south-central and western plains and Pacific coastal regions of North America (figure 8.2A). However, by the 1830s, the bird had expanded its breeding range into the Appalachians of the eastern United States, and by the 1920s

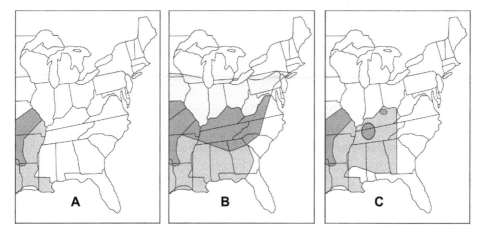

FIGURE 8.2   Bewick's Wren (*Thryomanes bewickii*) in eastern North America in 1800 (*A*), 1920 (*B*), and 1996 (*C*): *dark gray* = permanent resident portion of range; *light gray* = summer resident portion of range; *medium gray* = winter resident portion of range.

it was found as a partial migrant breeding throughout much of the northeastern United States and wintering in much of the southeast (American Ornithologists' Union 1886, 1895, 1910, 1931, 1957; Cole 1905; Todd 1940; Bent 1948; Kennedy and White 1997) (figure 8.2B). By the 1930s, the eastern breeding range had begun to collapse, and the bird is now rare at any time of the year east of the Mississippi (Bartgis 1986; Wilcove 1990; Sauer et al.1996; Kennedy and White 1997; American Ornithologists' Union 1998) (figure 8.2C). Kennedy and White (1997) suggest that extirpation of the Bewick's Wren from the eastern United States has occurred as a result of competition for nesting sites with the House Wren (*Troglodytes aedon*), which has been observed to destroy Bewick's Wren nests. This explanation seems unlikely for two reasons. First, House Wren distribution overlapped that of the Bewick's Wren during the entire period of Bewick's Wren range expansion in the northeast (American Ornithologists' Union 1931; Forbush and May 1939). Second, Bewick's Wren and House Wren distributions overlap extensively in the central and western portions of their ranges where no consistent changes in Bewick's Wren populations have been documented (Sauer et al. 1996). Other factors therefore must have been involved.

We suggest that disappearance of northeastern, migratory populations, which has been as rapid as their appearance (figure 8.2), has resulted from reversion of large amounts of scrub habitat to forest as farms were abandoned. Kennedy and White (1997) argue that this factor should not have caused range collapse because much seemingly suitable habitat still exists in the region. However, we contend that the fact that suitable seasonal habitat persists is not significant. What is important is relative success of the individuals either occupying the disappearing habitat or moving on (Adriaensen and Dondt 1990). If dispersing individuals are unsuccessful in locating breeding habitat or mates, they can go elsewhere because the movement pattern is not genetically fixed (Winkler 2005). Likewise, if migrants are unsuccessful in locating conducive conditions, they would most likely move on (Helms 1963; Gwinner and Czeschlik 1978; Terrill 1987).

• Bachman's Sparrow (*Aimophila aestivalis*). Bachman's Sparrow was first described as *Fringilla aestivalis* by Lichtenstein (1823) based on a specimen from Georgia. Audubon (1834), unaware of Lichtenstein's discovery, described the species again as *Fringilla Bachmanii*, Bachman's Pinewood-Finch, based on a specimen shot near Charleston, South Carolina. Audubon and his South Carolina host, Rev. John Bachman, found the bird to be quite common in "pine barrens" and "pine woods" with low scrub oak and tall pines—the mature pine savanna that was formerly a common habitat in the coastal plain of the southeastern United States. Notably, Audubon (1841:114) conducted an ad hoc roadside survey on his horseback journey from Charleston, South Carolina, to New York in June 1832 and "observed many of these Finches on the sides of the roads cut through the pine woods of South Carolina. At this time, they filled the air with their melodies. I traced them as far as the boundary between that State and North Carolina, in which none were seen or heard." As in the case of the Bewick's Wren, ornithologists working in eastern North America north of the Carolinas had not seen the bird in the early 1800s (Wilson and Bonaparte 1808–1814). Thus, it seems likely that its distribution was restricted to the mature open pine savanna of the southeastern United States from South Carolina to East Texas from the time of European settlement up until at least the 1830s as a year-round resident or partial migrant, with some birds withdrawing farther southward in winter (figure 8.3A). In the mid-1800s, distribution of the species was still understood to include only "southern states" (Coues 1872), but by the late 1800s, range expansion into the northeastern states was recognized (American Ornithologists' Union 1886, 1895). By the early twentieth century, migratory breeding populations were expanding northward rapidly (Brooks 1938) and were well established in second growth, scrub habitats associated with small farm holdings throughout much of the northeastern United States as far as Illinois, Ohio, and Pennsylvania by the 1920s, nearly 1,000 km north of the original range (American Ornithologists' Union 1931, 1957, Bent 1968) (figure 8.3B). At present, nearly the entire northeastern migratory population has disappeared (Dunning and Watts 1990; Dunning 1993; Sauer et al. 1996), and current distribution of the species looks very similar to what it must have been like in presettlement times (figure 8.3C). The process of habitat loss and abandonment of large sections of the migratory portion of the species' breeding range, described for Bewick's Wren, is also the likely explanation for disappearance of the Bachman's Sparrow from the northeastern United States (Dunning and Watts 1990).

• Loggerhead Shrike (*Lanius ludovicianus*). Wilson and Bonaparte (1808–1814, Vol. 1:107) described shrike distribution in the early 1800s as a resident bird inhabiting the warmer parts of the United States, specifically, "the rice plantations of Carolina and Georgia." By the 1920s, a migratory population had appeared, breeding in open farmland throughout northeastern North America north to

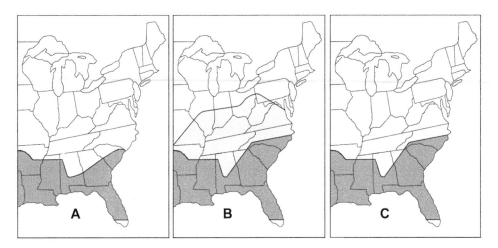

FIGURE 8.3   Bachman's Sparrow (*Aimophila aestivalis*) in eastern North America in 1800 (*A*), 1900 (*B*), and 1996 (*C*): *dark gray* = permanent resident portion of range; *light gray* = summer resident portion of range. Summer residents migrate to "permanent resident" portion of range for the winter. See text for discussion.

Maine and New Brunswick and wintering in the southeastern states; this population was considered sufficiently distinctive to be described as a separate subspecies (*Lanius ludovicianus migrans*) (Forbush and May 1939). As of the late twentieth century, this migratory population had largely disappeared from its northeastern breeding range (Yosef 1996).

• Snowy Egret (*Egretta thula*). Wilson and Bonaparte (1808–1814, Vol. 2:309) described Snowy Egret distribution in the early nineteenth century as follows: "This elegant species inhabits the seacoast of North America, from the Isthmus of Darien to the Gulf of St. Lawrence, and is, in the United States, a bird of passage; arriving from the south early in April and leaving the Middle States again in October." Audubon (1843:163) thought Wilson had overstated the northern distribution, finding that "it rarely proceeds farther than Long Island in the State of New York; few are seen in Massachusetts, and none farther to the east." He had no information on western distribution and thought the bird seldom traveled farther north up the Mississippi during the breeding season than Memphis, Tennessee. The American Ornithologists' Union check-list for 1886 (American Ornithologists' Union 1886) described the range as including temperate and tropical America, north along the East Coast to Long Island (New York) (figure 8.4A). In the late nineteenth century, the Snowy Egret, along with several other heron and tern species, was hunted intensively for its plumes, resulting in near extirpation of the species in the migratory portion of its range in eastern North America (American Ornithologists' Union 1910; Bent 1926; Ogden 1978; Parsons and Master 2000) (figure 8.4B). Proscriptions against hunting and other conservation measures, including creation of an extensive system of wildlife refuges, were instituted in the first half of the twentieth century in the United States, allowing recovery of many species. Currently,

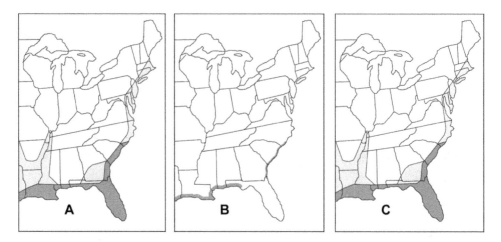

FIGURE 8.4   Snowy Egret (*Egretta thula*) in eastern North America in 1800 (*A*), 1900 (*B*), and 1996 (*C*): *dark gray* = permanent resident portion of range; *light gray* = summer resident portion of range. Summer residents may migrate into Central or South American portions of range for the winter.

the Snowy Egret has reoccupied most, if not all, of its former range, with a distribution that extends up the entire immediate coast of the United States eastern seaboard from Florida to Maine. Breeding Bird Survey and Christmas count data document that populations north of Georgia are mostly migratory, heading south by mid-September and returning in March or April (Sauer et al. 1996; Parsons and Master 2000) (figure 8.4C).

   • Cattle Egret (*Bubulcus ibis*). Members of this species, apparently derived from western Africa where populations are largely resident, first arrived in the Western Hemisphere along the coast of northeastern South America (Surinam) from 1877 to 1882 (Telfair 2006). The bird was not recorded outside this region until the 1930s when records began to accumulate from several Caribbean islands. The species was first observed in the United States in Florida in 1941, and the first breeding record, also from Florida, was recorded in 1953. From these initial dispersal events, populations rapidly occupied pastures, parks, and residential areas created largely by human activity (Telfair 2006). At present, breeding populations of the Cattle Egret are found throughout the eastern United States along the coastal plain from Texas to New Jersey. Most of these birds are migratory, departing their breeding grounds in September for wintering grounds in the Gulf States, Mexico, Central America, and the Caribbean, and returning in March or April (Sauer et al. 1996; Telfair 2006) (figure 8.5).

   In view of the historical evidence just outlined, we propose that the ultimate cause for the rapid range expansion and appearance of migratory populations, documented for the Bewick's Wren, Bachman's Sparrow, and Loggerhead Shrike was creation of extensive scrub habitat by farmers in formerly forested areas of

FIGURE 8.5   Cattle Egret (*Bubulcus ibis*) in eastern North America in 1996 (Telfair 2006): *dark gray* = permanent resident portion of range; *light gray* = summer resident portion of range. Summer residents may migrate into Central or South American portions of range during the winter.

the northeastern United States during the 1800s, which provided a new, seasonal environment for successful settlement by dispersing young individuals of these species. This cause has been endorsed by a number of authors (Brooks 1938; Bent 1948, 1968; Wilcove 1990; Dunning 1993; Kennedy and White 1997), but no mechanism has been suggested to explain how populations of principally or wholly sedentary species were able to occupy new breeding ranges hundreds of kilometers north of their original range as migrants in less than a century. Similarly, the Cattle Egret, newly arrived on the North American mainland in 1941, rapidly developed migratory populations occupying seasonal grassland habitats, mostly created by human activity, along the coastal plain of the eastern United States, and Snowy Egrets reoccupied as migrants in the 1900s the seasonal coastal habitats from which they had been extirpated in the late 1800s.

The continental changes in range and migration pattern for these five species are indicative of extensive changes that took place in the movement patterns for many North American species during the colonization and post-colonization period (table 8.2). They demonstrate the extraordinary flexibility in the avian dispersal/migration system, even in one that is newly developed. We contend that this flexibility results from the underlying physiologic and behavioral structure

TABLE 8.2   Changes in Movement Patterns for Selected North American Species

SPECIES	CHANGE
Greater Prairie-Chicken (*Tympanuchus cupido*)	Range of this resident grassland species expanded northward into areas cleared of boreal forest regions; populations of newly occupied regions were short-distance migrants (Cooke 1888).
Great Blue Heron (*Ardea herodias*)	Migratory populations were extirpated in the northeastern United States by the late 1800s; at present, migratory populations have reoccupied former range (Ogden 1978).
Great Egret (*Ardea alba*)	Migratory populations were extirpated in the northeastern United States by the late 1800s; at present, migratory populations have reoccupied former range (Ogden 1978).
Snowy Egret (*Egretta thula*)	See text for account.
Little Blue Heron (*Egretta caerulea*)	Migratory populations were extirpated in the northeastern United States by the late 1800s; at present, migratory populations have reoccupied former range (Ogden 1978).
Tricolored Heron (*Egretta tricolor*)	Migratory populations were extirpated in the northeastern United States by the late 1800s; at present, migratory populations have reoccupied former range (Ogden 1978).
Cattle Egret (*Bubulcus ibis*)	See text for account.
Loggerhead Shrike (*Lanius ludovicianus*)	Eastern United States populations of this species evidently were restricted to the southern portion of the region at the time of colonization; migratory populations developed in the northeastern region during the 1800s, which have now largely disappeared.
Bewick's Wren (*Thryomanes bewickii*)	See text for account.
Bachman's Sparrow (*Aimophila aestivalis*)	See text for account.

of migration, hinging on exaptations for migration, dispersal, and experience, and subject to genetic change. What causes a portion of a resident population actually to become migratory may not, therefore, entail an initial genetic change in capabilities, but rather competition, dispersal, and availability of a new, seasonal environment that is reachable by dispersing members of the population, presumably young birds for the most part, unable to compete with adults for breeding habitat within the existing, aseasonal range (Rappole and Warner 1980; Rappole et al. 1983; Rappole and Tipton 1992).

REACTION NORMS

Several lines of evidence indicate that individuals show significant flexibility in many aspects of their migratory habits, although the degree varies by species. For instance, differences in the precision of spring arrival times on breeding areas among birds have led to classification either as "calendar" or "obligate" migrants, in which departure timing from the wintering ground appears to be largely endogenous, as opposed to "weather" or "facultative" migrants in which departure timing from the wintering ground appears to be mostly under environmental control. These differences likely are in degree rather than kind. Even for calendar, obligate, long-distance migrants whose ultimate departure decisions are thought to be largely under the control of inherited migration programs (Gwinner 1986; Helbig 1996; Pulido and Berthold 2003), there is extensive evidence of flexibility in terms of the migration journey—for example, in the immediate timing of departure (known to vary according to the individual's physiologic state, local weather conditions, and presence of vocalizing conspecifics); direction of migration (known to vary by age and the individual's physiologic condition); duration of an individual migratory flight (varies according to physiologic state and, perhaps, presence of conspecifics); duration of the entire migratory journey (varies according to vicissitudes encountered along the way, and perhaps, age and sex). Thus, flexibility is built into the migration journey, even for individuals of species whose migratory behavior appears to be genetically programmed. What can account for such flexibility? Reaction norms have been proposed as providing an explanation—a theoretical construct defined as "a property of a genotype. It describes the phenotype formed by that genotype under different environmental conditions" (van Noordwijk 1989:455). For migrants, "environment" includes internal (e.g., age and physiologic condition) as well as external factors. Thus, reaction norms help to explain how individual members of a population with similar genotype can demonstrate markedly different responses to the same stimuli. They represent the mean value and variation for a particular trait in a given situation and are assumed to vary in their degree of flexibility by species.

The responsiveness for a particular migratory trait (e.g., immediate departure on fall migratory flight) to individual and environmental factors can be described by the reaction norm for a population (i.e., the mean value of a given trait for a given situation) (figure 8.6) (van Noordwijk 1989; Schlichting and Pigliucci 1998; van Noordwijk et al. 2006). The slope of the response over a set of conditions is a measure of phenotypic plasticity (i.e., the range of phenotypes expressed by the population over a set of conditions). A steep slope indicates that a population is highly responsive to different conditions, whereas a horizontal line reveals the absence of systematic variation in individual or environmental effects on migratory behavior. Individual birds vary around the population reaction norm and hence differ from mean migratory behavior. Such variation arises from genetic differences between individuals but also may reflect differences in age, sex, physiologic state, or experience.

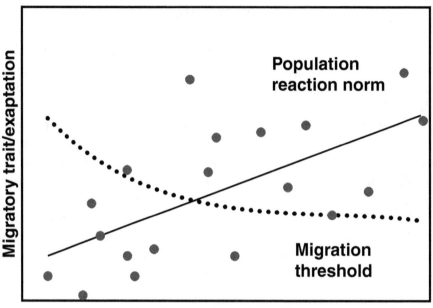

FIGURE 8.6   Schematic illustration of a population reaction norm. Expression of the migratory trait for immediate departure on fall migration (*y*-axis) varies over an environmental gradient that includes weather as well as individual status—for example, age, sex, or physiologic condition (*x*-axis). The slope indicates the mean departure response to environmental variation (i.e., the mean population reaction norm). Individual population members are shown by gray circles and scatter around the mean. Above a certain threshold value (*dotted line*), departure is initiated (based on ideas developed by B. Helm).

The power of the reaction norm is that it provides a theoretical basis for the behavior of actual migrants. As we have discussed extensively elsewhere (e.g., chapters 4 and 6), individual migrants show a great deal of flexibility in terms of their ability to respond to environmental variation according to their own specific circumstances. Its weakness lies in the fact that it is, for the present at least, difficult to explain based on single-gene control over physiologic and behavioral responses of the individual.

POPULATION DIFFERENTIATION

The chief impetus for the evolution of migratory species is the development of seasonal environments, which probably date back at least to the Eocene (37–48 mya), and perhaps longer. However, roughly 2.7 mya, the amount of seasonality increased sharply with the beginning of the Pleistocene (Molnar and Cane 2002), characterized by pulses of polar ice sheet advance and retreat, which continue to the present. Many modern species of migrants evolved over the past

2 million years or so during this time of greatly increased seasonality, and a number of workers have attributed this speciation to migrant breeding range fragmentation resulting from glacial advances (i.e., creation of habitat pockets called "refugia") (Mengel 1964, 1970; Bermingham et al. 1992; Bell 2000), although Zink and Slowinski (1995) have questioned the hypothesis of increased rates of Pleistocene speciation (for all species, not just migrant birds), and Zink and Klicka (1997) found that speciation events for several North American migrants appeared to predate Pleistocene glaciations. The beauty of the hypothesis positing Pleistocene refugia as the principal agents of recent migratory bird speciation is that it can be made to fit nicely with Mayr's concept of how species are formed, and it is clear from current distributions in some migrant groups that *something* related to Pleistocene events affected species formation (Mengel 1964; Rappole 1995:115–122).

Mayr (1982:273) defines a species "as groups of interbreeding populations reproductively isolated from other such groups." He further defines the process of species formation or speciation as the acquisition of reproductive isolation by a population or group of populations, where "reproductive isolation" is assumed to mean geographic isolation of populations, at least during the breeding season. Although evidence for actual breeding ranges for any migrant bird species or their ancestors during the various Pleistocene glacial advances and retreats is fragmentary or nonexistent, the extent and age of habitat fragments in refugia are better understood (figures 8.7 and 8.8). Thus, these reasonably well-established refugia, along with rather crude molecular dating of the timing of the speciation event, provide a theoretically satisfying explanation for how speciation could have occurred in migrants based on Mayr's allopatric model.

Mayr's allopatric model has tremendous explanatory power, but it may not be the only process at work in population differentiation. For instance, data on populations occupying neighboring, but strikingly different, ecological zones indicate that speciation may occur despite the potential for interbreeding between the populations (Rappole et al. 1994; Herder et al. 2008; Winker 2010a; Pfaender et al. 2010; Pfaender 2011). In addition, migrant life history presents some unique circumstances that may be relevant to the speciation process—for example, from where in the range (i.e., breeding versus wintering) the migrant population is derived, and how important mixing of metapopulation endogenous timers might be in terms of discouraging interpopulation breeding.

DERIVATION OF MIGRANT POPULATIONS

We have proposed that migrant populations derive mostly from ancestral populations resident in the aseasonal portion of the species' range (i.e., the "wintering" ground for the most part). Thus, it is possible, perhaps probable, that extant species could coexist with ancestral species, contrary to common thought among systematists (e.g., Zink and Slowinski 1995:5833).

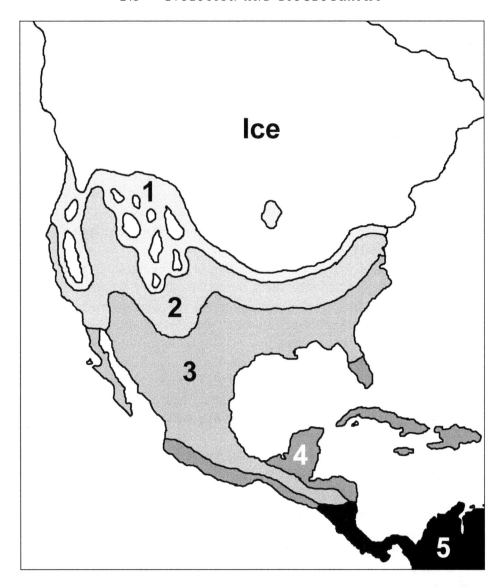

FIGURE 8.7   North American climatic conditions during the Pleistocene glacial maximum: (1) arctic; (2) subarctic; (3) temperate; (4) subtropical; (5) tropical (based on Mengel 1964).

Our hypothesis is based on the following information discussed elsewhere in the book:

• *Phylogenetic relationships.* Resident relatives (conspecifics or congeners) for most migrant species are found on the aseasonal portion of the range, not the seasonal portion (see chapter 5).

• *Community ecology.* If migrant species were derived from populations resident in the seasonal portion of their range, then the first migrants would be invading aseasonal or, at least, less seasonal environments with existing communities of

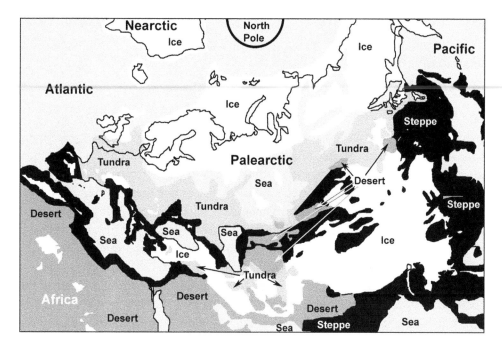

FIGURE 8.8 Palearctic habitats during the Pleistocene glacial maximum (based on Finlayson and Carrion 2007): *white* = ice; *light gray* = water; *medium gray* = tundra; *dark gray* = desert; *black* = steppe, woodlands, and savanna.

resident species. Unless resources were available throughout the wintering season in excess of what the resident birds could consume, migrants would have to exist as "fugitive" species by continually moving from one patch of superabundant resources to another, as predicted by MacArthur (1972). Observations of many migrant species, however, document that they exist as members of the communities they enter during the wintering period (see chapter 5).

• *Population ecology*. For those few migrant species that have been investigated, populations appear to be limited by food resource availability in the aseasonal portion of the range. In terms of niche theory, this finding is indicative that it is the aseasonal portion of the range in which critical aspects of the foraging niche evolved (i.e., where the individual can outcompete members of any other species for limiting food resources) (see chapters 5 and 7).

If the hypothesis of migrant origin from the aseasonal range is correct, then the first migrants must overlap with conspecific residents during the wintering period. Thus, for most species of migrants, differentiation would occur, whereas the migrant portion of the migrant and resident populations overlapped completely, albeit if only for a portion of the annual cycle. Such overlap is found for many species, with varying levels of differentiation in evidence (Rappole 1995:110–111). This speciation process may be considered to be a form of allopatric speciation in that the populations breed in geographically separate locations. However, it is not quite equivalent to Mayr's model because we do not know the degree to which individuals of the migrant population can elect to remain on the wintering ground to breed with resident conspecifics.

ENDOGENOUS PROGRAM DIFFERENCES BETWEEN METAPOPULATIONS

Much of this book has been devoted to review of adaptations migrants possess, among which are endogenous genetic programs governing control over migration to winter quarters and the timing of major life history events (e.g., molt, migration, and reproduction). Evidence from several recent studies indicates that there are negative fitness consequences for hybrids between members of neighboring breeding populations of migrants that may derive from mixing of these programs. For instance, research on eastern (*Setophaga coronata coronata*) and western (*Setophaga coronata auduboni*) populations of the Yellow-rumped Warbler has shown that individuals from the two populations interbreed in a long, narrow hybrid zone, the width of which has not changed in 40 years (Brelsford and Irwin 2009) (figure 8.9). They conclude, "Assuming that all selection maintaining the hybrid zone falls on hybrids, we find selection equivalent to a single-locus heterozygote disadvantage of 18% is necessary to account for the cline width and LD [linkage disequilibrium] we observe" (Brelsford and Irwin 2009:3057). Given that there is little evidence of assortative mating, this finding raises the question of precisely what the "heterozygote disadvantage" might be.

Hybrid zones exist for many migrant species (Mengel 1964, 1970; Price 2008:323–366), with the genetic closeness of the hybridizing populations running the gamut from complete species (e.g., Rose-breasted Grosbeak [*Pheucticus ludovicianus*] × Black-headed Grosbeak [*Pheucticus melanocephalus*]; Baltimore Oriole [*Icterus galbula*] × Bullock's Oriole [*Icterus bullocki*]) to those with barely discernible morphologic differences (e.g., Orange-crowned Warbler [*Oreothlypis celata*] [Foster 1967] and northern and southern populations of the Prairie Warbler [*Setophaga discolor*] [Nolan 1978]). Brelsford and Irwin (2009:3057) suggest three mechanisms by which selection might operate disproportionately against hybrids: "Differences in migratory pathway (e.g., Helbig 1991; Irwin and Irwin 2005), adaptation to different environments (e.g., Price 2008: chap. 15), and intrinsic genetic incompatibilities (e.g., Bronson et al. 2005) may all play a role."

For reasons cited in chapter 4, we think it unlikely that incompatible migration pathways play a critical role in maintenance in population differentiation, although we agree that life history factors likely are important. We propose two additional possibilities for disproportionate selection against hybrids deriving from these sources: (1) differences in endogenous timing programs that characterize each migrant population (e.g., molt) and (2) differences in genetic programs for wintering area location.

• *Differences in endogenous timing programs*. The timing of major life history events (e.g., reproduction, molt, and migration) can differ sharply among populations of

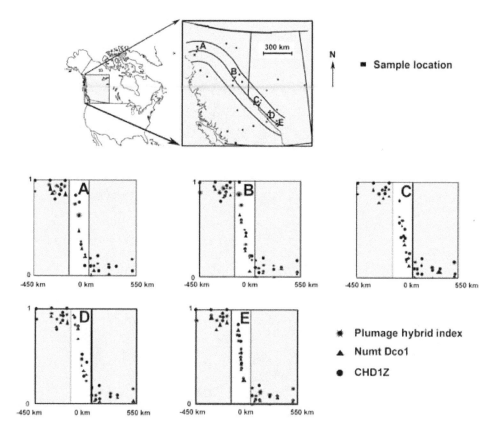

FIGURE 8.9   "Location of five hybrid zone transects and allopatric reference sample sites in British Columbia and Alberta. Parallel curves on map denote hybrid zone center and 100-km buffer. Plots of plumage pattern and two genetic markers along five transects show the consistent width of the hybrid zone between Myrtle and Audubon's warblers. Shaded areas of plots contain samples outside the 100-km hybrid zone buffer; these same samples are plotted on graphs for all five transects and were used in the cline analysis of each transect" (Brelsford and Irwin 2009:3052). Numt Dco1 = nuclear sequence of mitochondrial origin; CHD1Z = a nucleotide sequence (intron) of the chd1 gene on the avian sex chromosome, Z.

the same species occupying different portions of the range. For instance, Foster (1967) found that members of four different populations of the Orange-crowned Warbler (*Oreothlypis celata*) (figure 8.10) showed considerable within-population similarities and between-population differences in timing of molt. In fact, in one of the populations (*Oreothlypis c. sordida*), prebasic molt appeared to take place not only with different timing but in a different locality (i.e., on the wintering ground rather than on the breeding ground as in the other populations). Timing differences between populations probably are not unusual, although they require considerable intensive research to document. Additional examples are provided by major differences in timing of life history events between northern and southern populations of the Prairie Warbler (*Setophaga discolor*) and Purple Martin (*Progne subis*) (see discussion in chapter 4). Without some factor (e.g., assortative mating

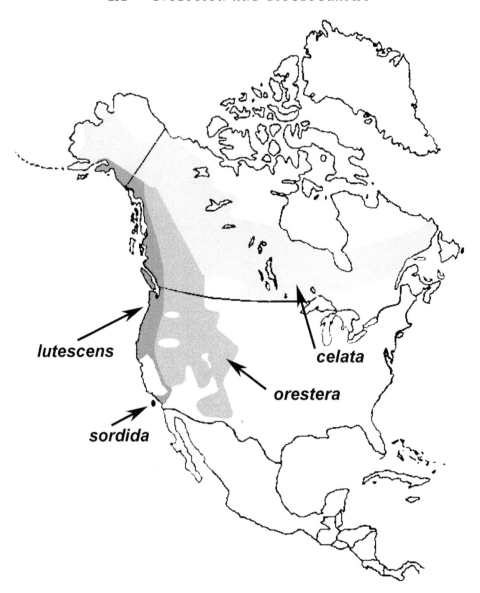

FIGURE 8.10 Breeding range of four populations of the Orange-crowned Warbler (*Oreothlypis celata*) (Foster 1967).

or hybrid disadvantage), it is difficult to see how such timing differences could be maintained in freely interbreeding populations.

• *Differences in endogenous migration route or wintering area programs.* Irwin and Irwin (2005) propose that if migrant populations differ in their genetic migration route programs, hybrids between the populations could suffer in terms of fitness. In chapter 4, we argue that migration route programs were not likely for most migrants because they allow insufficient flexibility to explain observed phenomena. As an alternative, we proposed a hypothesis stating that migratory birds possess a genetic program enabling them to locate the range of the resident ancestral population from which they were derived (i.e., the wintering area, in most cases). This hypothesis would allow the flexibility observed in navigation between breeding and wintering areas by naive (juvenile) birds. Nevertheless, the negative fitness consequences of the resultant mixture of two different wintering area programs in hybrids might have fitness consequences equally severe or worse than those proposed by mixing of migration route programs, as in those species in which neighboring populations go to completely different wintering areas ("migratory divide").

As an example of possible consequences of mixing migration programs, consider the hybridization of experiments performed by Helbig (1991) with eastern and western populations of the Blackcap (*Sylvia atricapilla*). Eastern populations of this species winter in East Africa; western populations winter in southern France, Iberia, and northwest Africa. *Zugunruhe* experiments with eastern populations indicate an initial heading of southeast whereas western populations have a heading of southwest. Hybrids have a heading of south, which indicates a direct crossing of the Mediterranean and the Sahara. Helbig argued that these experiments demonstrated genetic mixing of the migration route program in hybrids, but one could argue that it results from genetic mixing of the wintering area location program.

CONSEQUENCES FOR MIGRANT SPECIATION

Price (2008:324) assumes that hybrid zones result from recent contact between formerly allopatric populations (on the breeding ground for migrants), arguing that "there is good evidence that many species with currently parapatric distributions have initially diverged in allopatry, and there appears to be no good evidence that any have formed without a period of geographical isolation" (Price 2008:325). The first part of this statement certainly is true, at least for some species—for example, North American woodpeckers of the genus *Sphyrapicus* and some superspecies complexes of wood warblers in the genus *Setophaga*, including the earlier example of the Yellow-rumped Warbler (Mengel 1964, 1970; Weir and Schluter 2004; Rohwer et al. 2001). Retreat of the glacial ice sheet within the past 20,000 years likely brought representatives of previously isolated populations into contact after long isolation, at least in terms of breeding distributions.

The second part of Price's (2008:325) hypothesis—"there appears to be no good evidence that any [hybrid zones] have formed without a period of geographic isolation"—contains a circular argument rather than evidence. On the basis of the actual data on which this statement is formulated—that is, the various hybrid zones known to exist—Price could just as correctly have stated that "there appears to be no good evidence that all hybrid zones have formed as a result of recent contact between populations formerly geographically isolated." In other words, it is as difficult to prove the presence of a prehistorical separation no longer present as it is to prove the absence of one. This point is important because metapopulations of many migrant species differ in critical endogenous program timing, which may have important effects on hybrid fitness.

If migrant populations derive from resident populations in an aseasonal environment, as we have proposed, we see three major kinds of potential distributional relationships developing between migrant and resident conspecifics over evolutionary time (figure 8.11):

1. Overlap in winter between a derivative migrant population and a parent resident population. This scenario is likely the first step toward reproductive isolation of the resident parent and migrant derivative populations.
2. Overlap in winter with the parent resident population between two or more derivative migrant populations that have separate breeding areas. This scenario is a likely successor to scenario A over evolutionary time, particularly for migrant populations with broad breeding distributions. In this scenario, reproductive isolation begins not only between migrant and resident populations but between migrant populations of the same species that may have different breeding distributions.
3. Non-overlap in winter or breeding distribution for two derivative migrant populations.

Thus, migratory birds may present at least two differences from nonmigratory species in terms of species origin:

1. They derive from, and often overlap with, resident populations at least during the nonbreeding period, from which they ultimately form separate species either with or without the resident ancestral population remaining in place.
2. Differences in life history parameters between neighboring (i.e., sympatric or parapatric) populations may lead quickly to negative fitness for hybrids between either the resident populations from which they are derived or neighboring populations on their breeding areas.

Scenario C, in which hybrids would suffer from *both* differences in endogenous timing programs related to breeding and postbreeding molt *and* endogenous programs for location of the wintering area, would be the most likely to produce sibling species from formerly parapatric populations. Scenario C could easily apply to that proposed by Mengel (1964) for the Pleistocene evolution of several groups of Parulidae, the main difference with Mengel's argument being in the importance of simultaneous separation of both breeding and wintering areas (Rappole et al. 1983:51–58; Rappole 1995:115–122).

ORNITHOGEOGRAPHY AND MIGRATION PATTERNS

A central theme of this work has been that migration is a form of dispersal and therefore is nearly ubiquitous in both space and time wherever organisms that can move

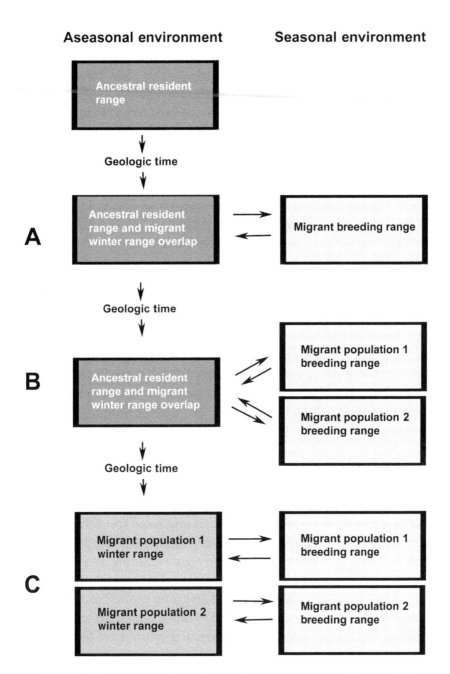

FIGURE 8.11 Three scenarios for evolution of migrant species: (A) from gradual accumulation of endogenous differences between migrant and ancestral resident populations; (B) from gradual accumulation of endogenous differences between sister populations that are allopatric or parapatric during the breeding season and sympatric during the wintering period; and (C) from gradual accumulation of endogenous differences between sister populations that are allopatric or parapatric during both the breeding and wintering periods.

confront environments that differ in their suitability for survival or reproduction. On the basis of this view, the phenomenon shares all the same basic components regardless of where in the world it occurs or the taxa involved. Nevertheless, many factors have potentially profound effects on characteristics of migration peculiar to each region. Exactly how these factors work to shape migration is a topic open to considerable debate (Wolfson 1948; Rappole and Jones 2002; Bell 2005; Rappole 2005a; Gauthreaux et al. 2006). Newton (2008:369–397), in his discussion of continental migration patterns, focuses on seasonality as the chief driver; that is, the more seasonality a particular portion of a continent experiences, the greater the percentage of its avifauna that is migratory. We agree: Breeding-period seasonality is a key issue, but it is not the only issue. Many of the major differences between migration systems have nothing whatsoever to do with differences in breeding-period seasonality.

The treatment presented here of the factors governing migration patterns, and subsequent discussion of the world's major migration systems, is cursory: It is meant to be indicative of the many factors affecting composition of any given migration system. Thorough review of the topic would require book-length treatment. Indeed, at least four such volumes exist of which we are aware on specific migration systems: Moreau's (1972) on the Palearctic–African migration system; McClure's (1974) on the South Asian systems; Rappole's (1995) on the Nearctic–neotropical system; and Dean's (1994) on the Australian nomadic system. Presumably, many other migration systems exist that deserve study (e.g., the Australo–Papuan system described by Dingle [2004]).

PHYLOGENY

Understanding the dynamics of any given migration system requires an understanding of where the migrants in the system originated from; that is, were they residents in the main breeding areas who migrated *from* increasingly seasonal *to* aseasonal environments or residents of the main wintering areas who migrated *from* aseasonal environments *to* seasonal breeding environments. Bell (2005) presents the former hypothesis in which most migrants evolved as residents in the Palearctic breeding area. According to Bell's theory, migration to relatively aseasonal tropical wintering environments evolved in populations of sedentary Palearctic residents as a result of gradual change in climate (over hundreds or thousands of years), which in turn caused change in the habitats in which they were resident from an environment in which members of these species could survive throughout the year to one in which they could not survive during the winter period. This theory differs from the one that we have presented throughout this book in which migration occurs as a form of dispersal, generally from the aseasonal environment to the seasonal environment. These different theories result in different predictions with regard to the phylogenetic origins of the species composition of migrants. For instance, the main landbird resident species in both the Old World and New World derive from some of the same families (e.g., Fringillidae, Paridae,

Picidae, and Corvidae). However, the main migrant families in these two regions are quite different from the temperate residents in terms of phylogeny, deriving largely from the tropical avifaunas (Mayr 1946; Rappole et al. 1983:2–3). If Bell's (northern home) theory were correct, we would expect that the closest relatives to migrants would be Temperate Zone residents, which is certainly not the case in any Holarctic migration system (Rappole 1995:128–129).

PALEOGEOGRAPHY

Finlayson (2011:31) states, "Ultimately, the extant Palearctic avifauna is the product of plate tectonics." This statement is clearly borne out by the composition of the Palearctic avifauna as a whole, and migrants in particular. For the most part, the families of the main landbird migrants of the Old World are quite different from those of the New World. For instance, Trochilidae, Tyrannidae, Parulidae, Vireonidae, and Icteridae are major families of migrants in the New World. These families do not occur in the Old World where Muscicapidae make up a large portion of long-distance migrants—a family that has few New World representatives (Rappole 1995:128–129). These hemispheric phylogenetic differences in migrant avifauna derive from the fact that the tropical portions of the Old World and New World avifaunas have been separated for greater than 100 million years (Cattermole 2000), resulting in very different evolutionary histories.

HABITAT AND THE MIGRANT NICHE

Following from the question of where the migrant niche evolved (i.e., from the aseasonal or the seasonal portion of the life cycle) is the related concept of what habitats are found in the breeding areas, wintering areas, and stopover areas. For instance, forest of any kind is very nearly absent from nearly 3,000 km along the potential route of a migrant from sub-Saharan Africa to southern Europe because of the intervening Mediterranean Sea and Sahara Desert (Rappole 1995:126). The absence of this habitat type appears to have had a profound effect on the number and species composition of migrants in the Palearctic–African migration systems, likely resulting in the very small number of Palearctic migrants that use forest habitats in the winter (three species) (Mönkkönen 1992). We suggest that the lack of Palearctic migrants drawn from the forests of Africa results from a lack of forest environments in the northern subtropics and tropics of this region (Rappole and Jones 2002).

The question of habitats en route and at the breeding area as determinant of what migrants will exist in any given system is related to the migrant niche and, ultimately, the phylogeny of the pool of potential migrants occupying the aseasonal portion of the system (i.e., birds with a certain phylogenetic history often share similar niches). An example is the Tyrannidae of South America. This group is composed largely of forest-related insectivores, and they make up nearly one-third of all known intra–South American migrants (Chesser 2005:169) (for further discussion see the "Intra-South American" portion of the section on "The World's Principal Migration Systems").

If migrants were derived largely from Temperate Zone resident species, one would expect that their main adaptations for foraging would reflect that derivation (i.e., that they would show tight adaptation to foraging in the temperate environments in which they breed) and they would behave as fugitive species (i.e., living off resource surpluses) in the tropical environments in which they winter. However, the niches for most long-distance migrants that breed in seasonal environments and winter in aseasonal environments simply do not exist during the winter period in the seasonal environment. In addition, many move to aseasonal environments during the nonbreeding period in which resources are not superabundant but rather in the lowest ebb for the annual cycle (Rappole 1995:58). Members of these migrant species must compete for stable resources during the nonbreeding period as members of the tropical communities in which they live (Rappole 1995:49–74). Thus, many migrants seem best adapted to the resources exploited during the nonbreeding season rather than the breeding season (Rappole 1995:58).

Some recent theories on the evolution of migration focus specifically on the niche of migrants as a major influence on development of migration, especially within the intra-tropical/subtropical systems (e.g., Levey and Stiles 1992; Chesser and Levey 1998).

In our view, the migrant niche, in combination with the predominant habitats of a region, plays a major role in terms of the numbers and kinds of birds composing any given migration system. This fact is well illustrated by a comparison of the landbird portion of the Palearctic–African system, which is composed mostly of open-country birds, with the landbird portion of the Nearctic–neotropical system, nearly half of which birds are forest-related (Rappole 1995:183–197). Essentially, there are few forest environments outside the equatorial zone in the tropics and subtropics of Africa, whereas such forests are widespread in the tropics and subtropics of the New World. Thus, niches for potential forest-related migrants are scarce or absent along routes to the Palearctic from Africa, whereas they are plentiful along routes to the Nearctic. In contrast, niches for open-country species are plentiful along routes to the Palearctic. Thus, landbird migrants of the Palearctic–African system likely originated as residents of the African, open-country habitats.

This origin for Palearctic migrants is also relevant to understanding "Moreau's paradox." In his examination of the Palearctic–African migration system, Moreau pointed out that the most important wintering area for Palearctic migrants (composing 40% of migrant species) is the open country (savanna, grassland, scrub) of sub-Saharan Africa north of the equator (Moreau 1972). The paradox is why should migrants from the Palearctic migrate to such habitat when it is seemingly quite inhospitable in terms of resources (i.e., it is during the dry season when resources are relatively scarce) (Jones 1996). Certainly, this observation presents a paradox, if one views migrants from the Palearctic from a "northern home" perspective (i.e., as "fugitive species" that can choose any environment they wish for the non-breeding period) (Newton 1995). However, if migrants derive from African open-country residents, then their return to such habitats in winter makes a great deal of sense, as it is within these habitats that their niches exist and for which their foraging adaptations originally evolved.

WEATHER EN ROUTE

Weather can exert strong influence on migration patterns (Rappole et al. 1979; Buskirk 1980; Butler 2000). In addition, weather probably affects the kinds of birds that can exploit seasonal environments. For instance, Beebe (1947) in Venezuela and McClure (1974) in Southeast Asia found evidence of seasonal movement in groups of tropical species (e.g., broadbills, pittas, and parrotbills). These birds of forest understory would experience significant losses attempting migration routes that included flights over water where their lack of adaptation for long-distance flight would be exposed when encountering adverse winds or storms. Thus, migration systems occurring entirely over land (e.g., the austral South American and austral African systems) could be expected to include different kinds of birds from those that exploit the seasonal habitats of the Holarctic, reaching most of which must involve long flights where winds and storms potentially serve as major factors affecting migrant success (Buskirk 1980).

LANDFORMS

As in the case of weather, the various landforms that occur in a migration system can influence the species composition of the migrant avifauna (Chesser 2005). Oceans and mountains can serve as obstacles or barriers to potential migrants depending on their position relative to possible routes between breeding and wintering areas. The most obvious such obstacle is the Himalayas, which exert an enormous effect on the species composition and evolution of Asian avifauna in general and migrants in particular (Rasmussen and Anderton 2005a, 2005b; Irwin and Irwin 2005; Price 2008)

SEASONS EN ROUTE

We have hypothesized that migration begins as a form of dispersal in which birds from an aseasonal environment disperse outward in attempts to locate habitats with superabundant resources (i.e., those that the resident species cannot completely harvest). Such resources generally occur in environments where seasonal variation results in periodic resource deficiencies and flushes. As mentioned earlier, Newton (2008:369–375) reports that migration is most prevalent (i.e., includes the largest portion of breeding avifauna) in places that show the sharpest seasonality. However, seasonal surpluses must also be available en route at stop-over sites between breeding and wintering areas in order for migration to be a successful strategy.

THE WORLD'S PRINCIPAL MIGRATION SYSTEMS

PALEARCTIC–AFRICAN

If Moreau (1972) was not the first to discuss continental patterns of seasonal move-ment as "migration systems," his treatment of the Palearctic–African migration system was certainly the most influential. In his review of the topic, he made a convincing case that major patterns were a result of the specific characteristics of seasonal rainfall and plant communities of the African continent (figure 8.12). These characteristics produce three of the salient attributes of the system: itiner-ancy, lack of forest-related species, and size (i.e., number of species in the system).

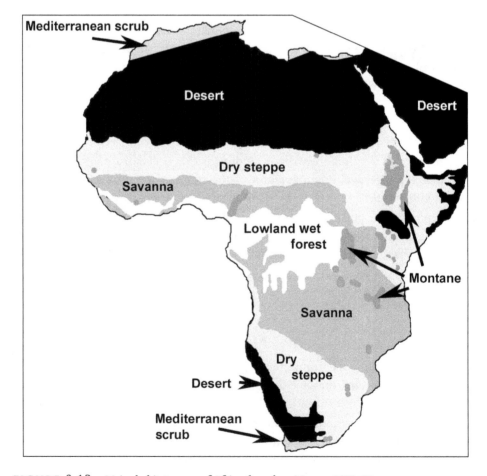

FIGURE 8.12   Major habitat zones of Africa (based on Moreau 1972:61).

• *Seasonality en route.* As discussed in chapter 1, "itinerancy" as reported by Moreau (1972) is a form of stepwise migration in which the bird moves southward in steps from its breeding (or hatching) area in Europe to migration stopovers sites in southern Europe or northern Africa from August to October, then to seasonally wet areas just south of the Sahara from November to December, then to a site farther south for the remainder of the wintering period, as has been documented by light datalogger data for the Nightingale (*Luscina megarhynchos*) (figure 8.13). Documentation of this migration pattern, pieced together largely by remarkable field efforts now confirmed by research using new technologies (e.g. dataloggers), has shown that as many as 19 species of Palearctic migrants undertake some form of this stepwise migration (Moreau 1952, 1972:x; Jones 1985, 1995, 1996; Jones et al. 1996; Newton 2008:708). Although there are data to indicate that a few western Nearctic migrants follow a similar strategy (e.g., Butler et al. 2002), no comparable portion exists in any other migration system of which we are aware (although, note that the Palearctic–South Asian system is poorly investigated, and there are indications that some species, e.g., the White Wagtail [*Motacilla alba*], may follow a similar strategy [Rappole et al. 2011a]). The presumed reason for the remarkable,

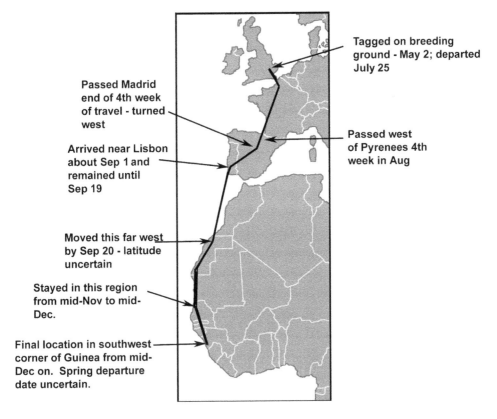

Tagged on breeding ground - May 2; departed July 25

Passed Madrid end of 4th week of travel - turned west

Arrived near Lisbon about Sep 1 and remained until Sep 19

Passed west of Pyrenees 4th week in Aug

Moved this far west by Sep 20 - latitude uncertain

Stayed in this region from mid-Nov to mid-Dec.

Final location in southwest corner of Guinea from mid-Dec on. Spring departure date uncertain.

FIGURE 8.13   Itinerancy in a Nightingale (*Luscina megarhynchos*) (based on British Trust for Ornithology 2011). See text for discussion.

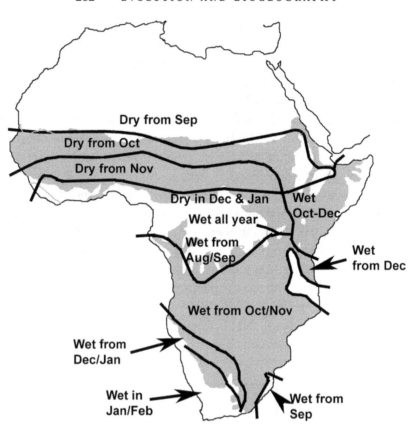

FIGURE 8.14   "Wet and dry season zones in Africa during the northern winter when Palearctic migrants are present. The shaded area represents seasonal savannas where migrants are most abundant" (based on Jones 1995:396).

stepwise movement so characteristic of the Palearctic–African system is the seasonality pattern of sub-Saharan Africa (figure 8.14), which produces seasonal surpluses in sub-Saharan savanna and scrub habitats sequentially during the winter period (Jones 1995).

   • *Lack of forest-related migrants.* Forest-related migrants (i.e., those that both breed and winter in broad-leaved forest habitats) constitute less than 2 percent of the Palearctic–African migration system (Mönkkönen 1992). This number stands in marked contrast to other Holarctic–tropical migration systems—for example, the Nearctic–neotropical system (35%) and Palearctic–Asian system (31%) (Rappole 1995:124–126; Rappole and Jones 2002). The presumed reason for lack of forest-related migrants in the Palearctic–African system is the absence of forest habitat en route between breeding and wintering areas (Rappole 1995:124–126; Rappole and Jones 2002).

   • *Total number of species.* Although the African, Asian, and South American tropics show comparable levels of diversity and presumably provide comparable numbers of candidates for invasion of Holarctic habitats, the Palearctic–African system is depauperate in terms of landbird migrants (185 species) compared with the Nearctic–neotropical systems (318 species) and the Palearctic–South Asian system (338 species) (Rappole 1995:124). We suggest that the relative paucity of migrants in the system is related to lack of diversity in stopover habitats available to potential migrants moving from the African tropics to the Palearctic presented by the Sahara Desert and Mediterranean Sea (Rappole 1995:127; Rappole and Jones 2002).

NEARCTIC–NEOTROPICAL

Some of the salient characteristics of this system were mentioned earlier in contrast with the Palearctic–African system (e.g., diversity of subtropical habitats, including forest, available as stopover habitat). Another striking characteristic of this system is the landform and major habitat positions relative to one another. Broad-leaved, evergreen tropical forest is located largely south and east of the Nearctic in Middle America (Central America plus Mexico) and South America and the Caribbean islands, whereas grassland, scrub, deciduous, and coniferous forest habitats are located in Mexico and the Pacific slope and mountains of Central America. These factors result in a sharp dichotomy in the breeding distribution of migrant land-birds in the Nearctic–neotropical system, with most broadleaved forest-related migrants deriving largely from Central America, South America, and the Caribbean migrating to breeding habitat in the forests of eastern or northern North America (e.g., American Redstart [*Setophaga ruticilla*]) (figure 8.15), whereas many grass-land, scrub, and coniferous-forest species derive largely from Mexico or the Central American Pacific slope or mountains breeding in temperate central and western North America (e.g., Townsend's Warbler [*Setophaga townsendi*]) (figure 8.16).

PALEARCTIC–SOUTH ASIAN

This system has not received the research attention of the other Holarctic migra-tion systems, and, as a result, key elements are not known or understood. Neverthe-less, those studies and general surveys that have been done are at least indicative of some of the important aspects (e.g., Smythies 1953; King et al. 1987; Lekagul and Round 1991; Robson 2000; Rasmussen and Anderton 2005a, 2005b; Rappole et al. 2011a).

An obvious difference between the Palearctic–South Asian system and other Holarctic systems is the relative lack of water obstacles for potential migrants between breeding and wintering areas in the Palearctic–South Asian system. Most species from the eastern half of the Nearctic–neotropical migration system make long over-water flights across the western North Atlantic, Caribbean Sea, or the Gulf of Mexico (see chapter 6), at least on their southbound journeys, whereas European-breeding species of the Palearctic–African system must pass over, or around, the Mediterranean. Many Palearctic species do not cross large waterbodies

FIGURE 8.15   American Redstart (*Setophaga ruticilla*) breeding (*light gray*) and wintering (*dark gray*) range.

FIGURE 8.16   Townsend's Warbler (*Setophaga townsendi*) breeding (*light gray*) and wintering (*dark gray*) range.

to reach their wintering areas (e.g., Red-breasted Flycatcher [*Ficedula parva*]) (see figure 5.8), although some do cross large water expanses in order to reach the Pacific island and northern Australian parts of the system (e.g., Brown Shrike [*Lanius cristatus*]) (figure 8.17).

A second critical aspect of the Palearctic–South Asian system is the extreme seasonality of many stopover and wintering areas of the region. Africa, of course, shows some of this same seasonality, which results in extensive intra-African migration as well as stepwise migration in some Palearctic species (Jones 1996).

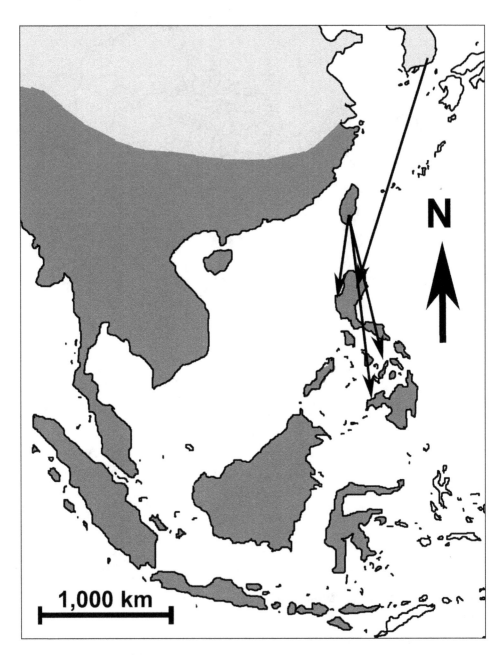

FIGURE 8.17 Brown Shrike (*Lanius cristatus*) over-water migration in East Asia, Southeast Asia, Indonesia, and the Philippines (based on McClure 1974:267): *light gray* = breeding; *dark gray* = winter. Arrows originate at banding sites and terminate at recovery sites.

It seems probable that similar levels of intratropical and tropical–subtropical movements occur in this system, but they are not so well documented (Rappole et al. 2011a). In addition, it is likely that intratropical and subtropical seasonality in South Asia has profound effects on the kinds of species that can make seasonal movements to the Palearctic.

HIMALAYAN–SOUTH ASIAN

Parts of this system are fairly well understood, thanks to recent summaries provided by Rasmussen and Anderton (2005a, 2005b). These summaries provide new breadth to the meaning of the term "altitudinal migration." Usually, this term is applied to the seasonal up- and down-slope movements observed in many of the world's mountain ranges, which usually involve only a few species in any given region (Rappole and Schuchmann 2003; Newton 2008:379). However, the Himalayan–South Asian migration system includes many species that migrate seasonally hundreds or thousands of kilometers from tropical wintering sites in India or Southeast Asia to highland breeding sites in the Himalayas—for example, the Bristled Grassbird (*Chaertornis naevia*) (figure 8.18). The eastern portions of this system are still poorly investigated but likely include movements of many species from the eastern Himalayas into Burma and Southeast Asia (Smythies 1953; King et al. 1987; Rappole et al. 2011a). The dominant feature of this system, obviously, is the Himalayas, which provide seasonally temperate environments during the Palearctic summer. However, the tropical portions of the system are subject to extreme seasonality in terms of rainfall, which certainly must have a profound effect on system structure as well.

INTRA-AFRICAN

Moreau (1966) estimated that 91 of the 998 open-country inhabitants of sub-Saharan Africa were intratropical migrants. Excellent field work by Elgood, Grimes, and others has greatly increased this number (Elgood et al. 1973, 1994; Grimes 1987). On the basis of their work, Jones (1996) estimates that as much as 40 percent of West African landbirds may be intratropical migrants. A major reason behind these extensive, seasonal movements is the extreme seasonality in terms of rainfall over large portions of the region (figure 8.13). However, other aspects are important as well—for example, phylogeny of the African species pool of potential migrants and the ecology of the various seasonal habitats available for exploitation by migrants (i.e., the kinds of foraging niches they provide). As in the case of the Palearctic–African landbird migrants, most intra-African migrants are open-country species as well (Jones 1996). The lack of forest-related intra-African migrants likely speaks to lack of seasonal forest environments for potential migrants to invade, a situation in direct contrast with that which occurs in both the intra–South American and intra–South Asian migration systems where forest-related migrants predominate (Harrison 1962; Chesser 2005).

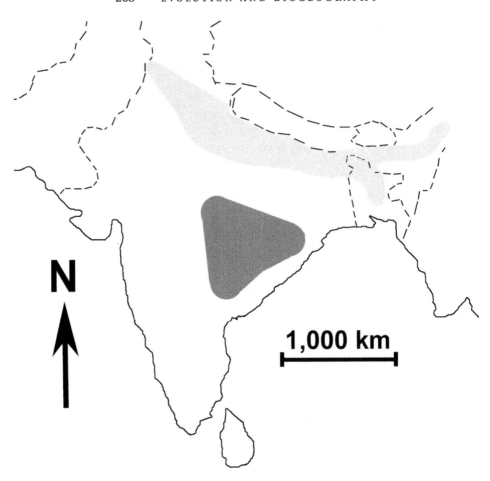

FIGURE 8.18   Range of Bristled Grassbird (*Chaetornis striata*) illustrating long-distance altitudinal migration (based on Rasmussen and Anderton 2005a): *light gray* = Himalayan breeding; *dark gray* = Indian wintering.

INTRA–SOUTH AMERICAN

At least 220 species compose the intra–South American system ("South American austral" of some authors): These migrate from less seasonal environments along the equator to more seasonal environments in the southern portion of the continent in South America (Azara 1802–1805; Zimmer 1938; Willis 1988; Hayes et al. 1994; Chesser 1994, 1997, 1998, 2005). Chesser (1998) has made the case that this movement is driven largely by seasonality (i.e., the more seasonal the environment, the greater the proportion of the avifauna that is migratory). He illustrates this concept with examples from the distribution of migrant and resident species of the genus *Myiarchus* in South America, in which members are almost entirely migratory in southern South America, but include an increasingly larger proportion of residents with decreasing latitude. This concept is correct, no doubt, but does not explain the composition of the austral South American migration system compared, say, with the Nearctic–neotropical migration system, although drawn from the same potential pool of tropical residents. As an example, assuming an aboriginal population of Swainson's Thrushes (*Catharus ustulatus*) resident in the foothills of the northern Andes, why did this species develop migratory populations to the boreal regions of the Nearctic rather than the southern portions of South America? We believe that the answer to this question lies in the fitness reward differences between the two regions, which, in turn, depend on many of the factors discussed earlier—for example, exaptations for migration of the resident population, contrasts between the different migration routes in terms of fitness costs, and contrasts between the different prospective breeding areas in terms of fitness benefits.

It is also of interest to compare the intra–South American migration system with the intra-African migration system. One major difference in particular stands out: the phylogenetic origins of the migrants in the two systems. As mentioned earlier, nearly one third of all intra–South American migrants are members of the New World flycatcher family, Tyrannidae, a group that does not even occur in Africa.

TROPICAL–SUBTROPICAL INTRA-ASIAN

This system is the least-studied of all the major systems. Only a few works have been published that contain information on the system (e.g., McClure 1974; Rappole et al. 2011a), which is likely to be as diverse and complex as the other intratropical–subtropical systems. The information that is available is intriguing, indicating many of the same types of migration seen in other systems—for example, apparently long-distance movements by tropical understory species like the Hooded Pitta (*Pitta sordida*).

INTRA-AUSTRALIAN

This system is relatively well understood thanks to extensive field investigations and excellent book-length summaries (e.g., Dean 2004). The tropical forests of extreme northern Australia contain a few landbird migrants to other systems and, of course, several species of waterbirds and shorebirds, but most of the movements take place within the continent's boundaries and are the result of extreme seasonal variation in rainfall. Unlike in other migration systems, the rainfall patterns are much less predictable, resulting in extensive nomadism among Australian bird species (Dean 2004).

CHAPTER 9

MIGRATORY BIRDS AND PATHOGEN MOVEMENT

T HE ECOLOGY, evolution, and life history of migratory birds are rather arcane subjects for the public at large until they present the potential of impinging upon human well-being. Then it becomes obvious that there are important connections between humans and migrant birds and that there are certain critical aspects of migrant biology that are important for us to understand. At the beginning of the West Nile virus outbreak in October 1999, I was contacted by the U.S. Centers for Disease Control and Prevention (CDC), the U.S. Department of Defense, and a staffer from a U.S. Senate committee, all of whom had questions regarding how West Nile virus was likely to have entered the United States and how it was likely to spread. Colleagues and I provided speculation on this topic based on what we knew about migrants and what was known about the virus (Rappole et al. 2000b). Based on this information, we concluded that the virus probably invaded the Western Hemisphere via legal or illegal imports of infected birds (not natural migration) and predicted that the virus would spread throughout the hemisphere in a matter of months. During 2000, 2001, and 2002, as the virus moved inexorably (but slowly) across the continent, fear among the general public regarding the role of migrants resulted in calls for government action to eliminate the threat by eliminating large flocks of obvious migrants (e.g., geese). Similar kinds of hysterical reactions occurred among the public in the United States during the HPAI H5N1 avian influenza virus outbreak. Indeed, on publication of our article expressing skepticism regarding the role of migrants and likelihood of HPAI H5N1 spread to the Western Hemisphere (Rappole and Hubálek 2006), a physician from New Jersey wrote to me expressing her desire that I be the first to die from the disease when it reached our shores. These experiences make clear the value to human society of knowledge concerning migrants and their pathogens, the serious lacunae in that knowledge, and the potential consequences for migrant conservation.

The fact that migratory birds can serve as hosts for many kinds of viral and bacterial pathogens is well known (Hubálek 2004). In particular, avian involvement in the epidemiology of both West Nile virus (Hubálek and Halouzka 1999) and avian influenza (Webster et al. 1992) has been well documented. Nevertheless, answers to important questions concerning specific aspects of the role of birds in two recent epidemics involving these pathogens remain unclear. Despite lack of conclusive data, a general consensus has been reached in the published literature that migratory birds were the principal agents responsible for introduction and spread of West Nile virus in the Western Hemisphere during the epidemic that began in August 1999 (McLean 2006) and for the introduction and movement of avian influenza subtype H5N1 across Eurasia and into Africa that began in 1996 (Morris and Jackson 2005; Gilbert et al. 2006; World Organization for Animal Health 2011) (although see Brown and O'Brien 2011).

West Nile Virus

West Nile virus (WNV), a flavivirus related to Japanese encephalitis virus and St. Louis encephalitis virus, was first described from the eponymous region of Uganda in East Africa in 1937 (Smithburn et al. 1940). Subsequent studies showed WNV to be endemic throughout much of northeastern Africa (Taylor et al. 1956). The virus was first reported in Europe from Albania in 1958, followed by sporadic outbreaks across the continent from widespread localities, mostly in eastern Europe, over the next four decades. By the late 1990s, the virus was widespread in Africa and Eurasia (Hubálek and Halouzka 1999).

Birds have long been known to play an important role in WNV epidemiology as the principal hosts for the virus: Members of a large number of Old World birds have been reported to have tested seropositive to antibodies for WNV (Rappole et al. 2000b). The virus has been isolated from 43 different mosquito species as well as from several other hematophagous arthropods (Hubálek and Halouzka 1999), but bird-feeding mosquitoes (e.g., several members of the genus *Culex* as well as some *Aedes, Mimomyia,* and *Coquillettidia*) appear to be the main vectors. These mosquitoes then transfer the virus to humans. Humans and most other mammals appear to be dead-end hosts for WNV because virus titers do not reach sufficient levels to allow transfer (Hubálek and Halouzka 1999).

In addition to serving as the main amplifying hosts for WNV, it has long been assumed that birds were the principal introductory and reservoir hosts, at least for European outbreaks (Hubálek and Halouzka 1999). According to this scenario, migratory birds wintering in sub-Saharan African wetlands, where WNV is endemic, contract the virus. They then carry it north in active form in their blood where they introduce it into mosquito populations in wetlands at stopover sites in the Middle East and southern Eurasia during spring migration. The virus circulates between the mosquito vector population and reservoir host bird populations until late summer when large numbers of birds (amplifying hosts) begin to congregate in wetlands either preparatory to or during southward migration. Outbreaks in humans thus occur only during a "perfect storm" situation in which introductory host (viremic northbound migrants), vector (mosquitoes), reservoir host (avian summer residents), amplifying host (large flocks of fall migrants), and susceptible human population interact under specific environmental conditions (Hubálek and Halouzka 1999).

Proving conclusively that migratory birds are, in fact, the main introductory hosts for WNV causing European outbreaks of the disease in humans is difficult. However, there is a significant amount of circumstantial evidence:

1. Outbreaks in European temperate regions occur only sporadically, often at sites widely separated from previous outbreaks.
2. The outbreaks usually occur in late summer or early fall, coincident with pre-migratory or migratory flocking by birds.
3. The outbreaks usually occur in wetlands or urban areas where large numbers of migrants congregate in places (e.g., parks, riparian areas, etc.) where there are large numbers of mosquitoes.
4. Most Old World birds that contract the virus do not appear to become ill when infected, indicating that they could be healthy enough to migrate while in a viremic state.
5. Members of many species of Old World birds captured during migration have tested seropositive for the virus indicating that they have survived infection.
6. Several Old World species exposed to WNV in the laboratory show high, long-term viremia (several days), and active virus can persist in organs of infected birds for weeks (Hubálek and Halouzka 1999).

These findings were taken to be a convincing case regarding the role of migratory birds in WNV movement in the Old World, although as pointed out by Hubálek and Halouzka (1999:648), the "migrants as principal introductory hosts for WNV" hypothesis is not the only one that fits the facts. Nevertheless, it is likely this explanation would not have been questioned any time soon had not WNV invaded the New World. Its sudden appearance in August 1999 in the New York City borough of Queens provided the first and only known introduction of the virus in the Western Hemisphere, essentially created a large-scale experiment allowing testing of existing hypotheses for WNV movement. Shortly after this occurrence, Rappole et al. (2000b), in response to a request by the CDC, considered the following hypothesis concerning the probable role of migratory birds in spread of WNV in the Western Hemisphere:

Migratory birds are the main introductory hosts for WNV, serving to move the virus from point of origin to new localities.

On the basis of this hypothesis and assuming that the Queens, New York, appearance of the virus was the first and only locality from which any subsequent records of WNV in the hemisphere could be derived, several predictions could be made. These predictions are provided in the text that follows, along with data relevant to their evaluation. Note that the references to "movement" of the virus actually refer to recovery of dead birds testing seropositive for the virus. Unlike WNV in the Old World, the New World form, apparently derived from the Middle East (Lanciotti et al. 1999), proved quite lethal, at least initially, to members of most Western Hemisphere avian species. Thus, each arrival of WNV at a new locality was signaled by the appearance of dead birds, often in large numbers (Steele et al. 2000).

• *WNV will move southward from New York, following the main migratory pathways used by fall passage migrants in the northeastern United States.* The virus moved nearly as far north (170 km) and east (230 km) as it did southwest (300 km) during the initial outbreak in 1999 before mosquito activity died out and the last dead bird seropositive for WNV was found (November 5, 1999) (figure 9.1) (Rappole and Hubálek 2003). No evidence of a north–south movement pattern has accumulated in the years since the initial outbreak. Instead, the dispersal pattern for WNV appears to be nondirectional (Rappole and Hubálek 2003).

• *WNV movement from the point of origin will be rapid during migration periods, covering hundreds of kilometers in a matter of days, in concert with migratory bird movement capabilities.* After years of data collection involving a huge collaborative effort between the CDC, U.S. Geological Survey, and the public health authorities for nearly every state plus the governments of Canada, Mexico, and several Caribbean and Central and South American countries, as well as a number of academic institutions, it is clear that this prediction is not supported. WNV did not move rapidly from the point of origin (U.S. Geological Survey 2003). While migrants move 30 to 70 km/h and 200 to 600 km/day (Cochran et al. 1967; Cochran 1987; Kerlinger 1995; Stutchbury et al. 2009), WNV spread gradually outward from the original epicenter at a rate of about 70 km/mo during the temperate mosquito activity period (April to October) for the first 2 years after its original occurrence until it reached the south temperate region when the annual rate of spread increased, perhaps due to additional months of mosquito activity (Rappole and Hubálek 2003).

• *During the winter of 1999–2000, WNV will become established in mosquito and resident bird populations at stopover and wintering sites in the southeastern United*

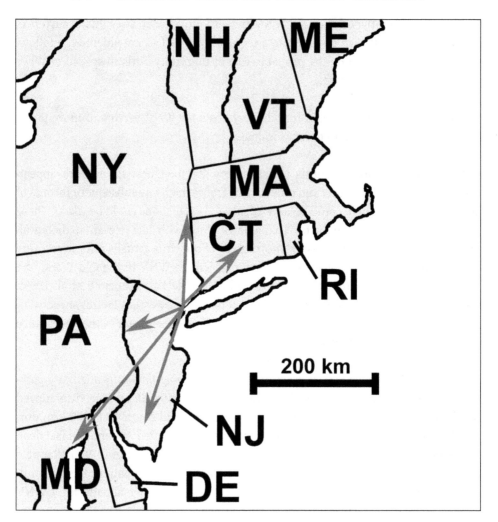

FIGURE 9.1   Northeastern United States. Arrows originate at the outbreak site (August 1999) for WNV in the Western Hemisphere and terminate at the most distant records recorded in 1999 (based on data from U.S. Geological Survey 2003).

States, the Caribbean region, Mexico, and Central and South America where tens of thousands of individuals of many species of migrants that pass through the New York region spend the winter. No records of WNV were found south of North Carolina in the year 2000 (Rappole and Hubálek 2003). WNV populations did not become established in the North American subtropics (southern Florida) until late 2001 and in the tropics (southern Mexico) until early 2003 (Estrada-Franco et al. 2003; U.S. Geological Survey 2003) and was not confirmed in South America until early 2006 (Bosch et al. 2007).

• *Failing southward movement out of the Temperate Zone, the virus will disappear because of failure to establish endemic populations in the New World tropics or subtropics.* The first record of WNV in the Western Hemisphere in the year 2000 was on May

22 less than 50 km from the 1999 New York outbreak site, followed shortly thereafter by many other records, all from within the region from which records for WNV were recorded during the previous fall (Rappole and Hubálek 2003).

• *The pattern of WNV outbreaks in 2000 in North America will be similar to that which has been observed in Europe, that is, sporadic outbreaks in late summer and fall in widely separate, Temperate Zone localities.* As the 2000 summer season progressed, records for WNV were reported in a gradually widening, and seemingly random, pattern centered on New York City, eventually extending 480 km to the north, 370 km to the east (stopped from further eastward movement by the Atlantic Ocean), 700 km to the southwest, and 570 km to the west before cold weather again shut down movement (Rappole and Hubálek 2003). No outbreaks occurred outside the eastern United States.

• *Large numbers of migratory birds of many species captured during migration or the wintering period in the southeastern United States will test seropositive for WNV, indicating their ability to survive infection under wild conditions.* Massive efforts involving capture and blood sampling of thousands of transient and wintering migratory birds of many species in the southeastern United States yielded very low numbers of migrants seropositive for WNV (McLean 2006).

Reports of the appearance of WNV in horses in Argentina confirmed that WNV had spread nearly throughout the Western Hemisphere by 2006 (Morales et al. 2006). Available data as summarized earlier indicate that migratory birds were not the main agents responsible for this spread. This finding begs the question of what did cause WNV movement across the hemisphere. Hubálek and Halouzka (1999), in their review of European outbreaks of WNV, hypothesized that overwinter persistence of the virus either in the mosquito population, through hibernation or transovarial transmission, or in resident bird populations (i.e., chronically infected reservoir hosts) could explain outbreaks. Three findings from the WNV movement across North America (1999–2004) provide support for this hypothesis:

1. Each year following the winter period of mosquito quiescence in the North American Temperate Zone (October to April), the virus reappeared in most sites where it had been documented the previous year (U.S. Geological Survey 2003).
2. Overwintering in mosquito populations has been documented for related viruses (Reeves 1974), and its occurrence in WNV has been documented in the laboratory and the field (Baqar et al. 1993; Peiris and Amerasinghe 1994; Turell et al. 2001). Furthermore, intensive work by entomologists in the New York City area revealed evidence of transovarial transmission, allowing the virus to overwinter in the mosquito population (Nasci et al. 2001).
3. Komar et al. (2003) documented that House Sparrows (*Passer domesticus*), House Finches (*Carpodacus mexicanus*), and Blue Jays (*Cyanocitta cristata*) appeared to be competent reservoir hosts for WNV.

Thus, mosquitoes and resident birds could explain persistence of the virus in areas previously infected. But how did the virus move? Rappole et al. (2006) compared known WNV movement based on CDC data with models of viral spread based on

- A migrant bird (Swainson's Thrush [*Catharus ustulatus*]) undergoing normal, long-distance, directional migration as the movement agent or introductory host
- A resident bird (House Sparrow [*Passer domesticus*]) undergoing normal, short-distance, random, nondirectional dispersal

They were able to demonstrate that the migrant bird model could not be made to fit the actual WNV movement pattern assuming normal migratory behavior, whereas the resident bird model could fit by varying the number of dispersing viremic individuals. In fact, a dispersing-mosquito model could also be made to fit the pattern of WNV spread across the hemisphere. Thus, mosquitoes and resident birds as overwintering reservoirs and/or dispersing introductory hosts provide viable alternative hypotheses for both persistence and spread of WNV across the Western Hemisphere.

August 2012 saw the largest and most widespread number of West Nile virus cases among humans in North America since the virus first appeared in the Western Hemisphere in 1999 (CBS News 2012). The sporadic nature of West Nile virus outbreaks, in both space and time, is typical of the disease in the Old World, as noted previously. Researchers speculated that this attribute resulted from the need of a particular confluence of environmental circumstances favorable to the virus; a confluence that only occurs in some places in some years. These circumstances included the following (Hubálek and Halouzka 1999):

1. The virus is introduced by a viremic host—a migratory bird, which would move the virus from the tropics, where it is endemic, to a temperate site, where it only occurs in summer.
2. A local mosquito (vector) becomes viremic by feeding on the infected introductory host migrant's blood.
3. The vector passes the virus on to reservoir hosts (summer resident birds), ensuring that the virus persists in the area by passing back and forth between host and vector populations through the summer.
4. Concentrations of large numbers of migratory bird (amplifying hosts) and mosquito vectors in the same site in late summer and early fall (e.g., a wetland) result in rapid "amplification" of the virus (i.e., increase in the number of viremic birds and mosquitoes;such concentrations probably occur in only some years and a few sites when temperature and rainfall are especially advantageous for mosquito reproduction).
5. When the wetland where the viral amplification has occurred is located in close proximity to human populations, then viremic mosquitoes infect humans as well.

When West Nile virus first invaded the Western Hemisphere, it represented the serendipitous initiation of a giant experiment, which provided a test of the speculative sequence of events hypothesized by Hubálek and Halouzka (1999). One of the results of this experiment was to demonstrate that no migrant introductory host was necessary for a West Nile virus outbreak to recur in a later year at a temperate site. If mosquitoes can maintain the virus through transovarial transmission from mother to daughter, as has now been demonstrated, there is no need to postulate an introductory host. This finding also obviates the need for migratory bird reservoir and amplificatory hosts as well; the mosquito population can serve as both. While birds, not necessarily migrants, may serve this purpose, all that is really needed for a viral outbreak among humans is a large population of infected mosquitoes, which could result solely from aspects of the physical environment (temperature, rainfall) that result in large variations in the size of mosquito populations from year to year and site to site.

Avian Influenza

Avian influenza is a group of influenza A viruses (members of the family Orthomyxoviridae) characterized by eight RNA segments of negative polarity that encode proteins composing the virion (Lamb 1989). Taxonomy for members of the group is based largely on the different forms of two major surface antigen proteins: hemagglutinin, of which 16 variants now are known, and neuraminidase, for which nine variants are known (U.S. Centers for Disease Control and Prevention 2007). Thus, an avian influenza virus possessing the fifth-described variant of the hemagglutinin protein and the first-described variant of the neuraminidase protein is a viral subtype of avian influenza referred to as H5N1. These viruses are further distinguished based on their effects on their hosts, with those variants that cause little host reaction being referred to as "low pathogenic avian influenza" (LPAI), whereas those that cause an intense host reaction are referred to as "highly pathogenic avian influenza" (HPAI). Most influenza A virus subtypes are found only in birds, but a few are found in humans and other mammals, especially pigs. H1N1, H1N2, and H3N2 are the only subtypes of influenza A viruses currently known to be circulating widely in human populations (U.S. Centers for Disease Control and Prevention 2007).

Avian influenza viruses representing nearly all subtypes are common among birds where they preferentially infect host cells lining the intestinal tract (Webster et al. 1992). Whereas most bird species appear susceptible to infection, the most commonly infected are waterbirds (e.g., ducks and shorebirds) because of the principal mode of viral interhost transfer, which is by means of excretion in high concentrations in fecal material. Such feces contaminate shared water bodies where the viruses readily infect new hosts by being consumed. Thus, avian influenza differs fundamentally from WNV in that no intermediate vector (e.g., the mosquito) is required for interhost transfer. Nevertheless, the avian influenza viruses and WNV share an important factor in their epidemiology: the apparent role of migratory birds in introducing the viruses to new, geographically distinct localities. However, whereas the role of avian migrants in movement of WNV is conjectural, at least for WNV spread in the Western Hemisphere, the role of migrants in movement of avian influenza is well known and documented for most subtypes (Webster et al. 1992).

Appearance of highly pathogenic avian influenza subtype H5N1 was first documented from samples collected from domestic geese (*Anser anser*) in Guangdong Province of southern China in 1996 (Xu et al. 1999). HPAI H5N1 was quickly recognized as a serious threat to domestic fowl, among which it spread rapidly causing high mortality (Morris and Jackson 2005). Thus, each new outbreak resulted in massive culling of exposed birds. Fears regarding the virus were greatly increased by the first human deaths in Hong Kong in 1997 where six people died among at least 18 that were clinically diagnosed (Morris and Jackson 2005). Sporadic outbreaks in domestic fowl occurred between 1997 and 2003, but in late 2003 and early 2004, HPAI H5N1 spread began to reach epidemic proportions, with occurrences in seven Asian countries in a matter of weeks (Morris and Jackson 2005). From November 2003 until June 2009, 433 human cases of H5N1 virus infection were reported from 15 countries, 262 of which resulted in the death of the patient (Uyeki 2009).

Despite the fact that migratory birds provide the normal mode of long-distance travel and introduction to new localities for avian influenza viruses, there was no indication of migratory bird involvement in HPAI H5N1 movement during the first years of the epidemic. Outbreaks were entirely restricted to domestic fowl and could be explained by known or suspected patterns of poultry marketing. The virus was first recorded from wild birds in samples collected in a Hong Kong park in 2002 (Ellis et al. 2004). Since then, HPAI H5N1 has been found in a number of wild bird species at localities across Eurasia (World Health Organization 2007). The consensus arrived at regarding HPAI H5N1 movement, at least among public health officials, virologists, and molecular geneticists, was that migratory birds were the principal agent (Morris and Jackson 2005; World Organization for Animal Health 2006), although, as noted by Normile (2006), "dissenters remain," most of whom are ornithologists or ecologists (Steiof 2005; Fergus et al. 2006; Yasué et al. 2006; Gauthier-Clerc et al. 2007).

HPAI H5N1 provides a situation similar to that for WNV in the Western Hemisphere in that it is essentially a new virus also originating from a single locality (Guangdong Province, China). Assuming that all subsequent records for this virus are derived from the original outbreak site, the following hypothesis can be formulated along with a set of predictions that should allow testing of the hypothesis concerning viral movement using observations of how the virus actually has moved over time:

Migratory birds are the main introductory hosts serving to move HPAI H5N1 from its evident point of origin in southern China westward across Asia and into Europe and Africa.

On the basis of this hypothesis and assuming, as all published data confirm, that the southern China appearance of the virus was the first and only locality from which any subsequent records for this particular viral subtype anywhere else in the world could be derived, several predictions can be made. These predictions are provided in the following text, along with data relevant to their evaluation. As in the case of WNV, most data documenting viral movement are derived from seroposi-tive samples taken either from dead birds or cloacal swabs of living birds.

• *One or more migratory species with continental distributions will be found to be common carriers of HPAI H5N1 in infectious form.* No such common carrier has been found to date despite extensive search efforts involving hundreds of research-ers and tens of thousands of birds in Asia, Europe, Africa, and North America (Munster et al. 2005; Spackman et al. 2005; Chen et al. 2006; Olsen et al. 2006; Ducatez et al. 2007; Gaidet et al. 2007; Wallenstein et al. 2007; Winker et al. 2007). Indeed, the entire direct evidence documenting that such wild carriers exist comes from two studies: L'vov et al. (2006) in western Siberia and Chen et al. (2006) in Hong Kong and Jiangxi, China. The Russian study reported that "Seven HPAI/ H5N1 strains were isolated from the tracheal/cloacal swabs of clinically healthy, ill and recently dead great-crested grebes (*Podiceps cristatus*), cormorants (*Phala-crocorax carbo*), balt-coots [sic] (*Fulica atra*), and common terns (*Sterna hirundo*)" (abstract translated from Russian from L'vov et al. 2006 as reported in PubMed). The Chinese study reported that of 13,115 migratory birds sampled, six apparently healthy, viremic individuals were found representing three species: Mallard (*Anas platyrhynchos*), Falcated Teal (*Anas falcata*), and Spot-billed Duck (*Anas poecilorhyn-cha*). In both studies, the wild migratory birds were sampled in areas where HPAI H5N1 outbreaks in domestic poultry were well under way, and the results do not link samples positive for the virus with specific individuals by species, nor do they describe how viremic migrants were determined to be healthy or for how long a period (hours? days?) they were known to have remained healthy.

• *Evidence documenting the presence of the virus (i.e., infected dead birds or swab samples) will occur in patterns reflective of the principal migration pathways of those migrant species most commonly infected.* Gilbert et al. (2006) presented overlays of continental migration pathways and HPAI H5N1 outbreak sites and found "dis-crepancies between the geographic spread of HPAI H5N1 virus and overall pattern of wild bird migration." Yet they concluded that migration and outbreak patterns in central Asia appeared to overlap, which, in their view, strongly implicated migra-tory birds, especially Anatidae, in the spread of the virus. However, these findings do not provide proof of migrant involvement as introductory hosts. In fact, given the broad distribution of outbreaks in this region and elsewhere across Eurasia, it would be curious if they did not find overlap with many migratory routes. More indicative of the failure to connect migrants with spread of the virus is the many examples in which major migratory routes do not coincide with outbreak sites. Figure 9.2 illustrates many such examples. This figure shows the distribution of HPAI H5N1 as of 2007 by areas where the outbreaks were mainly in domestic poultry versus those in which they were presumed to be mainly in wild migratory birds. Two observations can be made based on this figure. First, the vast majority of outbreak sites (>95%) involved domestic fowl. For many of these, the market connections resulting in viral spread are known. Second, there are vast holes in viral distribution that would not be expected were migrants serving as the principal introductory hosts. The most obvious of these occurs along the eastern shore of the Mediterranean Sea, perhaps the most heavily trafficked region of Eurasia in terms of both numbers of migrant species and individuals (Moreau 1972). Areas located north (e.g., Turkey) and south (Israel, Egypt, Sudan) of this region have had outbreaks; yet HPAI H5N1 was unreported from most parts of this corridor as of 2007. A second example occurs along the principal East Asian migration corridors. McClure (1974) and others (Morris and Jackson 2005:26) have identified major pathways for waterbirds (cranes, anatids, scolopacids). Comparing these pathways with the data in figure 9.2 on outbreaks, it can be seen that major gaps in outbreak distribution occur: for example, proceeding south from Korea and Japan (both infected) across Taiwan and the Philippines (not infected) to the Celebes (infected) (McClure 1974:8); similarly, passing from eastern China (infected) through Thailand, Laos, Cambodia, Vietnam, and northern Malaysia (all infected), on to southern Malaysia and Sarawak (not infected) to wintering areas in Sumatra, Java, Borneo, and New Guinea (all infected) to Papua New Guinea and Australia (not infected). Many other examples could be drawn from figure 9.2 in which migrant pathways as well as major breeding and wintering congregation sites fail to overlap outbreak distribution.

• *The timing of outbreaks will coincide with timing of major migration movements by the most commonly infected migrants.* HPAI H5N1 became an epidemic in February 2004 when major outbreaks occurred almost simultaneously in South Korea, China, Japan, Laos, Vietnam, Thailand, Cambodia, northern Malaysia, and Indonesia (Morris and Jackson 2005). No Asian migrant of any species, let alone waterbirds, has a migration schedule that would fit this pattern. Similarly, the timing for most other outbreaks since 2004 failed to coincide with any obvious schedule related to the arrival of migratory species (Rappole and Hubálek 2006:table 1).

• *Outbreaks will occur mainly at wetland sites where commonly infected migratory species congregate at staging areas, stopover sites, and wintering areas, separate from domestic fowl.* Jourdain et al. (2007) identified the wetlands in the Camargue region of southern France as a major wintering area for waterbirds from continental Europe as well as the tundra and taiga regions of Scandinavia and Siberia. No evidence of HPAI H5N1 had been identified from this region as of 2007 or indeed at hundreds of other sites where waterbird migrants are known to congregate (figure 9.2).

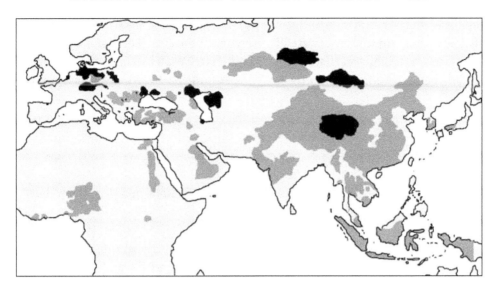

FIGURE 9.2   World distribution of HPAI H5N1 as of May 2007 based on data from the World Health Organization (2007): *dark gray* = distribution of outbreak sites mainly in domestic poultry; *black* = distribution of outbreak sites mainly among wild birds.

Using summaries provided by The World Health Organization, the Food and Agriculture Organization of the United Nations, and the World Organization for Animal Health Morris and Jackson (2005) documented that HPAI H5N1 outbreaks occurred predominantly in domestic fowl, a circumstance supported by subsequent summaries (e.g., World Organization for Animal Health 2006; The World Health Organization 2007, 2011). In fact, from the time of its first appearance in July 1996 until December 2002, outbreaks occurred solely in domestic fowl, with no evidence of migrant bird infection (Rappole and Hubálek 2006). Thereafter, of 3,095 outbreaks of HPAI H5N1 reported between December 2003 and February 2005, all involved captive birds or domestic fowl (Morris and Jackson 2005). During this period, no infected migratory birds were discovered at any locality where they were not in direct contact with infected domestic fowl (Rappole and Hubálek 2006). The first outbreak among migrants at a site where domestic fowl were thought to be absent occurred at Qinghai Lake and Xinjiang Province, China (April, May 2005) (Chen et al. 2005), followed thereafter by outbreaks at Lakes Erhel and Khunt in northern Mongolia (August 2005) and elsewhere in Asia and European Russia. However, complete separation from infected domestic fowl could not be confirmed for these occurrences because of the near ubiquity of free-ranging domestic fowl throughout these regions wherever human beings live (Feare and Yasué 2006).

In 2005 and 2006, there were reports from several European countries of sick or dead migrants, mostly waterfowl or hawks, at sites where no obvious evidence of the presence of infected domestic fowl or other exotic birds existed (Heneberg 2006; Rappole and Hubálek 2006). An interesting aspect of the appearance of sick and dead wild birds at sites in Europe apparently isolated from domestic fowl is that no healthy viremic birds were discovered near the sites (Feare and Yasué 2006) or, indeed, anywhere on the continent as of 2007 (Wallenstein et al. 2007). This finding has at least three explanations:

1. The viremic migrants that imported the virus were sick during their movement or became sick shortly thereafter.
2. The occurrence of healthy, viremic migrants is very rare, making capture of the introductory host individual improbable.
3. Some other agent (e.g., infected domestic fowl or exotic pet) was present but undetected.

The HPAI H5N1 virus epidemic appears to have nearly run its course in the human population. As of February 9, 2011, only Egypt (3) and Cambodia (1) reported new human infections for the current year, down from a world peak of 115 in 2006 , although new outbreaks continue to occur in poultry (World Organization for Animal Health 2011). No data have yet been found to provide direct support to the hypothesis that migratory birds served as the principal means of movement for HPAI H5N1 during the early years of the outbreak (1996–2004). Indeed, until late 2005, there was very little indirect evidence that migratory birds were involved at all as a means of long-distance movement for HPAI H5N1, let alone the principal means. The 2006 and 2007 appearance of dead infected wild birds at a number of sites in Europe where no other obvious source of infection existed changed that situation. Although these records do not confirm migrant introduction of the virus to new localities, they do provide the strongest circumstantial evidence to date. If migrant introduction ever were to be confirmed, it would not be a surprise. Tens of thousands of migrants have now been exposed to the pathogen across most of Eurasia and large parts of Africa. Obviously, many of these migrants *do* contract the virus. Available data indicate that nearly all birds, regardless of species, that come in contact with this virus contract it and get sick from it, and most die, as confirmed by massive cloacal swabbing efforts across Eurasia and Africa, which have failed as yet to find more than a few healthy, viremic individuals. Nevertheless, we can be sure that migrant survival in a viremic state after exposure to the virus is not zero. Some migrants will survive infection and be capable of introducing the virus to new localities. The broader the distribution of outbreaks, the more migrants are exposed, and the greater the number of potentially surviving introductory migrant hosts.

A second factor in operation that is likely to increase the probability of viremic, healthy migrants is that both the virus and its host species are evolving. In the case of the virus, natural selection is likely to favor decreased lethality, whereas selection should favor increased resistance among migratory bird hosts. Most avian influenza viruses are benign (Webster et al. 1992), and that is the likely fate for HPAI H5N1 over time as selection forces continue to operate on both virus and host.

If migrants were not the cause of the extraordinarily rapid spread of HPAI H5N1 across Eurasia and into Africa, then what was? The most obvious alternative is legal and illegal trade in poultry, uncooked poultry products, and exotic avian pets. In many cases, it is well known that this was the route followed by the virus to infect new regions (Morris and Jackson 2005). However, a major portion of this trade is illegal and therefore unreported, if not aggressively hidden. Thus, absence of records documenting the contamination of birds at a new site by imported, infected domestic or exotic pet birds or bird products does not constitute evidence that such an import did not take place.

CHAPTER 10

CONNECTIVITY AND CONSERVATION

A s our knowledge of the migrant life cycle has increased, it has become clear that understanding only one portion of that cycle will not necessarily allow us to determine what threats confront each migrant species. The study of migrant "connectivity" represents an attempt to confront this basic issue.

Connectivity

"Connectivity" is defined as "the degree to which individuals of populations are geographically arranged among two or more periods of the annual cycle (Webster et al. 2002; Marra et al. 2006)" (Boulet and Norris 2006a:1). The concept is pertinent to understanding several different aspects of life history, and, where appropriate, "connectivity" data of various types have been presented in preceding chapters (see chapters 5 and 6). Nevertheless, the main focus of connectivity studies has been on conservation (Crooks and Sanjayan 2006; Faaborg et al. 2010a, 2010b). The logic of this approach is impeccable: If a population is dependent for its survival on geographically separate breeding, postbreeding, fall transient, winter, and spring transient environments, as discussed in chapter 7, the starting place for preservation is to know where those environments are located, what threats exist at those sites, and where population limitation is occurring (Fretwell 1972; Terborgh 1974; Rappole and Warner 1976:210; Sillet and Holmes 2002; Rappole et al. 2007:4).

We have long had the ability to connect movements of migrant populations to the various parts of their breeding, transient, and wintering areas through the use of observational records (Cooke 1915; Moreau 1950) supplemented by work with specimens, especially those with subspecific plumage differences (Lack 1944b; Salomonsen 1955; American Ornithologists' Union 1957; Phillips 1986, 1991; Ramos 1988), and banding–recapture studies (Moreau 1972; McClure 1974; Bellrose 1976; U.S. National Bird-Banding Lab 2011). These methods have provided clear ideas of breeding, transient, and winter distributions of most migratory bird populations (Poole 2010; del Hoyo et al. 1992–2011); indeed, the basic structure of migratory connectivity was worked out by Salomonsen (1955) based on precisely these kinds of data (see chapter 1). Nevertheless, recent technological developments (e.g., satellite radio-tracking, light-level geolocators, stable isotope analysis [SIA], and DNA hybridization) now allow us to connect the various parts of the migrant annual cycle for at least some individuals of some species. The result provides the potential for a new perspective, upon which researchers have been quick to capitalize (Webster et al. 2002; Boulet and Norris 2006a).

SATELLITE RADIO-TRACKING

Cochran et al. (1967) were the first researchers, and among the only ones, to place a transmitter on a small migratory bird, the Swainson's Thrush (*Catharus ustulatus*), at a stopover site in-transit and to follow that bird for hundreds of kilometers. They accomplished this feat using a small (<3 g) transmitter mounted on the bird and a receiver mounted on a single-engine aircraft. Within the past two decades, the ability to follow migratory birds carrying transmitters has greatly expanded through the use of large transmitters mounted on the birds and receivers mounted on satellites (Brodeur et al. 1996; Fuller et al. 2003; Berthold et al. 2004; Bildstein 2006). To date, the utility of the technique is limited by three factors—(1) the size of the transmitter (only relatively large species can carry transmitters of a size necessary to have sufficient battery life and transmission power), (2) expense in terms of transmitters and satellite time, and (3) potential effects on behavior of birds to which transmitters have been attached (Phillips et al. 2003)—although technology is changing quickly to reduce or eliminate these problems (British Trust for Ornithology 2011a). Despite these limitations, the data quality in terms of connectivity establishment are unrivaled by any other method. Error estimates for satellite locality points are of the order ±1 km (Berthold et al. 2004).

LIGHT-LEVEL GEOLOCATORS

Global location sensor loggers, or geolocators, were developed by the British Antarctic Survey (BAS) (Afanasyev 2004; Phillips et al. 2004). Depending on requirements in terms of kinds of data to be recorded and battery life, they can be made quite small (1.0 g) so that even medium-size songbirds (30 to 40 g) can carry them on their migratory journeys (British Trust for Ornithology 2011b). These dataloggers can provide remarkable data on migration route and wintering area (Phillips et al. 2007; Guilford et al. 2009; Bächler et al. 2010; Bowlin et al. 2010; Robinson et al. 2010). The type used by Stutchbury et al. (2009) in their study of Purple Martin (*Progne subis*) and Wood Thrush (*Hylocichla mustelina*) movements recorded visible light intensity every 60 seconds and maximum light per 10-minute interval. Their geolocator dataloggers, holding data of variable accuracy on the bird's daily location throughout the period, were recovered by recapturing the marked birds (usually 9 months or so after original capture) after completion of a round-trip between breeding and wintering sites. The data on the geolocators were downloaded and read using the TransEdit program developed by the BAS. Stutchbury et al. (2009) estimated accuracy at the breeding site in northwestern Pennsylvania (41°N, 78°W) as ±300 km in latitude and ±100 km in longitude. Takahashi et al. (2008) used 7-g geolocators that recorded immersion in seawater and water temperature in addition to light intensity on two Streaked Shearwaters (*Calonectris leucomelas*) (figure 10.1) and with an error estimate of ±200 km.

The advantage of the geolocators over other connectivity-measuring techniques is that they provide large amounts of data on daily location throughout the non-breeding period at relatively low cost. Previously, these kinds of data could only be obtained by satellite radio-tracking. The accuracy of geolocator location data (±100 to 300 km) is nowhere near as accurate as location data derived from satellite radio-tracking (< ±1 km for a 5-g transmitter [Fuller et al. 2003:360]), and geolocator accuracy varies according to time of the year (latitude cannot be determined within ±15 days of an equinox) and proximity to the equator (Hill 1994), factors that do not affect satellite-tracking data. Given these limitations regarding location accuracy, it is surprising that journals publishing such studies have not required authors to show migration paths derived from the data as probability distributions hundreds of kilometers in width rather than as lines on maps, which have the same precision as lines resulting from satellite radio-tracking but differ by two orders of magnitude in terms of accuracy. As with other techniques requiring capture, handling, and attachment of a device, including radio-transmitter attachment, the geolocators can affect the behavior of the marked bird (Igual et al. 2004; Rodriguez et al. 2009).

STABLE ISOTOPE ANALYSIS

Ratios of the stable isotopes of certain elements have been found to have characteristic continental distribution patterns—for example, the occurrence of deuterium (D), the stable isotope of hydrogen that occurs in characteristic values in precipitation (figure 10.2). Stable isotope ratio distribution across continents and oceans has been a powerful tool with extensive applications in Earth and marine sciences, resource exploitation, environmental science research, and, within the past 30 years or so,

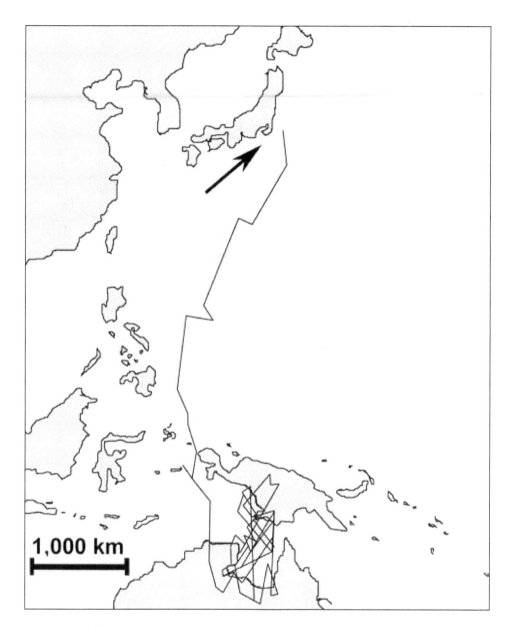

FIGURE 10.1 Tracking migratory movement of a Streaked Shearwater (*Calonectris leucomelas*) using a light-level geolocator. Individual points are ±186 km in accuracy. This bird was tracked from October 16, 2004, to January 13, 2005, during which period it traveled south from its breeding colony (*arrow*) on the island of Mikura (off the southern coast of Japan) across 60 degrees of latitude to the Gulf of Carpentaria off the north coast of Australia (Takahashi et al. 2008).

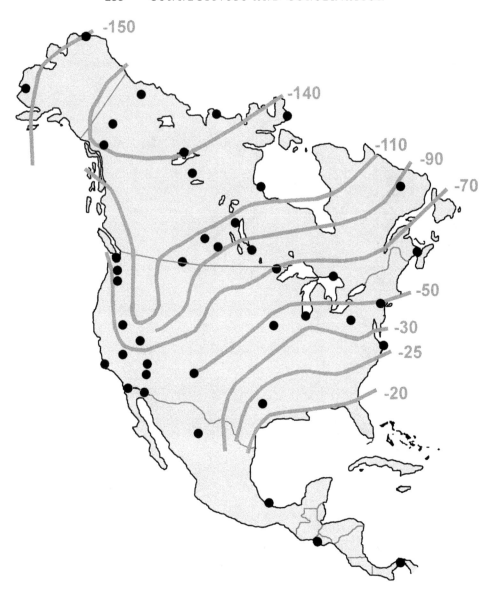

FIGURE 10.2   Patterns of stable hydrogen isotope ratios (δD) for growing-season average rainfall in North America (based on data from Hobson and Wassenaar 1997): *circles* = locations of sampling stations.

ecological investigations (Rau et al. 1983; Lajtha and Michener 1994). More recently, these stable isotope ratio patterns have been used extensively in studies of migratory bird connectivity (see reviews in Hobson and Wassenaar 2008; Faaborg et al. 2010a). Hobson (2005:1038) describes the concept as follows:

The principle behind the application of SIA of avian tissues to infer migratory connectivity is simply that isotopic patterns in nature can be variable; when a bird moves between one known "isotopic landscape" or "isoscape" (G. J. Bowen pers. comm.) and another, its tissues will retain isotopic information from the previous location for a period depending on the elemental turnover in that tissue. For feathers, isotopic information is usually "locked in," because feather keratin is metabolically inert following formation; so this material is particularly useful when molt chronology and general location (breeding vs. wintering grounds vs. en route) are unambiguous.

In the majority of migrant connectivity studies, feather samples are collected from birds from several known breeding areas for a particular migrant species, and an average is taken of one or more isotope ratios—usually carbon, nitrogen, or deuterium, but sometimes others (e.g., sulfur or strontium). Then feathers are collected from birds of the same species taken from some point along the migration pathway or on the wintering ground. The stable isotope ratios are compared with the averages from the breeding-ground samples, and a conclusion concerning the breeding-ground origin of the individual bird is reached. For example, Chamberlain et al. (1997) analyzed feathers from across the breeding range of the Black-throated Blue Warbler (*Setophaga caerulescens*) from which they measured the stable isotope ratios for carbon, hydrogen, and strontium and derived mean values for each element isotope ratio from each major breeding area (figure 10.3). They then collected feathers from wintering birds in the Caribbean (Jamaica, Dominican Republic, Puerto Rico). On the basis of the values of the Caribbean sample, they drew two conclusions concerning these birds:

1. Most derived from the northern end of the breeding range.
2. Birds from different breeding areas may use the same wintering area.

These are not surprising findings and appear to follow what might have been concluded based on banding data. However, a study using the same breeding-ground

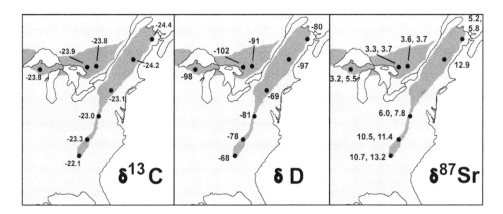

FIGURE 10.3   Mean stable isotope ratios for hydrogen (δD) and carbon ($\delta^{13}C$) in feathers and strontium ($\delta^{87}S$) in bones of Black-throated Blue Warblers (*Setophaga caerulescens*) from eastern North America (Chamberlain et al. 1997): *dark gray* = breeding range.

feather data raised concerns regarding the technique. Graves et al. (2002) found that if instead of averaging samples by breeding area, one were to look at the individual values, a large amount of variation in stable isotope ratios occurred in birds from the same breeding areas, indeed, as much as was found across large areas of the bird's range in the Chamberlain study. When this variation was examined, certain patterns appeared, including differences in ratio by age, altitude, and from year to year for birds taken from the same general breeding area.

Further work with carbon and nitrogen stable isotope values has shown that these ratios vary not only by year, elevation, and age but also by habitat type in which the bird molted and by proportion of diet that is animal versus plant during the molt period, leading Hobson (2005:1039) to conclude: "Indeed, a strong case can be made that ^{15}N and ^{13}C isoscapes in North America, and indeed throughout the world, are simply too complex to be used reliably to track migratory birds, and most researchers tend to shy away from interpreting these sorts of data." He goes on to recommend stable hydrogen isotope ratios as a more reliable source for establishing migrant population connectivity between breeding and nonbreeding range. However, Smith and Dufty (2005) found some difficulties in establishment of connectivity using this technique as well. In their analysis of stable hydrogen isotope ratios in Northern Goshawk (*Accipiter gentilis*) feathers, they examined variation in values based on climate, geographic location, age, and sex. In addition, they compared values for different individuals taken at the same places and for feathers taken from different parts of the same bird (intraindividual variation). They concluded:

> Variation among individuals was nearly eight times the magnitude of variation within an individual, although age differences explained most of this interindividual variation. In contrast, most variation in δD values between multiple feathers from an individual remained unexplained. Additionally, we suggest temporal patterns of δD in precipitation (δD_p) as a potential explanation for the geographic variability in age-related differences that has precluded the description of movement patterns of adult raptors using δD_f [i.e., stable isotope ratios of hydrogen in feathers]. Furthermore, intra-individual variability necessitates consistency in feather selection and careful interpretation of δD_f-based models incorporating multiple feather types. Finally, although useful for groups of individuals, we suggest that variability inherent to environmental and intra-individual patterns of δD_p and δD_f, respectively, precludes the use of stable-hydrogen isotopes to describe movements of individual birds. (Smith and Dufty 2005:547)

Similar concerns have been pointed out by others (Norris et al. 2006; Faaborg et al. 2010a:28).

Stable isotope analysis may have potential for future connectivity studies, but the issues raised here must be addressed. Two major categories of assumptions have been made in many of the "connectivity" analyses that have been performed to date: (1) that stable isotope ratios can be obtained from feather samples that are characteristic of the area in which a bird underwent prebasic molt; and (2) that the site where a bird undergoes its prebasic molt is the same (in terms of location and habitat) as the site in which it bred or was born. The findings of Graves et al. (2002) and Smith and Dufty (2005) demonstrate that the first category (i.e., characteristic regional variation in ratios in feather samples for a given species) cannot be assumed. It must be tested for experimentally. The work reported in chapter 3 ("Postbreeding Period") show that the second category (i.e., that birds whose feathers are used to make the isotope ratios averages molted in the same locality in which they bred or were born in) also cannot be assumed; it must be tested (with some exceptions—see discussion of prebasic molt patterns in chapter 3). A characteristic of many studies in which stable isotopes have been used is that the resulting data often form the basis for compelling conclusions, which make a great deal of sense from an ecological or evolutionary perspective. These conclusions, however, should be viewed as working hypotheses and subjected to additional testing using other methods.

CARRYOVER EFFECTS BETWEEN WINTERING AND BREEDING GROUND

Raveling (1979) was able to demonstrate a direct relationship between fat levels accumulated during the nonbreeding period and timing of incubation on breeding grounds for female Cackling Geese (*Branta hutchinsii*). The need for subcutaneous fat reserves derived from stopover sites on arrival at breeding areas has been established for other High Arctic–breeding species as well, particularly other waterfowl and waders (Drent and Piersma 1990).

Similar claims of carryover linkage between quality of winter habitat occupied and breeding ground fitness have been made for several passerines in which various purported measures of "physiologic condition" (i.e., in terms of individual survival or reproductive success)—for example, corticosterone levels, subcutaneous fat levels, or even habitat quality (as an indirect measure of physiologic condition)—have been performed on wintering populations of passerines, often contrasting males with females (Marra and Holberton 1998; Studds and Marra 2005; Johnson et al. 2006; Newton 2008:768–772). The proposed relationship suggested between wintering ground physiologic condition and breeding success is logical, but there are three difficulties with many of the studies on which it is formulated. First, the linkages between wintering habitat quality and breeding habitat occupation usually are based on stable isotope technology, the problems with which have been discussed earlier; second, the physiologic measures (or habitat measures) taken generally have not been demonstrated to be related to fitness for the species in which they were measured (see chapter 5); and third, the relationship between these measures and reproductive success has not been demonstrated. Thus, the difference between these studies and those performed by Raveling (1979), Drent and Piersma (1990), and similar work is that the although "condition" measures are assumed to have a direct effect on breeding success, no actual demonstrations or tests are provided. In this regard, it is important to note that recaptures and return rates, used in some studies as measures of fitness (e.g., Johnson et al. 2006), are measures of fidelity, which is not necessarily equivalent to survivorship (see chapter 2). Because the meaning of the term "condition" as used in this context is predicated on some unknown relationship to fitness, and in addition is assumed predictive of reproductive success, the reasoning is circular and invalid until the relationship has been demonstrated. See chapter 7 for additional discussion of the relevance of physiologic measures (e.g., corticosterone levels) to migrant population dynamics.

It is a fair question to ask: If differences in corticosterone levels between males and females are not related to condition, then what do they mean? Unfortunately, our knowledge of the relationship between hormones, condition, and behaviors in birds is crude (Holberton and Dufty 2005; Ramenofsky 2010). Nevertheless, we do know that the life history challenges confronting the different sexes in migratory birds are quite different. These differences are evident in various ways throughout the life cycle (see earlier chapters), but especially in spring. The male must arrive as early as possible on the breeding area, commensurate with survival, in order to obtain and defend the highest-quality breeding territory possible, whereas the female is focused on assessment of male quality and selection of highest-quality territory for raising her offspring. Males not only depart earlier, on average, than females from their wintering sites but also demonstrate evidence of increased testosterone production during migration (e.g., aggressiveness, singing, development of a cloacal protuberance, and even sperm production) long before their arrival on the breeding area (Quay 1985). Thus, although we do not know what the different levels of corticosterone between wintering male and female migrants might mean, it would be surprising if there were no such differences, given their differences in terms of their life history.

More recent work attempting to demonstrate carryover effects on fitness resulting from occupation of wintering habitats of apparently different quality has focused on within-sex comparisons, specifically among males. Reudink et al. (2009), for instance, found that adult (after second year) male American Redstarts (*Setophaga ruticilla*) that apparently wintered in lower-quality habitats arrived later and had lower breeding success than adult males that wintered in higher-quality habitats. Although based on the same kinds of flawed stable-isotope technology as many of the studies mentioned earlier, we believe these kinds of intra-sex/age group comparisons offer a much more promising line of investigation than the inter-sex/age group comparisons that involve a number of potentially confounding life history variables of unknown importance. Nevertheless, these are complicated issues, and there may be more going on in this relationship between carbon isotope signature and breeding success than carryover effects. For one thing, it appears that in these New Hampshire redstarts, we are looking at the process of change from a monogamous to a promiscuous mating system. In addition, age

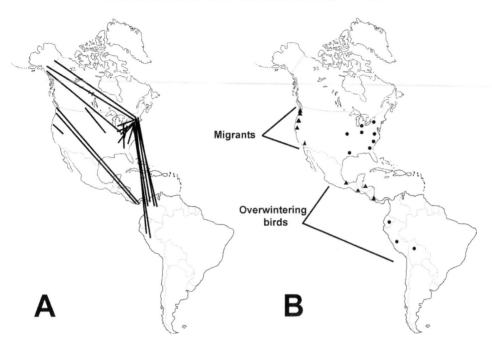

FIGURE 10.4   Swainson's Thrush (*Catharus ustulatus*) breeding areas, wintering areas, and migration routes based on banding data (*A*) and genetic data (*B*) (based on Ruegg and Smith 2002): *triangles* = coastal-breeding birds taken in winter; *dark circles* = continental-breeding birds taken on migration or in winter.

may still be an important factor in terms of wintering area and breeding success, even among after-second-year males.

GENETIC STUDIES

Subspecific plumage differences are essentially genetic markers that have been used for a long time to determine connections between breeding and nonbreeding populations of migrants (Lack 1944b; Salomonsen 1955; American Ornithologists' Union 1957; Phillips 1986, 1991; Ramos 1988). Recent advances have allowed increased sophistication in terms of technique, which now include mitochondrial DNA (Wenink and Baker 1996; Wennerberg 2001; Kimura et al. 2002; Ruegg and Smith 2002; Smith et al. 2005), random amplified polymorphic DNA markers (Haig et al. 1997), and DNA microsatellites (Clegg et al. 2003; Smith et al. 2005) for use in establishing connectivity between different portions of the annual cycle for populations (figure 10.4). The goal, of course, is to be able to identify the metapopulation source for any given individual, for which available techniques are, for the most part, not sufficiently fine-grained. However, the problem of connectivity identification using genetic markers is not a technological one alone as the degree to which markers are characteristic of a given metapopulation will depend upon the dynamics of individual populations in terms of mean dispersal distances within the species' winter range for juveniles and mean settling distances from natal territory for novice breeders.

CONSERVATION RELEVANCE OF CONNECTIVITY STUDIES

Technologically, connectivity studies are still under development and need considerable work to achieve the level of knowledge required to understand the conservation challenges of specific migrant populations. Nevertheless, the work emphasizes the practical importance of the equations developed in chapter 7, which demonstrate that, for any given declining population of migrants, the most likely cause is habitat loss during one phase of the annual cycle. Once connectivity studies have been refined and coupled with population studies, our ability to identify the exact nature of the problem confronting any given declining migrant will be greatly enhanced.

CONSERVATION

Conservation is what we think we should do when confronted with a particular set of ecological circumstances. People of good will can differ in their opinion concerning what is best to do depending upon the information they have and their basic value systems. Thus, conservation is not science, and much of what will be discussed in this section should be considered in that light.

Before human populations began to expand rapidly across the globe, extinction of species occurred as the result of environmental change and interspecific competition, including predation and diseases, in patterns established over millennia. Humans may have played a role in extinction of Pleistocene megafauna (Martin 1973) and caused prehistorical extinctions in some island faunas but probably did not begin to exert a global impact on other species until the age of exploration (sixteenth to eighteenth centuries) when improved technology allowed relatively rapid movement, as well as efficient harvest of resources for market purposes (although see Stinchcomb et al. 2011 for a contrary hypothesis). Experience has shown that in modern times (i.e., 1600 to the present), most extinctions occur in two ways: population destruction and habitat destruction. Conservation is a relatively recent concept, born of massive anthropogenic changes to the environment. As applied to migratory birds, it dates back only to the early 1900s, at least in the United States, when exponential growth in the human population in the eastern half of the country along with the plume trade and market hunting had caused actual extirpation of many migrant populations, arousing public concern and subsequent legislative action for the long-term welfare of disappearing species.

Migratory bird conservation depends on action in three major areas: research, national and international conservation policy collaboration, and management

(Rappole et al. 1983:75–107). Most countries now have policies in place for protection, conservation, and management of migratory birds. The series of policy and legislative actions and international treaties put in place in the late-nineteenth and early-twentieth centuries to promote wading bird and waterfowl recovery from market hunting in North America are exemplary regarding the efficacy of such an approach (Chandler 1986; U.S. Fish and Wildlife Service 2006). In addition, efforts by many nongovernmental organizations have been extremely important in helping to achieve migratory bird conservation goals (e.g., Convention on Migratory Species [United Nations Environmental Program 2004], Partners in Flight [Ruth 2006], Ramsar Convention on Wetlands [Ramsar 2011], International Union for Conservation of Nature 2011, BirdLife International 2011). Three programs that have been extremely important in recent efforts to preserve migratory bird populations are considered in the following.

SPECIES-SPECIFIC MIGRATORY BIRD CONSERVATION

The U.S. Endangered Species Act (U.S. Congress 1973) has had a large impact on the most threatened species of migratory birds in North America, although the approach suffers from a lack of pertinent knowledge. A major theme of this book is the complexity of the migrant life cycle and the incomplete state of our understanding of that complexity. This lack of understanding is an important aspect of migratory bird conservation. The conservation record is excellent for preservation of migratory bird populations suffering from factors that are known and understood (e.g., pesticides, hunting, or breeding habitat loss) (Nichols et al. 1995; Cade et al. 1997; Reynolds et al. 2001). When the causes are either not understood or misidentified, the record is much less encouraging. For example, causes of declines in forest-related migrants have been debated intensely in the literature (e.g., Rappole and McDonald 1994, 1998; DeGraaf and Rappole 1995:19–34; Rappole 1995:136–150; Latta and Baltz 1997; Faaborg et al. 2010a:17), and, at present, there is still no consensus among students of the issue, although declines continue. There are many other examples of threatened migratory species whose conservation is inhibited not by lack of interest or funds but by incomplete knowledge of the life cycle, two of which are discussed in the following. Nevertheless, knowledge is the foundation on which successful migratory bird conservation programs are built. The best will in the world is likely to fall short of conservation goals if the information on which it is based is faulty.

• Whooping Crane (*Grus americana*). The Whooping Crane is an example of a conservation success story for a migratory species, at least to date, and is illustrative of how the program can work when enough knowledge is available to confront the critical issues threatening populations. The Whooping Crane was formerly distributed across much of central, and perhaps eastern, North America, breeding in prairie marshes and wintering in coastal wetlands of the southeastern United

States (Lewis 1995). By 1941, the known migratory population was 15 to 16 birds, all of which bred at Wood Buffalo Park in northeastern Alberta and wintered on the Texas Gulf coast at Aransas. A small resident population persisted in coastal Louisiana but had disappeared by 1950. As of September 30, 2010, the world population of the species was 574 birds, 407 of which were in the wild (Whooping Crane Eastern Partnership 2011). The presumptive cause of decline was hunting, most of which probably occurred along the migration route. An obvious question is how a migratory population with a continental range could drop to such a low level and still recover. The answer probably lies in the species' social system. Adults and young remain together as family groups from the time they leave breeding territories in fall until their return in spring. Thus, a very small number of birds could serve as the nucleus for recovery once protection was in place because the entire population went to the same breeding, stopover, and wintering sites. Had family groups disbanded after breeding to migrate individually to disparate wintering sites, Whooping Cranes likely would have been extinct by the early 1900s. Recent efforts have resulted in expansion of both breeding and wintering areas (U.S. Fish and Wildlife Service 2011a). A large part of the success of this program likely is due to the fact that the Whooping Crane is a large charismatic species for which the funding was available to examine all aspects of the bird's life history—from its Alberta breeding marshes, its migration route across the Great Plains, and its winter quarters in salt marsh along the Texas coast.

• Golden-cheeked Warbler (*Setophaga chrysoparia*). The Golden-cheeked Warbler breeds in oak–juniper (*Quercus–Juniperus*) habitat of central Texas and winters in the pine–oak (*Pinus–Quercus*) habitat of the highlands of southern Mexico and Central America south to Nicaragua (figure 10.5). Steep declines in breeding population size prompted recognition of the species as "Endangered" by the U.S. Fish and Wildlife Service (1990). When the recovery plan was drafted and published (Keddy-Hector and Beardmore 1992), greater than 99 percent of all research, management activities, and funds were focused on the Texas breeding ground; and in fact most effort was directed to a small portion of the breeding area located at Fort Hood military base outside San Antonio (Jettj et al. 1998). Because almost no information was available on winter habitat for the bird at the time (<20 records in the literature), this focus was not surprising. However, a U.S. Fish and Wildlife Service–funded program largely rectified this problem, providing extensive information on winter habitat requirements (Rappole et al. 1999, 2000; King and Rappole 2000). In fact, remote sensing data on habitat amounts and bird densities for both the breeding and wintering area resulting from this work indicated that the population likely was controlled by lack of winter habitat (Rappole et al. 2003b, 2005). The pine–oak habitat required by the bird during the winter period has been recognized as a threatened environment (Perez et al. 2008; King et al. 2012). Nevertheless, current U.S. Fish and Wildlife Service programs continue to focus almost entirely on breeding ground issues (U.S. Fish and Wildlife Service 2009, 2011b).

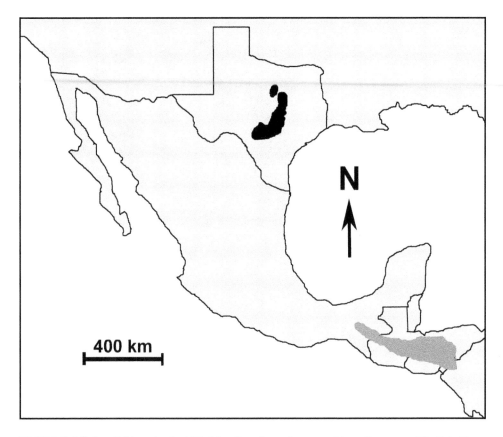

FIGURE 10.5   Golden-cheeked Warbler (*Setophaga chrysoptera*) breeding range (*black*) in Texas and winter range (*gray*) in southern Mexico and Central America (based on Rappole et al. 2003b).

• Kirtland's Warbler (*Setophaga kirtlandii*). The Kirtland's Warbler breeds in second-growth jack pine (*Pinus banksiana*) savanna in northern Michigan and winters in the Bahamas (figure 10.6). It has been rare since its first discovery and description in 1851 (Baird 1852; Mayfield 1992). The assumed reasons for its rarity were that breeding habitat was limited and productivity was low due to social parasitism by the Brown-headed Cowbird (*Molothrus ater*) (Mayfield 1992). Papers by Radabaugh (1974) and Haney et al. (1998) have raised questions concerning this hypothesis. Prior to their work, the winter habitat for the bird was assumed to be dense scrub, which has a rather broad distribution throughout the island chain (Mayfield 1992). Haney et al., however, found that most winter records for the species were from pine savanna, a much rarer habitat type than scrub, and one that has been subjected to decimation across the islands since the early 1900s. When these findings are considered along with the fact that jack pine savanna breeding habitat occurs after burns on sandy soil across a vast area of the North American continent (figure 10.7) (U.S. Department of Agriculture 2012) and that even apparently suitable breeding habitat in the northern Michigan range remains unoccupied,

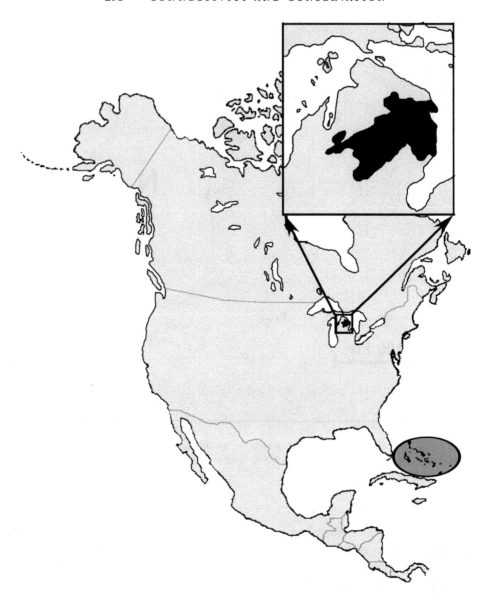

FIGURE 10.6   Kirtland's Warbler (*Setophaga kirtlandii*) range (based on Mayfield 1992): *black in inset* = main breeding area in Michigan; *dark gray* = wintering area in the Bahamas.

winter limitation of Kirtland's Warbler populations seems the likely conclusion. Nevertheless, as in the case of the Golden-cheeked Warbler, recovery efforts focus overwhelmingly on the Michigan breeding ground (U.S. Fish and Wildlife Service 2011c). It is interesting that these programs, which mostly involve massive Brown-headed Cowbird control and creation of large amounts of new breeding habitat, have met with success, despite the likelihood that winter habitat in the Bahamas is likely the principal controlling factor. Cowbird control alone had very little effect on the population until an accidental burn created extensive new areas of jack pine

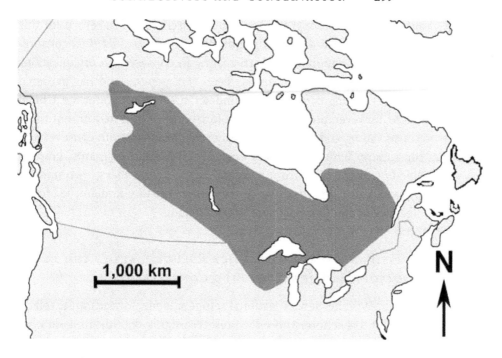

FIGURE 10.7   Jack pine (*Pinus banksiana*) distribution (*dark gray*) in northern North America (based on data from U.S. Department of Agriculture 2012).

in the general vicinity of the existing population. If one assumes that the cowbird control increased productivity, and the burn created new habitat in which young birds returning to their natal site could settle with the likelihood of finding a mate, then one can see how the breeding site programs could maximize population size despite the likelihood of overall control by winter habitat. Nevertheless, the long-term future of the species probably depends upon preservation and expansion of Caribbean pine savanna in the northern Bahamas.

COFFEE AND MIGRATORY BIRD CONSERVATION

One of the most successful nongovernmental programs in terms of public visibility and participation for conservation of migratory birds during the nonbreeding period is the promotion of shade coffee to provide habitat for migrants wintering in the neotropics. Conceptually, the idea behind the program was focused precisely on the main conservation issue for many migrant species: winter habitat loss. Many migrants rely on tropical forest habitats in winter, and traditional shade coffee, in which coffee plants are grown under a forest canopy, can provide such habitat (Greenberg et al. 1997). On the basis of this idea, programs and partnerships were put in place to encourage coffee growers to plant their crop in the traditional manner instead of using the highly mechanized processes involved in the growth of sun coffee (no canopy) (Philpott and Dietsch 2003). In return for this behavior, they would receive a market premium for their product (i.e., consumers were encouraged to pay more by participating in conservation organizations for bird-friendly coffee). This program has been very popular and has garnered widespread support among conservation-minded citizens throughout the United States. There are, however, problems. Market forces are a blunt instrument when it comes to conservation, and growers have different objectives from conservation-conscious consumers. With weak or nonexistent certification programs, growers have been able to market coffee from marginal habitats (e.g., that grown under a citrus canopy as "Shade Coffee") or, worse yet, convert existing undisturbed forest to shade-coffee production (Rappole et al. 2003c, 2003d).

THE U.S. FISH AND WILDLIFE SERVICE REFUGE SYSTEMS AND THE NEOTROPICAL MIGRATORY BIRD CONSERVATION ACT

The U.S. Fish and Wildlife Service national refuge system provides an excellent paradigm for the kind of action required to preserve migratory birds: find out what the threats are, and act to meet those threats. The chief threats to waterfowl conservation in the early twentieth century were hunting and migration stopover and wintering ground wetland habitat loss. Laws were put in place to control hunting, and a systematic program of wetlands purchase and preservation was established, which had a marked effect on conservation of migratory wetland species (Bellrose 1976; Nichols et al. 1995). The Neotropical Migratory Bird Conservation Act takes a similar approach for preservation of migratory birds traveling beyond the borders of the United States. The U.S. Fish and Wildlife Service established this competitive, matching grants program supporting public–private partnerships in carrying out projects throughout the Western Hemisphere in order to "promote the long-term conservation of Neotropical migratory birds and their habitats" (U.S. Fish and Wildlife Service 2011d), with greater than 75 percent of roughly US$24 million appropriated used to fund projects outside the United States.

MIGRATORY BIRDS AND PUBLIC POLICY

• *Crop damage.* Migratory birds can cause damage to crops during any portion of the life cycle, but most commonly the migrants have their greatest effects during migration and the wintering period when they can occur in large flocks. Bobolinks (*Dolichonyx oryzivorus*) so famously damage rice crops during migration and the winter period (Renfrew and Saavedra 2007) that "rice-eating" is part of their scientific name. Transient Red-winged Blackbirds (*Aegelius phoeniceus*) cause sufficient economic effects on corn and sunflower harvests to warrant major government control programs in the Great Plains areas of the United States (Wiens and Dyer 1975; Peer et al. 2003). Such damage to crops by migrants is found in many parts of the world—for example, European Starlings (*Sturnus vulgaris*) on cereal grains in Europe or Dickcissel (*Spiza americana*) consumption of rice in Venezuela (Avery et al. 2001). As Peer et al. (2003:248) note regarding economic losses from depredations by blackbirds, often losses caused by migrants would be inconsequential "if damage were distributed evenly [i.e., across all growers]; however bird damage is often localized around wetlands and can be economically debilitating to individual producers." Unfortunately, control efforts can pose potential damage to declining populations of some migrant species as well (Avery et al. 2001; Renfrew and Saavedra 2007).

• *Depredation.* Some migrants also have been involved in damage to livestock (e.g., Golden Eagle [*Aquila chrysaetos*]). Although it has been argued that lamb depredation due to this species is minimal across its range (Wiley and Bolen 1971), there are some reports of sites and incidents where it can be significant (Watte and Phillips 1994).

Fishermen throughout the world consider anything else that captures fish a competitor. In terms of the many migrants that consume fish (e.g., Common Merganser [*Mergus merganser*], Brown Pelican [*Pelecanus occidentalis*], and Great Blue Heron [*Ardea herodias*]), studies generally do not support claims of damage to local fisheries (Kushlan 1976; Nilsson and Nilsson 1976). Nevertheless, there are situations in which birds (e.g. Double-crested Cormorant [*Phalacrocorax auritus*] and White Pelican [*Pelecanus erythrorhynchos*]) can affect fish density, especially at hatcheries or other kinds of aquaculture sites (e.g., commercial Channel Catfish [*Ictalurus punctatus*] ponds) (Wires et al. 2001; King 2005). Also, several studies have shown that the Great Cormorant (*Phalcorcorax carbo*) can have important effects on populations of certain North Sea fish species, at least in some years (Leopold et al. 1998).

• *Aircraft collisions.* Collisions between birds and aircraft are a serious and expensive problem: worldwide. It is estimated that bird strikes cost roughly US$1.2 billion annually (Allan and Orosz 2001). Most of the damage is caused by low-flying birds (<150 m above ground) during aircraft takeoff and landing in the vicinity of airports during fall migration (July to October) (Dolbeer 2006). Migrants clearly play a major role in aircraft bird-strike issues.

• *Pathogens.* The ecology of migratory bird pathogens is a fascinating topic, involving as it does the mixture of complex life cycles of two completely different organisms: that of the disease agent and that of the migratory bird (e.g., Dickerman et al. 1972; Reeves 1974; Webster et al. 1992; Barraclough 2006). In recent years, association with pathogen movement has been the single most important public policy issue worldwide regarding migratory birds and has involved a great deal of public fear and funds (Epstein 2011). Unfortunately, as discussed in detail in chapter 9, our understanding of the role of migratory birds in pathogen movement is rudimentary.

CHAPTER 11

CONCLUSION

THIS CHAPTER represents a brief summation of major points concerning migratory bird ecology, life history, and evolution that have been addressed in this book. These points are presented in the following sections sequentially according to the chapters in which the points were raised. Because this discussion concerns matters of conjecture rather than research findings, at least in part, I switch from the first person plural to first person singular to emphasize the fact that the opinions expressed are my own.

CHAPTER 1: INTRODUCTION

I view migration as a behavior used to exploit geographically separate environments that differ seasonally in their value for survival and reproduction. I suggest that this behavior has origins in class Aves that are at least as old as flight and Earth's seasonality, although development of human understanding of the phenomenon is recent and still at an early stage. The various types of migration cover a broad spectrum of movements, ranging from a few kilometers to the breadth of the planet. A remarkable number of these different movement types can occur in different populations of the same species, from residency to long-distance migration. Birds are essentially "preadapted" for migration in that nonmigratory individuals possess many of the traits (exaptations) required for successful movement between separate environments (e.g., flight, navigation, and energy storage). Nevertheless, migratory behavior radically changes the selection pressures on a population, resulting in the evolution of an extraordinary variety of adaptations that affect every part of the migrant's annual cycle. These adaptations for a migrant lifestyle are the main focus of my treatment of the subject. In the remainder of the book, I examine each phase of the annual cycle, as well as key aspects of life history, from the perspective of how a migratory habit has molded individual migratory populations through the process of natural selection. In addition, I consider the extent of behavioral and physiologic responses (reaction norms) apparently built in to the migrant's genome that allow genetically identical individuals to respond differently to different environments.

CHAPTER 2: BREEDING PERIOD

Obvious differences between migrants and residents during the reproductive period include arrival time by sex and age, arrival condition (e.g., female arrival at breeding sites with fat reserves accumulated on wintering areas), site fidelity (i.e., annual return to breeding site), and breeding habitat selection. Clearly, these categories are unique to migrants because they involve finding, selecting, using, and returning to a breeding area that is geographically separate from nonbreeding areas—behaviors not required of residents after initial dispersal.

An additional difference between migrants and residents is the need to convert quickly from a nonbreeding (transient) mode to breeding mode in terms of physiology and behavior. Song rates (higher in migrants), reproductive organ size and growth rate (larger and faster in migrants), and male plumage differences (often "brighter" on average in male migrants than females) may represent adaptations resulting from a migrant lifestyle, at least in some taxa.

Differences between migrants and residents also include several aspects of the actual reproductive cycle, including the following:

- *Territory structure.* Most terrestrial, noncolonial migrants defend short-term type A territories, whereas related residents either hold long-term type A or year-round territories. Even in those resident species that hold type A territories for only a portion of the year, adults may associate as pairs during other times of the year.
- *Mating systems.* The majority of both migrants and residents are putatively monogamous. However, rates of successive polygyny and extra-pair fertilization are much higher among migrants than tropical residents.
- *Clutch size.* Clutch size is higher among temperate migrants than among tropical resident congeners.

- *Clutches per season.* Temperate migrants average more successful clutches per season.
- *Duration of incubation.* Shorter for migrants.
- *Nestling period.* Shorter for migrants.
- *Postfledging parental association with brood.* In temperate songbird migrants, postfledging parental care usually lasts 3 to 4 weeks, whereas among tropical residents parental association with the brood can last for months or even into the next breeding season.

Some of the reproductive-period differences between migrants and residents may result simply from the fact that reproduction for migrants occurs in a completely different environment from that in which reproduction occurs for resident counterparts. For instance, many migrants breed at much higher latitudes than resident congeners and thus are able to take advantage of longer summer daylengths to provision offspring, thus reducing nestling period. Similarly, higher rates of extra-pair fertilizations in temperate migrants could result from the increased breeding synchrony enforced by having a shorter breeding season available at higher latitudes. However, I suggest that high rates of extra-pair paternity in migrants with putatively monogamous mating systems may be a relatively recent development resulting from some aspect of environmental change.

In contrast with these potentially environmentally induced aspects of reproduction, others clearly are genetically based (i.e., they are adaptations for a migrant lifestyle that likely evolved since the populations became migratory). Among these is clutch size. Although environmental influences on clutch size have been documented for many species, genetic differences between temperate migrant and tropical resident populations of the same species have been proved experimentally. Greater differences in sexual dimorphism between migrant and resident congeners may also represent adaptations to migration in some species, as may a number of the other differences cataloged. Unfortunately, however, the relative roles of environment versus heredity for these have not yet been investigated. Nevertheless, the summary demonstrates the extraordinary potential for rapid modification through natural selection of reproductive cycle elements for migrants.

CHAPTER 3: POSTBREEDING PERIOD

In this chapter, I recognize the postbreeding period as a distinct and important segment of the migrant life cycle. For many residents, there really is no postbreeding period as a distinct interval between breeding and nonbreeding activities. The reproductive period often has no clear termination point; adults and their young often continue to associate in family groups for weeks or months after the young fledge, even during the prebasic molt. In addition, there is, of course, no preparation for departure in residents, at least by adults.

For most migrants, the postbreeding period has at least two major purposes: completion of the prebasic molt and preparation for departure on migration. Additional purposes have been suggested including exploratory movements by juveniles and future territory and mate prospecting by adults anticipatory of the next breeding season by males. These purposes are quite different from the main purpose of the reproductive period, which is maximization of number of offspring raised to independence. As a result, movements and habitat use for migrants can differ markedly between these two portions of the life cycle. Although postbreeding habitat use patterns are still not well known for many migrant species, there is evidence to indicate that change in habitat, or at least microhabitat, between the two periods is the norm, at least for species that breed in forested habitats. Duration of the postbreeding period is of the order 2 to 3 months for most calendar migrants, but there is considerable variation both within and among species.

Critical examination of the postbreeding period reveals a number of characteristics that appear to be unique to migrants in comparison with residents and are indicative of specific adaptations for a migratory lifestyle. These include

- Precise timing for cessation of breeding
- Precise timing for initiation of molt
- Precise timing for duration of molt
- Pattern of molt by age groups
- Geographic location of molting site
- Initiation of preparations for departure on migration
- Departure on migration to the wintering area

The pattern of postbreeding adaptations varies among migrant species and even subspecies and is extremely complex. Comparison of the different patterns among closely related groups composed of both migrant and resident taxa may be instructive in terms of understanding how long a given population has been migratory and in elucidating relationships.

Chapter 4: Fall Transient Period

The ultimate reason for migrant departure from the breeding area presumably is that probability of survival until the next breeding season is greater for birds that leave than for birds that stay. Evidence for how a migratory habit influences timing of other major life history events (e.g., timing of reproduction, molt, premigratory fattening, and departure) is extensive. These events overlap minimally or not at all in most migrants, and the degree to which they do may provide some indication concerning the length of evolutionary time for which a given population has been migratory. Both field and laboratory data demonstrate that timing of the initiation of preparation for migration, involving hyperphagia and laying down of subcutaneous fat reserves, are under endogenous control with photoperiod serving as the principal distal cue. The actual physiologic and behavioral processes are mediated proximally by hormones, although the exact nature of these controls is not well understood. Proximal cues governing timing of departure on migration are both physiologic (completion of preparations in terms of fat storage) and environmental—usually presumed to be light levels along with some aspect of weather (e.g., wind direction, temperature, or barometric pressure).

The main factors suggested as providing ultimate (evolutionary) control over timing of departure from the breeding ground are

- *Breeding latitude or elevation.* The higher the breeding latitude, the earlier the departure.
- *Wintering latitude.* The more distant the breeding from the wintering latitude, the earlier the departure.
- *Diet.* Migrants that depend on food sources sensitive to temperature (e.g., insects) leave earlier.
- *Molt.* Species in which molt is delayed until reaching the wintering ground leave earlier than those that molt on the breeding ground.
- *Sex and age.* The optimal timing of life history events differ by sex and age.
- *Balancing costs and benefits.* Trade-offs based on factors favoring early departure (e.g., deteriorating weather or early arrival on wintering areas) and those favoring later departure (e.g., raising of multiple broods or defense of next year's breeding territory [adult males]).

Precisely what aspects of migrant biology are exaptations for actual migratory flight versus adaptations is not clear: Many species that appear poorly adapted based on one or more aspects of their biology are migratory, whereas some that appear well adapted are not. Nevertheless, there is clearly a relationship between length of evolutionary time for which a species has been migratory, annual distance covered in migration, and certain types of morphology, physiology, and behavior affecting flight. Morphologic adaptations for long-distance flight include muscle fiber type (fast oxidative fibers) and wing shape (highly convex, pointed wing); physiologic adaptations include energy and water storage; behavioral adaptations include selection of optimal altitude for migratory flight.

When most migrants depart on migratory flight, they "know" where they are going. This can be explained for adults on the basis of the well-known attribute of both migrant and resident birds (as well as of many other organisms) of being able to home to places where they have already been. Juvenile migrants, however, must either accompany adults on their journey or possess a genetic program. Extensive experimentation has demonstrated that many juvenile migrants have a genetic program, although the precise nature of this program is not entirely understood. On the basis of observation and experimentation, two theories have been proposed regarding structure of this program:

- The "migration route" theory, in which each migrant has a genetic program for the specific compass directions to be followed for specific time periods
- The "destination" theory, in which each migrant has a genetic program for location of its ancestral population

Evidence regarding actual migration for a number of species is presented, which appears to favor the "destination" hypothesis over the "migration route" hypothesis for most species.

If the hypothesis regarding the origin of migrants as dispersing individuals from resident populations is correct, then orientation and navigation are exaptations for migration, as is indicated by the fact that many kinds of sedentary organisms, including sedentary birds like the Rock Pigeon (*Columba livia*), are capable of detecting and using a wide range of environmental cues for orientation purposes. Among those cues found usable by birds are topographical landmarks, the Sun's location in the sky, polarized light, star patterns, Earth's magnetic field, infrasound, smells, and wind direction. Other cues, not as yet detected, probably exist as well. Not all of these cues have been confirmed as being used by migrants in navigating from breeding to wintering areas—the main ones appear to be magnetic field, star maps, and polarized light, at least for nocturnal, passerine migrants. Even if the ability to detect and use these cues is a characteristic of resident bird species, refinement in migrant populations is a likely result over evolutionary time.

Stopover at various sites along the migration route for the purposes of resting, drinking, and rebuilding of fat reserves is characteristic of most long-distance migrants. Length of time spent at any given stopover point will depend on the experience and physiologic needs of the individual in the context of the environment (weather, competitors, predators).

CHAPTER 5: WINTERING PERIOD

Arrival on the wintering ground means different things to different species of migrants as well as to the different age and sex groups within a migrant population. Adult birds are capable of migrating to a specific, previously occupied wintering site or area based on experience. Juveniles, however, must depend upon some sort of genetic program to find their population's winter range, two types of which have been suggested: migration route or destination. Understanding of duration of the fall migratory flight and arrival at the wintering ground are quite different depending upon which hypothesis is followed. Under the migration route program, juveniles migrate in specifically programmed directions for specifically programmed times. At the end of the programmed time for migration, the bird has arrived at its winter range, and migratory flight stops. Under the destination program, there is no program for a specific duration of migratory flight. The bird follows its genetic program for location of the wintering area but may continue migratory movement even after arrival on the winter range if resources are not readily available; or it can stop along the way as long as resources are available and continue on to the winter range when they run out.

Sociality during the nonbreeding period is dictated by two aspects of fitness: reproduction and survival. Balancing of these can be seen in the different winter distributions of age and sex groups of partial or differential migrants in which adult males tend to remain on or near breeding territories, presumably enhancing future reproduction at some cost in survival, whereas adult females and juveniles migrate to a separate winter range.

For migrants in which all age and sex groups leave the breeding range entirely to follow an individual migration and wintering program (as opposed to pairs or family units), nonbreeding season sociality is largely a balance between resource harvest efficiency and predator avoidance, with the different forms of sociality (e.g., individual territoriality, wandering, participation in mixed-species or single-species flocks) being dictated by the kinds of resources being harvested, their amounts and distribution in space and time, and the individual's experience and competitive ability.

A major life history characteristic of the wintering period for at least one-third of migrants is conduct of a prealternate molt. This molt shows a remarkable amount of variation in the species in which it occurs, often resulting in different plumages not only among the various sex and age groups but also within groups as a result of balancing selection factors that include both breeding and nonbreeding period influences.

The winter range of migrants formerly was attributed in large part to interspecific competition and the temporal distribution of resources. Now it is recognized that many factors in addition to these shape the winter range including geography, climate, habitat, predators, disease, sex, age, and evolutionary history.

CHAPTER 6: SPRING TRANSIENT PERIOD

The purpose of spring migration is to place the individual in the optimal environment for reproduction. The ultimate factor that controls timing of departure from the wintering area is an endogenous program for calendar migrants and probably for most facultative migrants as well. An endogenous program governing the timing of breeding may have served as the ultimate factor governing spring movement even in the first migratory generation derived from a resident population, with intraspecific competition for breeding territories serving as the proximate factor. In any event, the greater the number of generations separating a migratory population from its resident ancestors, the greater endogenous control over departure timing is likely to be.

Timing of wintering area departure within the migration season (i.e., whether a bird is likely to be an early- or late-departing migrant) is affected by several factors including breeding area latitude, wintering area latitude, diet, molt, sex, age, whether or not the bird is a calendar or a facultative migrant, and balancing selection governing fitness trade-offs between the aforementioned factors and probably many others.

Proximate factors governing actual departure date may include nothing more than physiologic status (i.e., sufficient energy stores) or local food availability for migrants wintering at tropical latitudes, with endogenous control serving as the principal timer. However, photoperiod and weather are likely to play increasingly important roles the closer the wintering area is to the breeding area, at least for populations wintering in the temperate zone.

Routes between breeding and wintering areas are assumed to represent the most direct possible, depending on the location of en route obstacles and barriers, and fall and spring migration routes are assumed to be the same for most species. These assumptions may be true for some short-distance, continental migrants but are probably not true for the majority. The value of a route from a fitness perspective is to allow the bird to travel between breeding and wintering area in the safest way possible commensurate with timing needs (i.e., a particular spring arrival date may be important in terms of balancing survival and reproduction probabilities). Field data, where available, document that the greater the distance between breeding and wintering area, the less likely it is that fall and spring routes will be similar. For some species, like the Red Knot, there is evidence that specific fall and spring routes, and stopover points, are part of a genetic migration program.

Some short-distance migrants and even a few long-distance migrants (e.g., Bar-tailed Godwit [*Limosa lapponica*]) may make the movement from wintering to breeding area in a single flight. However most migrants interrupt the journey at places along the way called stopover sites. Stopover has at least four possible purposes, not all of which may be operative for a given individual at a particular site: (1) rest, (2) rebuild fat reserves, (3) wait for favorable flight conditions (usually wind), or (4) en route delay so that arrival on the breeding area is optimal for survival and reproduction (i.e., early arrival may prove fatal, and late arrival may decrease probability of breeding).

The physiologic/behavioral states of migrants at stopover sites probably reflect the different needs required by the different purposes stopover serves, but to date only three have been documented: a "flying" state (*Zugstimmung*), a "feeding" state (*Zugdisposition*), and a "resting" state. I suggest that at least one additional state is likely to occur during stopover, namely a "transit" state during which the bird is foraging and, perhaps, searching for more appropriate habitat either to feed intensively in or to rest.

Proximate factors governing when and where an individual decides to stop include

- Time of day (some birds migrate at night and stop during the day, whereas the reverse is true for others)

- Habitat availability for resting or feeding
- Food availability
- Optimization of refueling needs
- Weather
- Predation
- Position along the route (whether close to or distant from the breeding area)

CHAPTER 7: POPULATION ECOLOGY

The prevailing paradigm for migratory bird population limitation assumes density-dependent effects during the breeding season (mostly affecting natality) and density-independent effects during the nonbreeding period (mostly affecting mortality). However there is extensive field evidence documenting that intraspecific competition during the nonbreeding portion of the life cycle could play an important role in population limitation for many migratory species. The potential for density-dependent population limitation in multiple critical habitats in different locations occupied at different times over the course of the annual cycle requires new theoretical formulations. These formulations, developed by Alan Pine, are presented in appendixes A and B. They show that the critical habitat with lowest carrying capacity will exert the greatest influence over population size, regardless of when during the annual cycle it occurs. The findings of this model are considered in light of data on population movement and change, and two corollaries are proposed:

1. Density-dependent interactions can affect populations even when they are below carrying capacity for any critical habitat occupied over the course of the annual cycle.
2. Such interactions can have carryover effects in which competition in one critical habitat can affect a population after it has moved on to the next critical habitat.

Actual data on migratory bird population change are reviewed from the perspective of the potential for limitation by a critical limiting habitat encountered at any point during the annual cycle. Various factors causing large amounts of migrant mortality are shown to have no effect on overall population size except when the population is below carrying capacity for all critical habitats. Under that circumstance, all mortality factors contribute to population size.

Measurement of migrant populations requires long-term counts of numbers of breeding individuals over large geographic areas; such counts have been in place in the United States since 1966 and in parts of Europe for roughly the same period. These counts reveal trends for different migrant species, analysis of which is often done by various kinds of ecological groupings by continent (e.g., North American waterbirds and European grassland birds). Such groupings may be helpful in terms of indicating where a closer look is needed, but no clear understanding of the population dynamics of any given migratory species can be obtained without a detailed investigation of the species' populations in all critical habitats occupied over the course of the annual cycle.

Suggested causes for migrant population change fall mainly into two categories, density independent and density dependent, although the same factor can be density independent under some circumstances and density dependent in others. Five major, density-independent factors that have been responsible for population limitation in one or more migrants species are discussed: hunting, pesticides, disease, accidents during migration, and climate change (both short and long term).

Density-dependent factors that have been suggested as potentially limiting one or more migrant populations are discussed from the perspective of Pine's multiple carrying-capacity model. These include breeding habitat loss, breeding habitat fragmentation, postbreeding (molt and premigration) habitat loss, stopover habitat loss, and wintering habitat loss.

The difficulty of determining whether population limitation is density independent versus density dependent is discussed, and I suggest that, where detailed information on the entire life cycle is lacking, predictions can be made based on the likely characteristics of populations controlled by different kinds of factors or during different portions of the annual cycle.

The problems involved in understanding where and how any given population of migrants is controlled are illustrated by an examination of the population dynamics of the White-winged Dove (*Zenaida asiatica*), a species that breeds in the southwestern United States and northern Mexico and winters in southwestern Mexico and along the Pacific slope of Central America. Because the species is hunted, there is extensive, detailed information on breeding density by habitat and hunting mortality dating back more than one-half century, at least for the United States portion of the population. These data, when considered in light of the theoretical findings presented by Pine in appendixes A and B, suggest the possibility that the population is limited by wintering-season factors (e.g., availability of tropical dry deciduous forest, the principal wintering habitat). This example illustrates the potential problems associated with determination of where and how any given population of migrants may be limited over the course of an annual cycle.

CHAPTER 8: EVOLUTION AND BIOGEOGRAPHY

Migration derives from postnatal dispersal of resident birds, and all of the basic phenotypic characters of structure, physiology, and behavior required (e.g., the power of flight, energy storage, homing, and orientation) are possessed by resident bird species as exaptations for migration. Nevertheless, migration places the population into a different selective environment from residents, likely to favor rapid evolutionary changes to many aspects of the phenotype, especially those governing the timing of major events in the annual cycle (e.g., migratory movement, reproduction, and molt). Comparison of timing of these events between migratory populations and their resident relatives may be instructive with regard to when migration and divergence began.

Laboratory experiments with onset and duration of migratory restlessness (*Zugunruhe*) demonstrates heritability aspects typical of single-gene control over migration. The meaning of these experiments, however, is unclear. The selection regime imposed may be on some trigger mechanism (e.g., temperature) related to onset of migration, not migration itself. Nevertheless, these and similar findings have led to the proposal of a "migratory syndrome" in which all migration traits form an ancient, integrated composite. I question the "migratory syndrome" hypothesis and suggest that modification of the migrant's genome through natural selection likely is profound, affecting most structural and physiologic systems.

The rapid appearance or reappearance of migration in formerly sedentary or extirpated populations of several species of birds in eastern North America presents powerful support for origin of migration through exaptations, competition, and dispersal.

Migrant populations demonstrate remarkable short-term flexibility. Reaction norms are a theoretical construct built to reflect integration of genetics, environment, ecophysiology, and developmental physiology in order to explain extensive information on the behavior of actual migratory populations. They represent the mean value of expression for a migratory trait in a given environmental situation and are assumed to vary in their degree of flexibility by species. They provide a powerful tool for understanding flexibility of migrants in responding to different environments based on their own immediate status.

The "origin of migration through dispersal" hypothesis requires neither a "northern" nor a "southern" home for the ancestral point of origin, only competition and a reachable, seasonal environment. Nevertheless, there is evidence for many migrant species (based on phylogenetic relationships and community and population ecology) that argues strongly in favor of origin in the aseasonal (i.e., winter) rather than the seasonal portion of the range.

Much work has been done on population differentiation in migrants, including exciting new research into hybrid zones demonstrating that hybrids between members of related populations of migrants experience decreased fitness despite no evidence of assortative mating. This finding has led to the formulation of a hypothesis in which the lowered fitness results from mixing of endogenous programs specific to each population, in particular those involving the migration route. I agree that the likely cause of decreased hybrid fitness likely results from mixing of population-specific endogenous programs but suggest that the programs compromised likely have more to do with timing of reproduction, molt, migration, and location of the wintering area.

These findings, along with observation of migrant taxonomy and distribution, lead to a model peculiar to migrants in which speciation occurs mainly in two ways:

- Between the migrant population and its resident ancestor
- Between populations that are parapatric or allopatric on both their breeding and wintering grounds, resulting in gradual accumulation of significant differences in important endogenous programs

Migration systems (i.e., the migrant species composition of major regions) present a fascinating overview of the mixture of evolution, biogeography, and ecology that goes into shaping the avifauna of a particular continent or hemisphere. Landforms, weather, seasonality, phylogeny, habitat, and paleogeography all play important roles in determining what species will be involved in migration in any given region. This interplay is discussed with reference to seven of the world's major migration systems: Palearctic–African, Nearctic–neotropical, Palearctic–South Asian, Intra-African, Intra–South American, Intra–South Asian, and Australian.

Chapter 9: Migratory Birds and Pathogen Movement

Migratory birds are directly involved in several kinds of pathogen transport, posing potential damage to human health and economies. Two recent pathogen outbreaks are considered: West Nile virus (WNV) in the New World and highly pathogenic avian influenza (HPAI) H5N1 in the Old World. In both cases, immense resources were wasted based on poor understanding of the role of migrants and other agents in virus movement. These examples highlight the value of knowledge concerning migrant biology and how it can interact with that of pathogens.

In April 2005, the Food and Agriculture Organization (FAO) of the United Nations published a report on the epidemiology of HPAI H5N1 based on the findings of a panel of experts in which it was concluded that "the 2003–5 epidemics appear to have arisen due to the establishment of H5N1 infection in wild birds" (Morris and Jackson 2005:1). Similarly, McLean (2006:44) concluded that WNV spread across the Western Hemisphere was the result of migratory bird movements. As summarized earlier, neither of these conclusions appears justified (Brown and O'Brien 2011). In both cases, alternative agents provide explanations more in line with available data: dispersal by resident bird or mosquito reservoir hosts for WNV and legal and illegal trade in poultry, poultry products, and exotic birds for HPAI H5N1.

WNV poses a threat to human and equine health at outbreak sites, in addition to avian populations that are small and confined (e.g., many Hawaiian birds). HPAI H5N1 had the potential to devastate local, regional, and even national poultry industries and could have posed a serious threat to human populations in the unlikely event of its evolution into a virus that was both lethal to, and readily transmissible among, humans. Thus, a clear understanding of how these viruses moved is critical. We do not have that understanding as yet. Despite this fact, major programs and initiatives were put in place based either on unsubstantiated assumptions or faulty logic. Particularly in the case of HPAI H5N1, huge amounts of money were spent confronting a threat whose likelihood was vanishingly small (i.e., introduction of the virus to the New World in a form devastating to human populations that would be moved throughout the hemisphere by migratory birds). Unfortunately, many of the programs that were put in place to address the potential problems posed by this virus in the United States were media and market driven (Epstein 2011; Rappole 2011). Obviously, rational evaluation based on known biology of migrants and viruses is the preferable alternative.

CHAPTER 10: CONNECTIVITY AND CONSERVATION

Conservation is not science; it is public environmental policy. As such its design is dependent upon the values of those making and implementing the policy and the quality of information that they possess. Migratory birds have been a major focus of conservation for more than a century, at least since the realization in the late nineteenth century that market forces were causing extirpation of large numbers of species in North America. Most initial efforts in the United States were focused on wading birds and waterfowl and were successful in no small degree because of the excellent biological information on habitat needs through the entire portion of the life cycle, which served to focus management actions on the most critical segments.

Recent research on migratory species has attempted to increase knowledge of all aspects of the annual cycle through what have come to be known as "connectivity" studies. The concept is not new, dating back to at least the late 1800s when banding programs and distributional studies of migrants based on plumage differences were initiated. Nevertheless, the past decade has seen the development of new or improved technologies for conduct of connectivity work, including satellite radio-tracking, geolocators, stable isotope analysis, physiologic measures, and genetic studies. Each of these new techniques holds promise but requires refinement to produce the kind of information necessary to make important advances of migrant conservation needs.

Conservation action to preserve migratory bird populations requires work in three main areas: (1) research, (2) national and international conservation policy collaboration, and (3) management. The fundamental importance of research is to identify where in the cycle the threat to a particular migrant species exists and to suggest the kinds of actions that might ameliorate the threat. However, research alone, without an understanding by the general public and policy makers, is wasted.

MIGRATION AS A LIFE HISTORY STRATEGY

During a few beautiful fall days in late October 1977, a symposium on migratory birds on their wintering grounds in the neotropics was held at the Smithsonian Institution's Conservation and Research Center in rural northern Virginia. Gene Morton was the principal organizer of that meeting, which brought together some of the leading ecological thinkers of the time along with graduate students and post-docs in an environment where there was very little else to do besides consider the topic at hand. The result was that young researchers at work on various projects concerning migratory birds in tropical environments could discuss at length what they were actually finding with leading theoreticians and field biologists like Stephen Fretwell, John Terborgh, Peter Grant, John Emlen, Gary Stiles, E. O. Willis, Doug Morse, and Dan Janzen.

Morton and his co-convener, Alan Keast, originally had the idea that they would publish not only the papers that were given but also the discussions that occurred afterward, as had been done during the previous Smithsonian symposium on migratory birds (Buechner and Buechner 1970). To this end, they had a recording setup in place. However, the discussions quickly outstripped capacity; indeed, entire impromptu papers were given from the audience, such as R. F. Whitcomb's presentation on island biogeography and habitat island use by migratory birds in the eastern United States. Had these extemporaneous comments been published, the resulting work would have run to several volumes instead of the one that came out (Keast and Morton 1980).

The most heated discussions occurred over the issue of migrant origins; that is, are neotropical migrants temperate birds pushed south in winter by inclement weather into tropical habitats or tropical birds that move north in summer to exploit seasonally superabundant resources in temperate regions? I recall a late-night interchange during the symposium involving myself, Pete Myers, and John Fitzpatrick on the topic: "*Resolved*—Where migrants originated from is not important; it is only important that we know how they are living at present." I took the position then, as I do now, that whether the resident ancestors of a migrant population originated from a place where there was little seasonal change in resources (tropics) versus one in which there were profound seasonal changes (temperate region) was of extreme importance in understanding every aspect of migrant life history, ecology, and evolution. In essence, this book, *The Avian Migrant*, represents a continuation of that interchange.

Fundamentally, this debate is the same one that I have carried out in public forums for the past three decades, including published interchanges in the *Auk* (Rappole and McDonald 1994; Latta and Baltz 1997; Rappole and McDonald 1998), the *Journal of Avian Biology* (Zink 2002; Rappole et al. 2003a), and *Ardea* (Rappole and Jones 2002; Bell 2005; Rappole 2005a). The critical issue on which the entire question depends is whether or not resident birds possess all of the behavioral and physiologic attributes (exaptations) necessary for successful completion of a round-trip journey between the area where they are hatched and the area where they breed. This hypothesis is stated in the opening paragraph of chapter 1. I am well aware that I have not proved this hypothesis with the information presented in the intervening pages. However, I hope that I have convinced the careful reader of the issue's importance and the need for continued investigation without prejudice.

APPENDIX A

POPULATION DYNAMICS OF PERIODIC BREEDERS

ALAN S. PINE

W E BEGIN our discussion of population dynamics with the Malthusian principle that the change in population dN over an infinitesimal time dt is proportional to the existing population N, or

$$\frac{dN}{dt} = rN$$

The proportionality constant r is usually associated with the difference between the birth and death rates, $b - d$, in a closed environment. These rates are the average number of births or deaths per capita per unit time. If r is constant in time, this results in exponential growth or decay, $N = N_0 e^{rt}$, of the initial population N_0 depending on the sign of r—that is, whether there are more or fewer births than deaths. Of course, if some critical resource such as food is limited, r cannot be constant. If the resource is consumable and nonrenewable, then once the resource is exhausted, the death rate would rapidly dominate births and the population would collapse. In principle, the resource could vary explicitly with time, such as seasonally or randomly or gradually, or it may depend on the population density itself. In this appendix, we will discuss the influence of both time and density factors on the population.

In the case that the resource is purely density dependent but renewable at some finite rate, then an equilibrium population carrying capacity K may be reached for a given habitat.

This requires that the birth *and/or* death rates adjust to the limited resource. For example, the per capita birth rate may be reduced due to reproductive stress, to limited nest sites, or to territorial expansion. The reduction could result from fewer progeny per individual or from a smaller percentage of breeders in the whole population. Density-dependent mechanisms that may affect the death rate include competition for food or epidemic disease.

The concept of an environmental carrying capacity was introduced by Verhulst (1845, 1847) through the "logistic" differential equation,

$$\frac{dx}{dt} = rx(1-x) \tag{A.1a}$$

Here, $x = N/K$ is the population density relative to the carrying capacity of the habitat, and r is now the "intrinsic" growth rate moderated by the density-dependent factor, $F(x) = (1 - x)$. Clearly, dx/dt vanishes when the population reaches the carrying capacity. The classical logistic equation has the analytical solution (Pearl 1927; Maynard Smith 1968; Pielou 1969; Acheson 1997),

$$x = 1/\left[1 - (1 - x_0^{-1})e^{-rt} \right] \tag{A.1b}$$

where $x_0 = N_0/K$ is the initial population density. For $x_0 \ll 1$, this solution starts out with nearly exponential growth but eventually saturates exactly at the carrying capacity, $x = 1$, exhibiting the well-known "sigmoidal" shape. Pearl (1927) demonstrated the utility of the logistic curve in fitting the growth of laboratory-cultured yeast, but similar logistic projections of the U.S. population (saturating at ~200 million in year ~2000) from historic census data (Pearl et al. 1940) have not been realized. Though extrapolation of any model fit to limited, noisy data is always risky, such failure likely results from inadequacies of the model.

The Verhulst logistic, equation A.1, is a deterministic, single-species, single-isolated-habitat, single-mechanism, age-independent, spatially homogeneous, instantaneous-response, first-order density-dependent model. These assumptions may restrict its applicability, and the literature is replete with extensions for statistical fluctuations of the parameters, predator–prey and competitive species interactions, age-specific birth and death rates, spatial diffusion, delayed response, and nonlinear density dependencies (Lotka 1956; Maynard Smith 1968, 1974; Pielou 1969; Royama 1992; Krebs 1994; Kot 2001; Caswell 2006; and references therein). In this work, we primarily consider generalizations of the logistic for multiple habitats and mechanisms encountered by migratory or resident seasonal breeders. We also examine threshold effects and possible instabilities due to the delayed response inherent in substituting discrete difference equations for differential equations, such as equation A.1a. We discuss the requirements for the validity of overall density-dependent population dynamics models for periodic breeders that exhibit age-structured birth and death rates in appendix B.

GENERALIZED SINGLE-HABITAT LOGISTIC MODELS

The Verhulst logistic model, equation A.1, has only two parameters, r and K, available to fit data, and K is just a scale factor. For increased flexibility, Richards (1959) and Gilpin and Ayala (1973) introduced an additional "asymmetry parameter," p, to the density dependence, in the form, $F(x) = (1 - x^p)$. Here, we refer to this p parameter as the "saturation power," as it affects the abruptness of the population saturation, as shown later. However, this modification, as well as the standard $p = 1$ case, may have some difficulties for populations that exceed the carrying capacity because of the negative-going $(1 - x^p)$ density-dependent factor. In particular, if the birth rate is less than the death rate, then extinction is predicted for $x_0 < 1$, as expected, but anomalous growth occurs for $x_0 > 1$. Indeed, for $r < 0$ and $x_0 > 1$, the denominator in equation A.1b vanishes at $t = r^{-1} \ln(1 - x_0^{-1})$, leading to a population singularity (Kuno 1991; Royama 1992). For this reason, the range of validity is often restricted to $0 \le x \le 1$. However, for multiple sequential habitats as considered here for periodic breeders, the population out of one habitat may exceed the carrying capacity of the next. To eliminate such anomalous growth for nonbreeding seasons, we modify the logistic by simply allowing for separate density dependencies for the birth and death rates,

$$\frac{dx}{dt} = x\left[b(x) - d(x)\right] = x\left[b_0 F(x) - d_0 G(x)\right] \tag{A.2}$$

Here, the birth-rate density-dependent factor, $F(x)$, is constructed to be unity for $x = 0$ and to vanish or be greatly reduced at the carrying capacity, whereas the death-rate density-dependent factor, $G(x)$, increases with density from unity at $x = 0$. Separation of the density dependencies of the birth and death rates is not a new idea of course (Williamson 1972; Sutherland 1996), but it is critical to the periodic breeder problem. Fretwell (1972) avoided the anomaly for "short generation" species by ignoring the seasonal change in the intrinsic growth rate, r. Kuno (1991) and Royama (1992) invoked a negative effective K for $r < 0$, which is unphysical, but reflects the dominance of $G(x)$ over $F(x)$ when deaths exceed births.

Some illustrative functional forms for $F(x)$ and $G(x)$ are considered here,

$$F(x < q) = [1 - (x/q)^p], \quad F(x \ge q) = 0 \tag{A.3a}$$

$$F(x) = \exp[-(x/q)^p] \tag{A.3b}$$

$$F(x) = 1/[1 + (x/q)^p] \tag{A.3c}$$

$$F(x \le q) = 1, \quad F(x > q) = (q/x)^p \tag{A.3d}$$

and

$$G(x) = 1/\delta(x), \quad \delta(x) = [1 - (x/q)^p] \text{ or } \delta(x) = \epsilon < 1 \text{ if } \delta(x) < \epsilon \tag{A.4a}$$

$$G(x) = \exp[(x/q)^p] \tag{A.4b}$$

$$G(x) = [1 + (x/q)^p] \tag{A.4c}$$

$$G(x \le q) = 1, \quad G(x > q) = 1 + p[1 - (q/x)] \tag{A.4d}$$

In these expressions, q is the ratio of any particular birth or death carrying capacity to an arbitrary K (usually the minimum value), and p is again the saturation power. Equation A.3a is basically the Richards–Gilpin–Ayala expression, with the physical provision that the birth rates cannot be negative. Equations A.3b and A.3c are continuous positive functions for all $x \geq 0$ and have been studied previously in discrete-time, single-habitat population dynamics models (Leslie 1948; Ricker 1954; Beverton and Holt 1957; May and Oster 1976; Thomas et al. 1980; Royama 1992). Equation A.3d is a hyperbola for $x > q$, piecewise continuous with unity for $x \leq q$, ensuring the physical requirement that $0 \leq F(x) \leq 1$. Functions A.4a–c for $G(x)$ are essentially the reciprocals of the corresponding functions A.3a–c for $F(x)$, with prevention of a vanishing or negative denominator in equation A.4a. Although equations A.3d and A.4d appear to be artificial constructs, they actually have a fairly straightforward biological interpretation. For example, if there are $K/2$ nest sites (assuming mating pairs) and any excess population is nonproductive, but benign, then A.3d with $p = 1$ simply represents the reduction of the per capita birth rates. Higher values of p represent a faster reduction of birth rates, perhaps due to competition for the available nest sites or territories. If the excess population leaves in search of a better habitat, then $G(x)$ can be represented by equation A.4d with p interpreted as the ratio of the emigration rate to death rate. Note that emigration in this case is equivalent to death for the habitat of interest. By contrast, immigration cannot be treated similarly to birth, as it need not depend on the existing population. The other functions may represent a distribution of quality or availability of nest sites or resources (Royama 1992). The $(1 - x)$ factor in the Verhulst model is sometimes referred to as the "available space" or "unutilized opportunity" (Krebs 1994) and in the analogous "epidemic" model (Acheson 1997) as the "uninfected."

Equation A.2 can be integrated for the population density at any time, provided an initial value, x_0. In general, the integration can be carried out numerically, very accurately using the fourth-order Runge–Kutta method (Press et al. 1992; Acheson 1997). For the special case when $F(x) = [1 - (x/q)^p]$ and $G(x) = 1$, and when the parameters are constant for a given time, t, there is an analytical solution to equation A.2 in the form,

$$x = q/\{a - [a - (x_0/q)^{-p}] \exp(-prt)\}^{1/p} \tag{A.5}$$

where $r = b_0 - d_0$ and $a = b_0/r$. This solution follows by integrating the partial fraction $1/[x(1 - ax^p)] = 1/x + ax^{p-1}/(1 - ax^p)$ after substituting $y = 1 - ax^p$ and $dy/dx = -apx^{p-1}$. The standard analytical solution to the Verhulst logistic, equation A.1b, is a special case of equation A.5 for $a = p = q = 1$. In the lower panel of figure A.1, we plot the time development of the population from equation A.5 for several values of p with $q = 1$ and $x_0 = 0.1$ and the corresponding density-dependent $b_0 F(x)$ and d_0 shown in the upper panel. Note that the saturations occur where the net growth rates vanish at the crossings of the respective birth and death rates in the upper panel, below the $x = N/K = 1$ nominal carrying capacity in this case. As p increases, the saturation approaches the carrying capacity, K, with a more abrupt inflection, and the growth portion more closely follows the density-independent exponential Malthusian curve. From the long-time behavior of equation A.5, we see that the saturation value in this case is $x_s = q/a^{1/p} = q[1 - d_0/b_0]^{1/p}$.

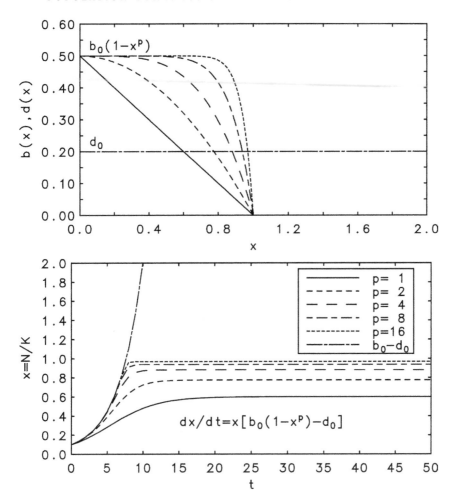

FIGURE A.1   *Upper panel*: Density dependence of birth and death rates for $F(x) = (1 - x^p)$, with $p = 1, 2, 4, 8, 16$ as noted, and $G(x) = 1$. *Lower panel*: Time development of population for single habitat for birth and death rates in upper panel, compared with Malthusian ($b_0 - d_0$) growth.

Numerical integration for a birth density dependence given by equation A.3d and death by equation A.4b yields the results shown in figure A.2. In this case, the saturations occur at the crossings above the $x = 1$ nominal carrying capacity, but again an increase in the saturation power, p, more closely approaches both the Malthusian growth and the $x = 1$ limits with a more abrupt transition. For the $F(x)$ factors given by equations A.3b and A.3c, saturation may occur above or below the carrying capacity, depending on whether $d_0 G(x)$ crosses the birth rates below or above b_0/e or $b_0/2$, respectively. The saturation value for a single habitat occurs for $b_0 F(x_s) = d_0 G(x_s)$, so the effective or combined carrying capacity depends not only on K but also on contributions from the distinct birth and death factors, F and G.

As indicated earlier, there may be more than one density-dependent mechanism operating simultaneously in the same habitat, controlling either the birth or the death rates. For example, birth rates may be limited by nesting sites and territorial behavior; death rates by food, water, shelter, predators, and communicable disease. Each of these mechanisms may

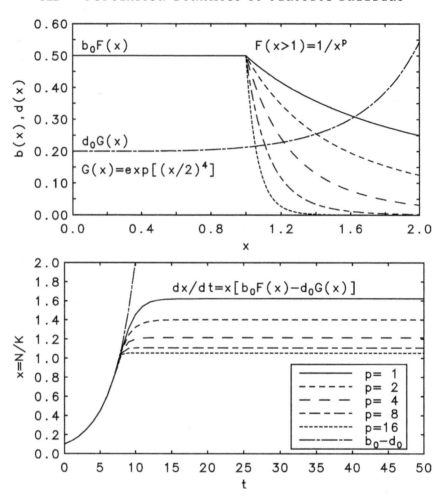

FIGURE A.2 *Upper panel*: Density dependence of birth and death rates for $F(x \le 1) = 1$, $F(x > 1) = 1/x^p$, with $p = 1, 2, 4, 8, 16$ and $G(x) = \exp[(x/2)^4]$. *Lower panel*: Time development of population for single habitat for birth and death rates in upper panel, compared with Malthusian $(b_0 - d_0)$ growth.

be characterized by a different functional form, such as equations A.3a–d or A.4a–d, with specific carrying capacities and saturation powers. The overall effective density dependence is then given by the product of the various functions of the operative mechanisms. As might be expected, the mechanism with the minimum carrying capacity dominates, but the other mechanisms generally reduce the saturation population and may inhibit the growth rate. A reduced saturation value is evident from the upper panels of figures A.1 and A.2 as additional limiting resources increase the upward (downward) curvature of the death (birth) rates, thereby lowering the crossing density. Slower growth may be inferred from figure A.1 where it is seen that the population deviates from the Malthusian curve well below saturation for any saturation power, the anticipation being greater for lower p. For the mechanisms represented by equation A.3d shown in figure A.2 or by equation A.4d, there is no deviation from Malthusian growth for $x < q$, where $F(x)$ or $G(x) = 1$, so there is no anticipation, and any other simultaneous mechanism would be manifest only very near saturation.

MULTIPLE-HABITAT MODELS

In the preceding discussion, the parameters representing the birth and death rates, carrying capacities, and saturation powers are assumed to be constant in time. In general, though, they may vary explicitly with time, perhaps randomly due to weather, gradually due to habitat deterioration, or periodically due to seasonal breeding or migration. If the time dependence is known or is predictable, then it may be incorporated easily into the numerical integration of equation A.2 (Press et al. 1992; Acheson 1997).

In the case of periodic breeders, the time dependence of the parameters generally may be expressed as a Fourier series. For example, Skellam (1966) has discussed a "periodic normal distribution" of the form, $b(t,T)/b_0 = 1 + 2\sum_{n=1}^{\infty} \exp(-n^2 T/\tau) \cos(2\pi n t/\tau)$, which has a single peak at $t = 0$ of width scaled by T in a total period of τ. However, here we take the parameters to be a sequential series of steps, being constant throughout the seasonal occupation of a particular habitat. This latter approach also permits a seasonal change of the density dependence functional form—for example, equations A.3 and A.4—if desired. In this case, we give a seasonal index label, m, to the parameters, b_m, d_m, q_m, p_m, and functions, F_m and G_m, and integrate equation A.2 over each season of duration τ_m separately, with an initial seasonal value obtained from the final value of the previous season (using x_0 for the first season). If there are M seasons in the total period time, τ (e.g., 1 year), then $\tau = \sum_{m=1}^{M} \tau_m$, and the m label is cyclical, $M + m = m$. Here, we refer to a "season" as the dwell time in a habitat of substantially constant parameters, such habitats being geographically separate for migratory species, but which may coincide or overlap for resident species. We assume that the entire population either migrates or not, in order to avoid the complications associated with partial migration (Kaitala et al. 1993).

In figure A.3, we show the time development of the population using the generalized logistic, equation A.2, for the case of two sequential habitats with $\tau_1 = \tau_2$, but only one breeding season. Here, we have taken the birth rate to have the controlling density dependence (smallest carrying capacity) with the functions in equations A.3a–d as noted; equation A.4b is chosen for the density dependence of the death rates in both seasons. The sawtooth curves given by the solid lines represent the seasonal variations of the populations. Here, we see that saturation occurs over a wide range below and above the nominal carrying capacity, depending on the form of $F(x)$. If the density-dependent mechanism controlling the growth is dominated by the death rates, then the functional form for the birth rates makes little or no difference. If the death rate carrying capacity is lower in the breeding season, the seasonal population variations are proportionally smaller than shown in figure A.3, whereas if death in the nonbreeding season dominates, the variations are larger.

As noted for a single habitat, saturation occurs at the crossing of the birth- and death-rate density-dependent curves, such as shown in figures A.1 and A.2. For multiple habitats, we define an analogous equilibrium population, x_{eq}, by adding the contributions of the various seasons, m, weighted by their duration, τ_m, according to

$$\sum_{m=1}^{M} \left[b_m F_m(x_{eq}) - d_m G_m(x_{eq}) \right] \tau_m = 0 \tag{A.6a}$$

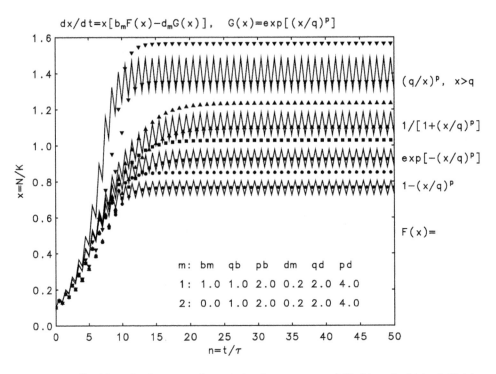

FIGURE A.3   Time development of population for two sequential habitats for $b(x) = b_0 F(x)$ in first season only, with $F(x)$ noted at right margin, and $G(x) = \exp[(x/2)^4]$ for both seasons of equal duration. Parameters are tabulated on the figure. Solid lines are used for continuous differential model, and symbols are used for discrete difference model.

An analysis of a number of growth curves, such as in figure A.3, for arbitrary parameters and M yields x_{eq} nearly equal to the average population over the yearly cycle once a steady-state oscillation has been reached. Equation A.6a is a key result, as it allows us to estimate the response of the ultimate population to change in any of the habitat parameters. As expected, a reduction in the habitat with the controlling (smallest) birth or death carrying capacity will have a greater effect on the population than a proportionate change elsewhere. Generally, the solution to equation A.6a may be obtained to arbitrary accuracy by numerical methods, such as root-finding bisection (Press et al. 1992). An analytical solution may be found in the special case when $F_m(x_{eq})$ is in the form of equation A.3a and $G_m(x_{eq})$ is given by equation A.4c, and if all $p_m = p$ are the same, whereby,

$$(x_{eq})^p = \left[\sum_{m=1}^{M} \left[b_m - d_m \right] \tau_m \right] \Big/ \left[\sum_{m=1}^{M} \left[b_m/(q_{bm})^p + d_m/(q_{dm})^p \right] \tau_m \right] \tag{A.6b}$$

Sutherland (1996) has considered the restricted, but informative, circumstance of two seasons of equal duration with breeding in just one. His net breeding and death slope parameters are related to ours by the respective derivatives, $b' = d[b_1 F_1(x) - d_1 G_1(x)]/dx$ and $d' = d[d_2 G_2(x)]/dx$, both evaluated at x_{eq}.

DISCRETE MODELS

The described age-independent differential equation models imply that the progeny have the same fertility and mortality at birth as their progenitors. This is unrealistic for most complex organisms, and Fretwell (1972) has also considered the equilibrium populations for "long-generation" species where the newborns are infertile during their birth season. In this case and for many species that abandon their eggs, the birth rate depends more on the adult population at the beginning of the breeding season than the total density at birth. In such circumstances, it may be appropriate to replace the derivative in equation A.2 with a seasonal difference relation, $dx/dt \rightarrow (x_{m+1,n} - x_{m,n})/\tau_m$, yielding an iterative expression,

$$x_{m+1,n} = x_{m,n}[b_m F_m(x_{m,n})\tau_m + s_m(x_{m,n})] \tag{A.7a}$$

where s_m represents the seasonal survival,

$$s_m(x_{m,n}) = 1 - d_m G_m(x_{m,n})\tau_m \tag{A.7b}$$

Here, $x_{m,n}$ is the density at the beginning of season m in breeding cycle n. Again, if there are M seasons per cycle, then $x_{M+1,n} = x_{1,n+1}$. The density-dependent factors, F_m and G_m, may still be given by equations A.3 and A.4, by substituting the seasonal carrying capacities, q_m, and saturation powers, p_m, and the $x_{m,n}$ for x. This approximation simplifies and speeds up the computations dramatically over the integration of the differential equation A.2. Of course, this model also implies that the death rate depends on the initial seasonal population, rather than the existing density, which is more difficult to justify biologically. Also, equation A.7b may lead to unphysical negative populations if $d_m G_m(x_{m,n})\tau_m > 1 + b_m F_m(x_{m,n})\tau_m$. These difficulties may be alleviated if we replace equation A.7b with

$$s_m(x_{m,n}) = \exp[-d_m G_m(x_{m,n})\tau_m] \tag{A.7c}$$

which represents the probability of survival and is always positive. Equation A.7c approximates A.7b if the probability of dying during the season is small—that is, if $d_m G_m(x_{m,n})\tau_m \ll 1$. Moreover, this approximation is compatible with the well-known discrete age-structured single-habitat population model of Leslie (1945, 1948). For age-dependent breeders in multiple habitats, equations A.7 are not valid for the total populations unless the age distribution takes certain stable forms as discussed in appendix B.

In figure A.3, we superimpose the population dynamics for the discrete model, equations A.7a and A.7c, for comparison with the continuous differential model, equation A.2, for the same seasonal parameters and functions. The unconnected symbols show the discrete results. Generally, the discrete points lag the changes in the comparable continuous curve, which tends to slow the early growth, but which may overshoot and exceed the continuous values somewhat. These lags and enhanced fluctuations result from the inherent delay in births and deaths in response to initial seasonal populations. Thus, the equilibrium

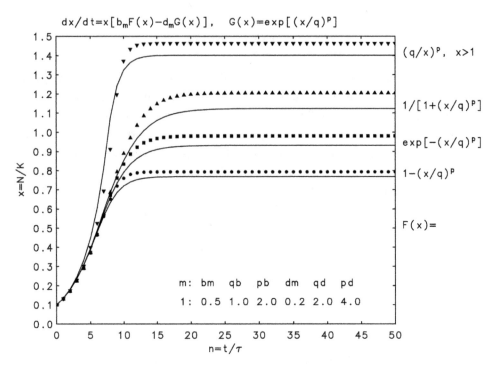

FIGURE A.4   Time development of population for single habitat for $b(x) = b_0 F(x)$, with $F(x)$ noted at right margin, and $G(x) = \exp[(x/2)^4]$. Parameters are tabulated on the figure. Solid lines are used for continuous differential model, and symbols are used for discrete difference model.

value, x_{eq}, found for the continuous model using equation A.6a is slightly lower than the average value for the discrete model. If we make the same comparison for the single habitat case, shown in figure A.4, we find that saturation values for the discrete models are still slightly higher than for the differential models owing to the approximation of equation A.7c. However, such smooth monotonic behavior in a single habitat is not necessarily the consequence of this discrete model.

We show in figure A.5 the dynamics predicted for a four times larger birth rate with the same density factors as in figure A.4. Here we obtain steady, complex oscillation when $F(x)$ is given by equation A.3a, damped oscillation for equation A.3c, and aperiodic fluctuations for equation A.3d. We note that the seasonal variations observed for multiple habitats, as seen for example in figure A.3, have a completely different physical origin than the inductive oscillations seen in figure A.5. These inductive fluctuations are a consequence of overshoot due to the delayed response in the discrete model and have been well studied for non-overlapping generations in a single habitat (q.v. May 1973, 1976; May and Oster 1976; Krebs 1994; Acheson 1997; Weisstein 1999). Here, we see similar behavior for overlapping generations (May and Oster 1976; Kot 2001), with a significant qualitative dependence on the functional form of the density dependence.

The single-habitat fluctuations seen in the upper trace of figure A.5 for $F(x > 1) = (q/x)^p$ are not entirely chaotic because the displacements above and below the differential saturation

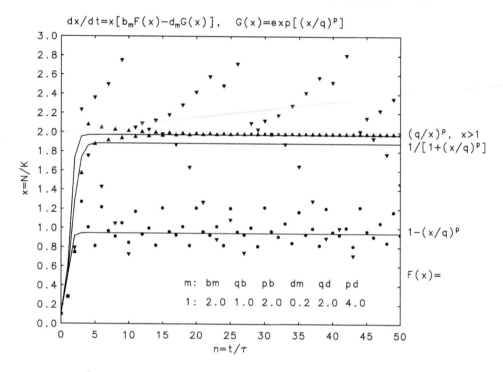

FIGURE A.5   Same as figure A.4 with birth rate four times higher, indicating instabilities from discrete model delays.

value show a high degree of symmetry. These fluctuations require not only a high birth rate but also the strong density dependence of the death rate. A density-independent death rate for the same high birth rate produces monotonic saturation. Also, such inductive instabilities may be repressed for multiple sequential habitats because of the shorter delays involved with seasonal updates. In fact, a large number of very short seasons would better approximate the continuous differential model in equation A.2. Here, we are ignoring explicit delayed density regulation effects (Maynard Smith 1968, 1974; Royama 1992; Kot 2001), which may arise, for example, for species with gestation, egg incubation, or seed dormancy longer than the cycle period. Arbitrary delay due to age-structured fertility is considered in appendix B.

METAMORPHIC AND SPAWNING SPECIES WITH NON-OVERLAPPING GENERATIONS

Equations A.7a and A.7c can also represent discrete non-overlapping generations for metamorphic or spawning species if we let m count the life stages, such as larva, pupa, adult, and set $s_m = 0$ for the breeding phase. Here, the prodigious output of many spawners is typically balanced by the low probability of survival of the pre-adult phases. This is consistent with the prior single-habitat, non-overlapping generations models, where the survival of the nonbreeding phases is incorporated into the effective birth rate (sometimes denoted

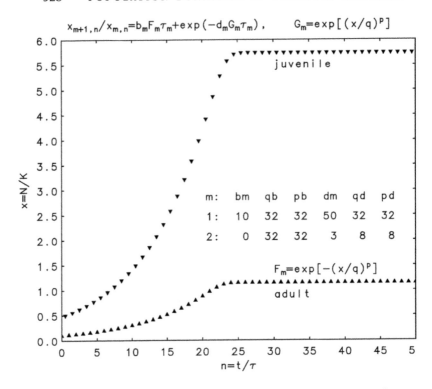

FIGURE A.6   Discrete model time development of metamorphic species with two stages of equal duration. Only adults breed and then die off; enough juveniles survive eventually to approach their carrying capacity.

as the biotic potential) in the adult phase. Splitting out the pre-adult phases, as done here, allows us to apportion the carrying capacities and mechanisms to the appropriate phases. In figure A.6, we show a hypothetical two-stage metamorphic species where the high death rate of the adults is sufficient to eliminate breeding survivors, but enough juveniles survive to allow overall growth until a saturation is reached near the nonbreeder carrying capacity. We note that the differential model, equation A.2, with a higher death than birth rate, just leads to extinction, so is not valid for metamorphic species. The life stage problem has been addressed previously using matrix formulations (Leslie 1945; DeAngelis et al. 1980; Caswell 2006), but this is not necessary if all births transition to the next stage in the cycle, as shown here for metamorphic species.

THRESHOLD EFFECTS

Some species may require a critical density, x_c, to survive or thrive, as discussed by Allee et al. (1949). This critical density could arise from the necessity to find a mate, establish a breeding colony, hunt cooperatively, school or flock, and the like. Such threshold effects can be incorporated into the multihabitat models described earlier if the appropriate birth or death rates, b_m or d_m, are respectively multiplied or divided by another density-dependent factor, $\Phi(x)$, that increases from a small residual value, $\epsilon < 1$, to 1 as x transitions through x_c. ϵ should be chosen so that the death rate exceeds the birth rate, for example, $\epsilon = 0$ for breeding thresholds or $\epsilon = d_m/(b_m + d_m)$ for survival thresholds. Some standard functions with this property are given by

$$\Phi(x \leq x_c) = \epsilon, \quad \Phi(x > x_c) = 1 \tag{A.8a}$$

$$\Phi(x) = 1 - (1 - \epsilon)\,\exp[-(x/x_c)^{\eta}] \tag{A.8b}$$

$$\Phi(x) = \frac{1+\epsilon}{2} + \frac{1-\epsilon}{\pi}\arctan\left[\frac{x - x_c}{\gamma}\right], \tag{A.8c}$$

$$\Phi(x) = \frac{1+\epsilon}{2} + \frac{1-\epsilon}{2}\,\mathrm{erf}\left[\frac{x - x_c}{\sigma}\right] \tag{A.8d}$$

Equation A.8a is a step function at $x = x_c$, representing a sharp, definite threshold. This modification does not change the dynamics discussed earlier except when the populations dip below x_c, whereupon they eventually go extinct. Equation A.8b represents a more gradual transition, characterized by the η parameter. Equations A.8c and A.8d are more probabilistic descriptions of this viability, representing a spread of effective critical densities given by a Lorentzian distribution, $\{1/[1 + (x - x_c)^2/\gamma^2]\}$, or a Gaussian, $\{\exp[-(x - x_c)^2/\sigma^2]\}$, respectively. Because the modifying functions in equations A.8b–d extend beyond the critical density, they may have a quantitative effect on the population dynamics, but the qualitative features discussed earlier remain. Such threshold effects may select against species whose population fluctuates wildly due to overbreeding (May and Oster 1976; Thomas et al. 1980).

Summary

We have shown, by the expediency of separating the density dependencies of the birth and death rates, that we can avoid the anomalous, unphysical behavior of the customary logistic model, equations A.1a and A.1b, for populations in excess of the carrying capacity. This procedure is essential for modeling the effects of sequential habitat variation encountered by seasonal breeders. The habitat with the minimum fertility or mortality carrying capacity generally controls the equilibrium or average population size. The biological interpretation of the density-dependent birth and death mechanisms is more explicit, even in a single habitat. Simultaneously operative limiting mechanisms, including threshold effects, can be treated by multiplicative density-dependent functions. The disadvantages include numerical integration in place of analytical solutions and a great increase in the number of model parameters, some of which may be indeterminate in fitting population dynamics measurements. Some parameters can be specified independently by methods such as sampling, tagging, and radio tracking. We have also demonstrated the deviations between continuous differential and discrete difference models, leading to time lags and possible inductive instabilities in the latter case.

APPENDIX B

AGE-STRUCTURED PERIODIC BREEDERS

ALAN S. PINE

I N T H I S appendix, we examine the conditions under which density-dependent over-all population dynamics models are justified for species that exhibit reproductive age structure. We find that a discrete generalized logistic model, strictly valid only for age-independent birth and death rates, is compatible with age-structured breeders under restricted circumstances involving particular youth-oriented stable or quasi-stable population age distributions and modified density dependencies. In particular, we discuss the effects of multiple sequential habitats and population-limiting mechanisms encountered by age-structured periodic breeders. We demonstrate that the conventional Leslie matrix analysis must be modified for species with more than one breeding season.

It is known that species with age-dependent birth rates may experience population waves on the maturation timescale (Leslie 1945, 1948; Maynard Smith 1968, 1974; Pielou 1969; Caswell 2006). However, this complication is usually neglected in overall population dynamics models, often due to the lack of age-specific demographic data. Nevertheless, it is of some interest to determine the age-structure requirements justifying density-dependent overall uniform population models. In particular, we are concerned here with the applicability of age-independent models developed for periodic breeders in appendix A to those species that exhibit age-dependent birth *and/or* death rates.

The concept of density-dependent population control was introduced by Verhulst (1845) through the differential "logistic equation,"

$$\frac{dx}{dt} = rx(1-x) \tag{B.1}$$

Here, $x = N/K$ is the population density, N, relative to the carrying capacity, K, of the environment, and the Malthusian growth rate factor, r, is usually given in terms of $b - d$, the birth minus death rates. The logistic r and K parameters are assumed to be constant in time, and the $(1 - x)$ factor represents the density dependence that limits the ultimate population to the carrying capacity. If the parameters are time dependent, say due to periodic breeding, we have shown in appendix A that separate birth and death density dependencies are required in order to avoid possible anomalous growth or singularities predicted by the Verhulst logistic in nonbreeding seasons. This modified logistic may be written,

$$\frac{dx}{dt} = x[bF(x) - dG(x)] \tag{B.2}$$

where the environmental factors, $F(x)$ and $G(x)$, represent the density dependencies of the respective birth and death rates. Because different limiting mechanisms usually pertain to births (e.g., nesting sites or territorial behavior) and deaths (e.g., food or water), we distinguish their functional forms. Here, the birth rate density-dependent factor, $F(x)$, is constructed to be unity for $x = 0$ and to vanish or be greatly reduced at the carrying capacity, whereas the death rate density-dependent factor, $G(x)$, increases from unity at $x = 0$.

Some illustrative functional forms for $F(x)$ and $G(x)$ have been considered in appendix A,

$$F(x < q) = [1 - (x/q)^p], \; F(x \geq q) = 0 \tag{B.3a}$$

$$F(x) = \exp[-(x/q)^p] \tag{B.3b}$$

$$F(x) = 1/[1 + (x/q)^p] \tag{B.3c}$$

$$F(x \leq q) = 1, \qquad F(x > q) = (q/x)^p \tag{B.3d}$$

and

$$G(x) = 1/\delta(x), \; \delta(x) = [1 - (x/q)^p] \text{ or } \delta(x) = \epsilon < 1 \text{ if } \delta(x) < \epsilon \tag{B.4a}$$

$$G(x) = \exp[(x/q)^p] \tag{B.4b}$$

$$G(x) = [1 + (x/q)^p] \tag{B.4c}$$

$$G(x \leq q) = 1, \; G(x > q) = 1 + p[1 - (q/x)] \tag{B.4d}$$

In these expressions, q is the ratio of any particular birth or death carrying capacity to an arbitrary K (usually the minimum value), and p is the "saturation power." Some of these $F(x)$ factors have been proposed previously by numerous workers (Leslie 1948; Gilpin and Ayala 1973; May and Oster 1976) to generalize and circumvent some difficulties with the standard Verhulst model. Functions B.4a–c for $G(x)$ are essentially the reciprocals of the corresponding functions B.3a–c for $F(x)$, with prevention of a vanishing or negative denominator in B.4a. In equations B.2–B.4, the parameters b, d, p, and q may depend explicitly on time, such as gradually, periodically, or randomly.

AGE-INDEPENDENT PERIODIC BREEDERS

Let us first summarize the results in appendix A for age-independent periodic breeders with an overall time period interval, τ (e.g., 1 year), divided into m sequential habitats or seasons. Then, if τ_m is the duration of season m, we have $\sum_{m=1}^{M} \tau_m = \tau$. The m label is cyclical such that $M + m \rightarrow m$. We allow each of the birth, death, carrying capacity, and saturation power parameters to vary seasonally, but we assume they are constant during a given season and can be labeled with the subscript index m. If we denote $x_{m,n}$ as the total population density at the beginning of season m in year n, we can solve equation B.2 for the density at the end of the season, which is then used to initiate the next season, $m + 1$. Note the periodicity condition that the last season of one year is followed by the first season of the next year, so $x_{M+1,n} = x_{1,n+1}$.

To simplify the mathematical computations for comparison with discrete age-structured models, we approximate the derivative in equation B.2 with a seasonal difference relation, $dx/dt \rightarrow (x_{M+1,n} - x_{m,n})/\tau_m$, leading to an iterative expression of the form,

$$x_{M+1,n} = x_{m,n}[b_m F_m(x_{m,n})\tau_m + s_m] \tag{B.5a}$$

$$s_m = \exp[-d_m G_m(x_{m,n})\tau_m] \tag{B.5b}$$

Here, s_m is the seasonal survival probability, given by equation B.5b rather than $s_m = 1 - d_m G_m(x_{m,n})\tau_m$ in order to avoid unphysical negative populations if $d_m G_m(x_{m,n})\tau_m > 1 + b_m F_m(x_{m,n})\tau_m$ and to account for deaths during the season, as noted previously in appendix A. This discrete approximation implies that the births during a season depend only on the initial seasonal populations, which may be appropriate for infertile newborns or species that abandon their eggs. The resulting delay in response to very high birth rates may lead to instabilities (appendix A), which are somewhat reduced from those for non-overlapping generations (May 1973, 1976; May and Oster 1976) due to the shorter seasonal durations.

AGE-STRUCTURED MODELS

These models ignore the effect of age structure on the birth and death rates, implying either that b and d are independent of age or that the population age distribution is nearly constant

(Pielou 1969), as we will show. Lotka (1922, 1956) has considered age-dependent fertility and mortality without density or seasonal dependence, demonstrating that Malthusian growth with a stable age distribution will ultimately be realized. Leslie (1945) has given a thorough description of a matrix representation for discrete age-dependent fertility and survival probabilities. The Leslie age-structured model was formally generalized for periodic birth and survival rates by Skellam (1966) using a matrix chain multiplication procedure, which we follow here and give an explicit form for the seasonal matrices. An alternative approach to a seasonal generalization of the Leslie scheme was presented by Gourley and Lawrence (1977) for species whose fertility and survival rates depend on both the current and the original birth season. Though Leslie (1948) discussed density-dependent control factors, neither Skellam (1966) nor Gourley and Lawrence (1977) considered this ramification, which we emphasize here.

The Leslie (1945, 1948) matrix formulation of discrete age-dependent population dynamics was for a single habitat and particular survival density dependence of the form of equation B.3c, mimicking the logistic sigmoidal growth. Here, we generalize this model for periodic breeders sequentially occupying multiple habitats with distinct birth and death rates and density dependencies. Let $y_{m,n,i}$ be the fractional population of age i at the beginning of season m in year n. The total population at the start of any season is the sum of the age-dependent fractional populations,

$$x_{m,n} = \sum_{i=0}^{\infty} y_{m,n,i} \tag{B.6a}$$

Now we define the density-dependent annual birth and death rates for season m and age i by $b_{m,i}(x_{m,n})$ and $d_{m,i}(x_{m,n})$, respectively. Newborns are assigned to the $i = 0$ age class for the next season. Thus, for $m < M$,

$$y_{m+1,n,0} = \sum_{i=0}^{\infty} b_{m,i}(x_{m,n})\tau_m y_{m,n,i} + s_{m,0}(x_{m,n})y_{m,n,0} \tag{B.6b}$$

and for $i > 0$,

$$y_{m+1,n,i} = s_{m,i}(x_{m,n})y_{m,n,i} \tag{B.6c}$$

where the terms

$$s_{m,i}(x_{m,n}) = \exp[-d_{m,i}(x_{m,n})\tau_m] \tag{B.6d}$$

represent the probability of survival of the $y_{m,n,i}$ from one season to the next. For the last season, $m = M$, the year is incremented by 1, as is the age for those born prior to the last season, so

$$y_{1,n+1,0} = \sum_{i=0}^{\infty} b_{M,i}(x_{M,n})\tau_M y_{M,n,i} \tag{B.6e}$$

and for $i \geq 0$,

$$y_{1,n+1,i+1} = s_{M,i}(x_{M,n})y_{M,n,i} \tag{B.6f}$$

Note that a finite cutoff age limit, i_{max}, on the sums implies that $s_{m,imax} = 0$, that is, no survivors above the cutoff. This cutoff will have little consequence if the death rate above the oldest initial population is sufficient to kill off the elders by i_{max}.

The relationship between seasonal age-dependent birth and death rates in the low-density limit and the average annual values is $\bar{b}_i = \sum_{m=1}^{M} b_{m,i}(0)\tau_m/\tau$ and $\bar{d}_i = \sum_{m=1}^{M} d_{m,i}(0)\tau_m/\tau$, and the average annual survival is the product of the seasonal survivals, $\bar{s}_i = \prod_{m=1}^{M} s_{m,i}(0)$, according to equation B.6d. Because the age class is incremented irrespective of the season of birth (much as racehorses have a common January 1 "birthday") and the index m refers to the current season, there is no memory of the original birth season in this model. Gourley and Lawrence (1977) have developed an alternative model that keeps track of the true age (in seasons), which is necessary for those species whose fertility or survival also depends on the original birth season. This is most important for species with more than one breeding season and which are fertile within their first year of life.

In figure B.1, we illustrate some hypothetical age-dependent population dynamics based on equation B.6 for an initial cohort, y_0, of 10-year-olds. The population distributions, y_C, y_G, y_L, after 100 cycles are plotted assuming constant ($B_C = b_{C0}$), Gaussian ($B_G = b_{G0}\exp[-z^2]$), and Lorentzian ($B_L = b_{L0}/[1 + z^2]$) age-dependent birth rates and a constant death rate ($d_i = 0.2$). Here, we choose a human time scale with $z = (i - 25.0)/5.0$, and we take $i_{max} = 100$ and birth and death density dependencies of the form, $F(x) = \exp(-x^2)$ and $G(x) = \exp(x/2)^4$. Two seasons of equal duration ($\tau_1 = \tau_2 = \tau/2$) are assumed, with breeding only in the first season and the

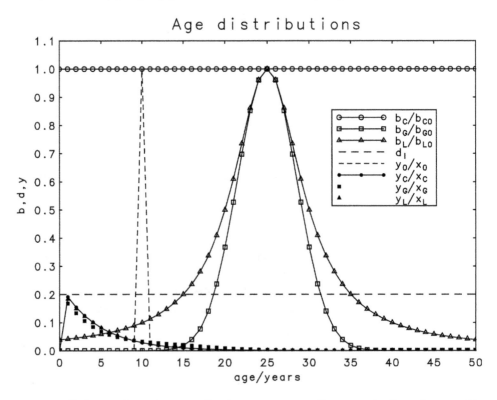

FIGURE B.1   Final population age distributions, y_C, y_G, y_L, for an initial cohort of 10-year-olds assuming constant, b_C, Gaussian, b_G, or Lorentzian, b_L, birth rates and a constant death rate, d_i. See text for definition of variables, normalizations, and model parameters.

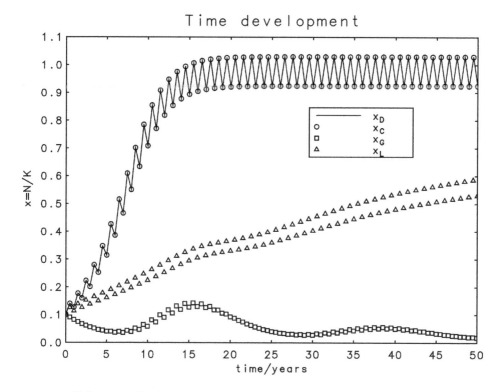

FIGURE B.2   Time development of total populations, x_C, x_G, x_L, for the three birth rates of figure B.1 compared to the discrete age-independent model, x_D.

same death rate in both. The birth rates $B_{C0} = 1$, $b_{G0} = 11.397$, $b_{L0} = 7.009$ are selected to yield equal average birth rates over the life span and are very high to demonstrate rapid changes. The population distributions in figure B.1 are all normalized to the total populations at the time. We see that the initial population spikes are all transformed into similar youth-oriented, nearly exponentially declining, age distributions after sufficient cycles.

Figure B.2 gives the time development of the total populations for the three fertility distributions in figure B.1, compared with the age-independent discrete model, x_D, in equation B.5. The sawtooth fluctuations depict the seasonal variations. The Gaussian case, x_G, exhibits strong maturation waves, taking about 15 years for the initial cohort of 10-year-olds to reach peak breeding, but declines overall because the progeny have such low fertility. The Lorentzian case, x_L, shows weaker maturation waves but steady growth because the fertility wing extends down to newborns. The constant birth rate, x_C, reproduces the age-independent model, as expected, for the initial spike or for any starting distribution that does not survive to i_{max}. Otherwise, it is clear that age-structured breeding generally cannot be described by age-independent models. However, in this appendix, we examine conditions for which the age-structured models of equations B.6 can be adjusted to reproduce the overall density-dependent population dynamics models in equations B.1–B.5.

Case 1. *Birth and death rates are independent of age.* For age-independent birth and death rates, we have $b_{m,i}(x_{m,n}) = b_m F_m(x_{m,n})$ and $d_{m,i}(x_{m,n}) = d_m G_m(x_{m,n})$, so $s_{m,i}(x_{m,n}) = s_m(x_{m,n}) = \exp[-d_m G_m(x_{m,n})\tau_m]$. For $m < M$, equations B.6a–d then yield,

$$x_{m+1,n} = Y_{m+1,n,0} + \sum_{i=1}^{\infty} Y_{m+1,n,i}$$

$$= \sum_{i=0}^{\infty} b_m F_m(x_{m,n}) \tau_m Y_{m,n,i} + \sum_{i=0}^{\infty} s_m(x_{m,n}) Y_{m,n,i} \qquad \text{(B.7)}$$

$$= x_{m,n} \left[b_m F_m(x_{m,n}) \tau_m + s_m(x_{m,n}) \right],$$

as the age-independent terms can be taken outside the sum over ages. This is identical to equation B.5. Also, when $m = M$, equations B.6e and B.6f similarly result in equation B.5, noting the cyclical condition, $x_{M+1,n} \to x_{1,n+1}$. As expected, this justifies the overall discrete population model in equation B.5 for age-independent birth and death rates, even though the rates may vary according to the seasonal habitat. For this reason, the density-dependent population models, such as in equations B.1–B.5, are sometimes referred to as age-independent dynamics. We note, however, that the assumption of an age-independent birth rate implies that newborns are fertile in their very next breeding season and that an age-independent death rate is technically inconsistent with a finite cutoff age, i_{max}, unless there are no survivors from the initial population distribution. The agreement of x_C with x_D in figure B.2 illustrates these points graphically.

Case 2. *Age-structured birth rates but age-independent death rates and density dependencies.* If the birth rates depend on age, but their density dependencies do not, then the age and density dependencies are "separable" according to $b_{m,i}(x_{m,n}) = b_{m,i} F_m(x_{m,n})$. Assume also that the death rates and hence the survivals are age independent, so, as before, $s_m(x_{m,n}) = \exp[-d_m G_m(x_{m,n})\tau_m]$. In this case, equations B.6a–d for $m < M$ yield

$$x_{m+1,n} = Y_{m+1,n,0} + \sum_{i=1}^{\infty} Y_{m+1,n,i}$$

$$= \sum_{i=0}^{\infty} b_{m,i} F_m(x_{m,n}) \tau_m Y_{m,n,i} + \sum_{i=0}^{\infty} s_m(x_{m,n}) Y_{m,n,i} \qquad \text{(B.8)}$$

$$= x_{m,n} \left[\langle b_m \rangle_n F_m(x_{m,n}) \tau_m + s_m(x_{m,n}) \right],$$

and similarly for $m = M$ using periodicity. Here, we define the age-averaged birth rates as

$$\langle b_m \rangle_n = \frac{1}{x_{m,n}} \sum_{i=0}^{\infty} b_{m,i} Y_{m,n,i} \qquad \text{(B.9)}$$

Thus, equation B.8 will reproduce equation B.5 if these average birth rates are constant, independent of n. This would require a seasonal scaling of the birth rates unless the population age distribution for a particular season m is stable from year to year. The "stable" distribution is characterized by the scaling relation, $Y_{m,n+1,i} = \varrho_{m,n} Y_{m,n,i}$, for all i where the scale factor is obtained from the yearly overall population gain ratio, $\varrho_{m,n} = x_{m,n+1}/x_{m,n}$. From equation B.9, such a stable distribution yields $\langle b_m \rangle_{n+1} = \langle b_m \rangle_n = \langle b_m \rangle$, which justifies the overall discrete population dynamics model under this condition. To have the average birth rates be equivalent to the age-independent values in equations B.5 and B.7, the peak birth rates are usually considerably higher as only a portion of the stable age distribution will be fertile. We now discuss how to obtain such a stable age distribution for the population.

Case 3. *Stable age distribution.* For a single habitat and density dependence, Leslie (1945, 1948) has shown that the stable age distribution may be obtained conveniently using a matrix formulation. We extend his approach here for multiple habitats and density dependencies. If we use the short-hand notation, $b_{m,n,i} = b_{m,i}(x_{m,n})\tau_m$ and $s_{m,n,i} = \exp[-d_{m,i}(x_{m,n})\tau_m]$, and restrict the ages to a finite cutoff value, i_{max}, then we may write equations 6b–d in matrix form,

$$
Y_{m+1,n} = \begin{bmatrix} y_{m+1,n,0} \\ y_{m+1,n,1} \\ y_{m+1,n,2} \\ \cdots \\ y_{m+1,n,i_{max}} \end{bmatrix} = A_{m,n} \cdot Y_{m,n}
$$

$$
= \begin{bmatrix} b_{m,n,0} + s_{m,n,0} & b_{m,n,1} & \cdots & b_{m,n,i_{max}-1} & b_{m,n,i_{max}} \\ 0 & s_{m,n,1} & \cdots & 0 & 0 \\ 0 & 0 & \cdots & 0 & 0 \\ \cdots & \cdots & \cdots & s_{m,n,i_{max}-1} & \cdots \\ 0 & 0 & \cdots & 0 & s_{m,n,i_{max}} \end{bmatrix} \begin{bmatrix} y_{m,n,0} \\ y_{m,n,1} \\ y_{m,n,2} \\ \cdots \\ y_{m,n,i_{max}} \end{bmatrix},
$$

(B.10a)

for $m < M$. Here, $Y_{m,n} = [Y_{m,n}]$ is a column vector of dimension $(i_{max} + 1)$, and $A_{m,n}$ is a square matrix of order $(i_{max} + 1)$. The nonzero elements are restricted to the first row and the diagonal. For the last season of the year, $m = M$, equations B.6e and B.6f yield a matrix expression of the form,

$$
Y_{1,n+1} = \begin{bmatrix} y_{1,n+1,0} \\ y_{1,n+1,1} \\ y_{1,n+1,2} \\ \cdots \\ y_{1,n+1,i_{max}} \end{bmatrix} = A_{M,n} \cdot Y_{M,n}
$$

$$
= \begin{bmatrix} b_{M,n,0} & b_{M,n,1} & \cdots & b_{M,n,i_{max}-1} & b_{M,n,i_{max}} \\ s_{M,n,0} & 0 & \cdots & 0 & 0 \\ 0 & s_{M,n,1} & \cdots & 0 & 0 \\ \cdots & \cdots & \cdots & 0 & \cdots \\ 0 & 0 & \cdots & s_{M,n,i_{max}-1} & 0 \end{bmatrix} \cdot \begin{bmatrix} y_{M,n,0} \\ y_{M,n,1} \\ y_{M,n,2} \\ \cdots \\ y_{M,n,i_{max}} \end{bmatrix}
$$

(B.10b)

This matrix consists of the first row and first subdiagonal and is the same mathematical form as given by Leslie (1945, 1948) for a single habitat, namely $M = 1$.

Propagation of the population distribution through sequential habitats is accomplished by multiplying the seasonal matrices, as indicated by Skellam (1966) without specifying the form of equation B.10a. For example, from the beginning of one year to the next, we have $Y_{1,n+1} = \widehat{A}_n \cdot Y_{1,n}$, where

$$\widehat{A}_n \triangleq A_{M,n} \cdot \prod_{m=1}^{M-1} A_{m,n} \tag{B.11}$$

Skellam (1966) and Caswell (2006) have given a formal discussion of the mathematical properties of such a chain of periodic matrices. Here, all the seasonal and annual matrices are square of order ($i_{max} + 1$). For species whose age-structured seasonal fertility and survival rates depend also on the initial birth season, Gourley and Lawrence (1977) provided an alternative description, counting age in seasons, resulting in much larger square matrices of order ($M_B \times M \times i_{max}$), where M_B is the number of breeding seasons. In equations B.10 and B.11, the product of matrices for $m < M$ retains zeroes everywhere except the first row and diagonal, just as for equation B.10a, whereas the product of $A_{M,n}$ with any combination of prior seasons within the year yields a matrix with zeroes everywhere but the first two rows and the first subdiagonal. As an illustration, we take $M = 2$ and $i_{max} = 4$, yielding

$$\widehat{A}_n = \begin{bmatrix} b_{20}\left(b_{10}+s_{10}\right) & b_{20}b_{11}+b_{21}s_{11} & b_{20}b_{12}+b_{22}s_{12} & b_{20}b_{13}+b_{23}s_{13} & b_{20}b_{14}+b_{24}s_{14} \\ s_{20}\left(b_{10}+s_{10}\right) & s_{20}b_{11} & s_{20}b_{12} & s_{20}b_{13} & s_{20}b_{14} \\ 0 & s_{21}s_{11} & 0 & 0 & 0 \\ 0 & 0 & s_{22}s_{12} & 0 & 0 \\ 0 & 0 & 0 & s_{23}s_{13} & 0 \end{bmatrix} \tag{B.12}$$

where we have suppressed the index n and commas in the matrix elements for brevity. We note that this form for more than one breeding season differs from the conventional Leslie matrix by the nonzero $s_{20}b_{1i}$ elements in the second row. The biological correspondence with Leslie's matrix elements will be discussed in the next section in connection with a single breeding season. Multiple breeding seasons may occur, for example, for birds that double clutch with varying success for the second brood or even for species with relatively uniform breeding throughout the year but with seasonal death rates.

In any case, we can find a stable solution such that $y_{1,n+1,i} = \lambda_n y_{1,n,i}$ when the λ_n are eigenvalues of the matrix $\widehat{A}_n$ and $[Y_{1,n}]$ are the corresponding eigenvectors. Leslie (1945) has discussed the eigenproperties of matrices of the form of equation B.10b, most of which apply to any matrix with non-negative real elements, such as equation B.12, or general population matrices as reviewed by Caswell (2006). The eigenvalues may be obtained either by diagonalizing the matrix numerically or by setting the determinant $\left|\widehat{A}_n - \lambda_n \mathbf{1}\right| = 0$ and solving for the roots of the characteristic polynomial equation. In general, there are $i_{max} + 1$ eigenvalues, of which at least one is purely real if i_{max} is even and two if i_{max} is odd. The maximum positive real eigenvalue, denoted the "dominant" eigenvalue, Λ_n, gives the most persistent population. The remaining eigenvalues, which occur in complex conjugate pairs (required because the matrix elements and, thereby, the coefficients of the characteristic polynomial are purely real), correspond to complex conjugate eigenvectors. Because any arbitrary initial age distribution can be written as a linear sum of the complete set of eigenvectors, the complex components give rise to population waves that are usually damped relative to the dominant stable form (Leslie 1945; Pielou 1969; Caswell 2006). Now we need to determine the circumstances under which the dominant eigenvalue tracks the population changes for the age-independent dynamics models, that is, when $\Lambda_n \to \varrho_{1,n} = x_{1,n+1}/x_{1,n}$.

In general, the matrix elements on the subdiagonal contain products of the seasonal survivals $\sigma_{n,i} = \prod_{m=1}^{M} s_{m,n,i}$, which we denote as the annual probability of survival for age i. The components of the eigenvectors for the dominant eigenvalue are then given by the ratios,

$$\gamma_{m,n,i+1}/\gamma_{m,n,i} = \sigma_{n,i}/\Lambda_n \tag{B.13}$$

for $i \geq 1$. For a stable population distribution, this ratio needs to be independent of n. This occurs, for example, when the population saturates, as then $x_{m,n+1} = x_{m,n}$, so $\Lambda_n = \varrho_{1,n} = 1$ and $\sigma_{n,i}$, which depends on n only through the density, is constant. Also, if the birth and death rates are density independent, then Λ_n is constant, and the population exhibits geometrical (Malthusian) growth with a stable age distribution. However, if the population gain, $\varrho_{1,n}$, changes from year to year, then the age distribution can be stable only in the unlikely circumstance that the n dependence of the annual survival compensates. If the death rates, and hence the survivals, are independent of age, equation B.13 represents an exponential decline with age for the stable population distribution. In fact, most realistic age dependencies for the death rates result in a highly youth-oriented stable age distribution.

If the density dependencies of the seasonal survivals given by equation B.5b do not compensate for the changing population, then we can still find a stable age distribution solution that tracks the age-independent dynamics. This may be accomplished by taking the density dependence as a common factor, $H_n = \varrho_{1,n}/\varrho_{1,0}$, multiplying all the matrix elements of $\hat{A}_n$, in place of the separate birth and death factors, $F_m(x_{m,n})$ and $G_m(x_{m,n})$. This is the approach originally taken by Leslie (1948) for introducing a density dependence into his single-habitat age-structured population model, except that he chose a simple generic function for the matrix factor, namely $H_n = 1/[1 + (\Lambda_{00} - 1)x_n]$, rather than trying to fit any particular age-independent dynamics model. Here, Λ_{00} is Leslie's "intrinsic" dominant eigenvalue in the limit $x_n \to 0$.

We first scale all of the seasonal birth rates in the first year, $b_{m,i}F_m(x_{m,0})$, with a single factor B_0 to yield the density ratio $\varrho_{1,0} = x_{1,1}/x_{1,0}$. This may be accomplished, for example, by numerical diagonalization of $\hat{A}_0$ with an estimated B_0, which can then be refined to any desired precision using bisection techniques (Press et al. 1992). Faster, more accurate methods are available under certain circumstances, as discussed in the next section for a single breeding season. The diagonalization procedure provides the initial stable eigenvector, $Y_{1,0}$, corresponding to $\Lambda_0 = \varrho_{1,0}$, which launches the age-structured dynamics model. The scaled age-dependent births, $b_{m,n,i} = B_0 b_{m,i}F_m(x_{m,0})\tau_m$, and survivals, $s_{m,n,i} = \exp[-d_{m,i}G_m(x_{m,0})\tau_m]$, are then held fixed at the initial values during the population evolution to eliminate the separate n dependence for the matrix elements.

The annual H_n factor for multiple habitats actually represents a composite of seasonal matrix factors; that is, $H_n = \prod_{m=1}^{M} h_{m,n}$, where $h_{m,n} = (x_{M+1,n}/x_{m,n})/(x'_{m+1,0}/x'_{m,0})$. Here, the primes designate solutions for the first year seasonal matrices, $A_{m,0}$, starting with the initial stable distribution, $Y_{1,0}$, and noting $x'_{1,0} = x_0$ and $x'_{M+1,n} = x_{1,n+1}$. All the $A_{m,n} = A_{m,0}$

elements need to be multiplied by $h_{m,n}$ in order to track the age-independent seasonal variations as well as the annual changes given by the H_n factor. If the birth rates and survivals for a particular season are density independent (i.e., $F_m = 1$ and $G_m = 1$), then $h_{m,n} = 1$. All the necessary H_n and $h_{m,n}$ factors multiplying the age-dependent matrices may be determined from the $x_{m,n}$ precalculated from the presumed age-independent population model, equation B.5.

Figure B.3 illustrates the initial, y_0, and final, y_F, stable age distributions for the same Lorentzian age-structured birth model discussed for figure B.1. Here, the initial stable distribution corresponds to the dominant eigenvalue, $\Lambda_0 = \varrho_{1,0} = 1.267$, yielding $B_0 = 19.809$. Note that $B_0 > b_{L0}$, as the fertility for the youthful stable age distribution is much lower than the lifetime-averaged birth rate. The final stable age distribution is obtained by the diagonalization procedure for $\Lambda_n = 1$, namely at saturation. Figure B.4 gives the time development of the corresponding total population, x_m, compared with the age-independent discrete model, x_D, for the seasonal matrix factors, $h_{1,n}$ and $h_{2,n}$, shown. Here, $h_{2,n} \approx 1$ because the second season is nonbreeding, but it is not exactly unity because the death rate is slightly density dependent over the range of x_m. The final distribution propagated here is given by y_m in figure B.3, and it is identical to the initial distribution, y_0, not y_F. Though we have illustrated a single-breeding season with constant death rates here for comparison with models that follow later, this procedure is applicable to multiple breeding seasons and arbitrary age-dependent death rates or survivals.

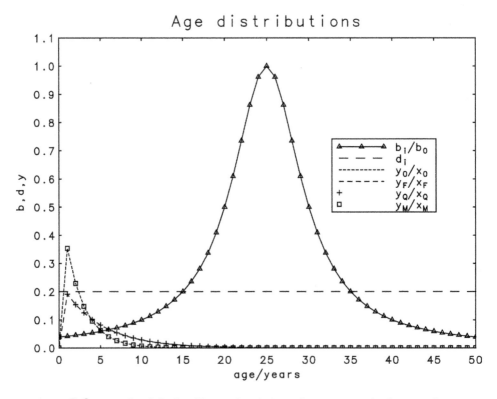

FIGURE B.3   Initial and final stable age distributions for Lorentzian birth rate and constant death rate model of figure B.1. See text for definitions and parameters.

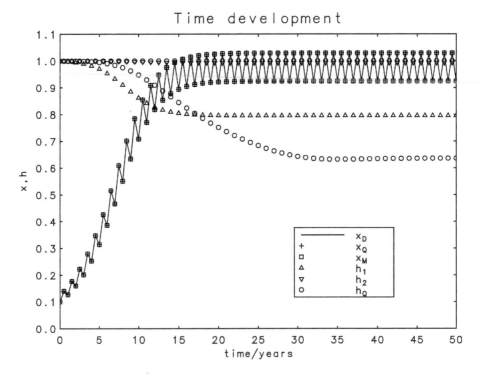

FIGURE B.4 Time development of total populations with stable, x_M, and quasi-stable, x_Q, distributions for the Lorentzian birth rate model of figure B.3 compared with the discrete age-independent model, x_D. The density-dependent fitting parameters, h, are discussed in the text.

Leslie (1945) has pointed out that the diagonalization procedure fails if any postreproductive population is included. That is, if $b_{m,i} = 0$ for all m for $i > i_r$ where i_r is the highest fertile age, then the matrix, $\hat{A}_n$, is singular because its determinant vanishes. Thus, Leslie (see also Pielou 1969; Caswell 2006) truncated the matrix at i_r, noting that the postreproductive individuals are not represented by descendants after $i_{max} - i_r$ years. However, these infertile individuals still consume resources and may affect the density-dependent controlling mechanisms. We find that this elder population can be included with negligible error if their birth rates are assumed to be very small (say 10^{-6} of the peak fertility), but not zero.

Case 4. *Single breeding season.* It often happens that breeding occurs in only one season. This affords considerable mathematical simplification, as the remaining habitats or seasons can be combined into a single nonbreeding season with an effective survival rate being the product of all the nonbreeding season survivals. Here, we ignore the stepwise seasonal declines, which may be significant if the death rates vary considerably, as with migratory species. Nevertheless, the contraction to two effective seasons, as seen in the example in equation B.12 with either all the b_{1i} or b_{2i} vanishing, reduces the annual matrix to one row and one subdiagonal, just as for the single habitat model of Leslie (1945, 1948).

Consider first the case that breeding occurs only in the first season, so $b_{2i} = 0$, which nulls the entire first row and results in one of the eigenvalues being zero. We then need to diagonalize the $i_{max} \times i_{max}$ block obtained by striking out the first row and column, yielding a matrix of the form (for $i_{max} = 4$),

$$\tilde{A}_n = \begin{bmatrix} \beta_1 & \beta_2 & \beta_3 & \beta_4 \\ \sigma_1 & 0 & 0 & 0 \\ 0 & \sigma_2 & 0 & 0 \\ 0 & 0 & \sigma_3 & 0 \end{bmatrix},$$

where the effective annual birth and survival rates are given by $\beta_i = s_{20}b_{1i}$ and $\sigma_i = s_{2i}s_{1i}$. Here, β_i involves the survival of the newborns for the rest of the year. Such sparse matrices have characteristic polynomials of the form (Leslie 1945, 1948; Pielou 1969)

$$\lambda^k - \lambda_{k-1}\beta_1 - \lambda^{k-2}\beta_2\sigma_1 - \lambda^{k-3}\beta_3(\sigma_1\sigma_2) - \cdot \cdot \cdot - \beta_k(\sigma_1\sigma_2 \cdot \cdot \cdot \sigma_{k-1}) = 0 \qquad (B.14)$$

where we have set $k = i_{max}$ and suppressed the index n. The roots of the polynomial result in k eigenvalues, with the maximum positive real root corresponding to the dominant eigenvalue, Λ_n, which we again associate with the population ratio, $\varrho_{1,n}$, and the stable eigenvector. Here, $\Lambda_n \gtreqless 1$, corresponding to population gain, saturation, or loss, according to whether the sum of the coefficients in equation B.14, $\sum_{j=0}^{k-1}\beta_{k-j}(\sigma_1\sigma_2 \cdots \sigma_{k-j-1}) \gtreqless 1$.

In the case that breeding occurs only in the last season, the results are similar, except that $k = i_{max} + 1$ and the i index starts at $i = 0$. Here, the effective birth rates in the first row of the matrix are $\beta_i = s_{1i}b_{2i}$, involving the probability of survival of the parent during the year up until birth. Mid-year births have survival contributions from both. The effective birth rates, β_i, correspond to the first-row "fertility" matrix elements given by Leslie (1945, 1948), who discussed random births throughout the interval in a single habitat and an ad hoc density factor applied to the overall matrix. This may be regarded as an alternative formulation of the Leslie matrix, showing more explicitly the approximations involved in the development from density-dependent birth and death rates in the generalized logistic, equation B.2. Of course, the single-habitat Leslie model does not exhibit seasonal variations.

For a single breeding season, the coefficients of the characteristic polynomial in equation B.14 are linear in the effective birth rates, so by substituting the initial eigenvalue, $\Lambda_0 = \varrho_{1,0}$, for λ in equation B.14, we may solve for the birth rate scale factor B_0 without numerical diagonalization or iteration as discussed in the previous section. This B_0 factor yields the density-independent matrices that, when multiplied by the combined density-dependent factors, $h_{m,n}$, reproduces the age-independent populations, $x_{m,n}$, for the stable population age distribution characterized by $y_{1,0,i+1}/y_{1,0,i} = \sigma_i/\Lambda_0$. This method is unaffected by a postreproductive population. Again, the overall factor, $H_n = h_{1,n}h_{2,n}$; and if the births and deaths are not density dependent for habitat m, then $h_{m,n} = 1$. If the breeding season is the first season, there are no newborns at the start of the new year, (i.e., $y_{1,n,0} = 0$), and the distributions start with a finite value at age $i = 1$, as shown in figures B.1 and B.3. This characteristic polynomial root method is much more efficient than the numerical diagonalization procedure discussed earlier—a factor of several thousand times faster in the example of figures B.3 and B.4 with a 101×101 dimension matrix.

Case 5. *Quasi-stable age distribution.* In the previous section, we have seen that the age-independent dynamics can be reproduced for age-dependent birth and death rates and a stable age distribution by multiplying the annual matrices with a combined density-dependent factor, H_n, in place of separate birth and death density dependencies. Of course, if there is only one breeding season, the H factor may be attributed to one or all of the remaining seasons, thus modifying only those survival rates. Notably, this logic does not work as a birth-only modifier for a hypothetical season with no deaths, as the H factor will also affect the unity survivals in that season. However, if the annual population changes are not too rapid and the death rates are independent of age, we can find a quasi-stable solution with a density dependence for the birth rates alone.

The quasi-stable solution begins just as for the stable solutions described earlier, with a birth scale factor, B_0, found to produce the first population gain ratio, $\varrho_{1,0}$, by the diagonalization-iteration procedure (for multiple breeding seasons) or the characteristic polynomial method (for a single breeding season). This yields the initial eigenvalue, $\Lambda_0 = \varrho_{1,0}$, and the corresponding stable eigenvector, $\mathbf{Y}_{1,0}$. We then propagate this population age distribution using equations B.6a–f with a factor $B_0 h_{m,n}$ augmenting the $F_m(x_{m,n})$ density dependence. The death density dependences, $G_m(x_{m,n})$, are not modified. The $h_{m,n}$ factor is adjusted (by bisection, for example) to reproduce the seasonal population gain, $x_{M+1,n}/x_{m,n}$ (or $x_{1,n+1}/x_{M,n}$ for the last season), precalculated for the age-independent dynamics model. In this situation, the age distribution of the population may vary slightly from year to year, but, if the total population saturates, it will eventually reach a new stable distribution corresponding to equation B.13 with $\Lambda_n = 1$ and σ_{ni} constant. Figure B.4 demonstrates that the time development of the quasi-stable total population, x_Q, can fit the discrete age-independent model, x_D, exactly, using the seasonal birth rate factors, h_Q, shown. In this case, however, the final quasi-stable population distribution, y_Q, is the same as the saturated value, y_F, shown in figure B.3, not the initial value, y_0. It should be noted that this final distribution, y_F, is identical to the distribution, y_C, in figure B.1 for an age-independent birth and death rate, no matter what the initial distribution. We also emphasize that propagating the initial stable distribution without readjusting the birth rate density dependence (i.e., setting $h_Q = 1$) generally does not reproduce the age-independent dynamics.

Because the $h_{m,n}$ factors are applied only to the birth rates, they are indeterminate for non-breeding, or even weakly breeding ($b_{m,i} \ll d_{m,i}$), seasons, and no corrections are made. If the death rates depend on age, then the uncorrected nonbreeding seasonal losses may vary from the precalculated model because the average death rates may differ from the age-independent model values. Because we are fitting population ratios rather than absolute values, any such discrepancies are cumulative. If there is only a single habitat (i.e., $M = 1$, $\tau_1 = \tau$), there are no seasonal variations, so the death rates can depend on age, and any variations in the $\langle d_1 \rangle_n$ averages can be absorbed in the birth h factor.

SUMMARY

Density-dependent overall population dynamics models, such as equations B.1–B.5, are convenient for assessing the mechanisms that control and regulate populations, including species sequentially occupying seasonal or migratory habitats. They may even be useful in fitting selected laboratory or field data. These models are strictly valid, though, only for species that do not exhibit age-structured fertility and mortality. However, if breeding maturity and death rates depend on age, it is still possible to realize the same dynamical behavior if the population age distribution is stable or nearly so. Some additional conditions are required for these stable age-dependent models to be applied to the age-independent dynamics. First, the distinct birth and death density dependencies, F and G, for each season is replaced by a single h factor multiplying the combined rates. Thus, the biological interpretation for h is not as transparent as for F and G, except in nonbreeding seasons where it can be related to G. A quasi-stable solution can also be found for a modified density dependence to the fertilities alone if the death rates for the nonbreeding seasons are independent of age.

Here, we have taken the approach of season-by-season exact fitting (one h parameter for each "data" point) of a particular discrete, density-dependent, age-independent model, equation B.5, for age-structured birth and death rates. However, any reasonable age-independent model can be fit in the same manner, even differential models, such as equations B.1–B.4, evaluated at seasonal intervals. Alternatively, one could parameterize the h factors by some simple analytical function of density, such as equations B.3a–d, and least-squares fit the predictions of an age-independent model or real data directly.

The requirement of a stable age distribution for age-structured breeders would seem to be a severe restriction on the applicability of the age-independent dynamics model. Newly colonizing species would be unlikely to start out with such a youth-oriented distribution. Nevertheless, any arbitrary initial distribution would eventually devolve into a stable distribution corresponding to a "dominant" eigenvalue, Λ_n, which may lead to steady population growth, saturation, or extinction. Even age-independent breeders will ultimately produce such a stable age distribution (exponentially decreasing with age in this case). Stability is approached at a rate dependent on the magnitude of the other eigenvalues, λ_n, of the $\hat{A}_n$ matrix relative to Λ_n and on the components of their corresponding eigenvectors in the initial distribution, as discussed by Leslie (1945, 1948) and Caswell (2006). The Leslie matrices must be generalized for more than one breeding season.

REFERENCES

Able, K. P. 1972. Fall migration in coastal Louisiana and the evolution of migration patterns in the Gulf region. *Wilson Bulletin* 84:231–242.

Able, K. P. 1980. Mechanisms of orientation, navigation, and homing. Pages 283–373 in *Animal migration, orientation, and navigation* (S. A. Gauthreaux, Jr., ed.). Academic Press, New York.

Able, K. P. 1982a. Field studies of avian nocturnal migratory orientation 1. Interaction of sun, wind, and stars as directional cues. *Animal Behaviour* 30:761–767.

Able, K. P. 1982b. Skylight polarization patterns at dusk influence migratory orientation in birds. *Nature* 299:550–551.

Able, K. P. 1996. Large-scale navigation. *Journal of Experimental Biology* 199:1–2.

Able, K. P., ed. 1999. *Gatherings of angels*. Cornell University Press, Ithaca, NY.

Able, K. P., and J. R. Belthoff. 1998. Rapid 'evolution' of migratory behaviour in the introduced house finch of eastern North America. *Proceedings of the Royal Society B* 265:2063–2071.

Acheson, D. 1997. *From calculus to chaos: An introduction to dynamics*. Oxford University Press, Oxford.

Adkisson, C. S. 1996. Red Crossbill *Loxia curvirostra*. In The birds of North America online (A. Poole, ed.). Cornell Laboratory of Ornithology, Ithaca, NY. http://bna.birds.cornell.edu. bnaproxy.birds.cornell.edu/bna/species/256/articles/introduction. Accessed September 8, 2012.

Adriaensen, F., and A. A. Dondt. 1990. Population dynamics and partial migration of the European robin (*Erithacus rubecula*) in different habitats. *Journal of Animal Ecology* 59:1077–1090.

Aebischer, A., M. Perrin, M. Krieg, J. Struder, and D. R. Meyer. 1996. The role of territory choice and arrival date on breeding success in the Savi's Warbler *Locustella luscinioides*. *Journal of Avian Biology* 27:143–152.

Afanasyev, V. 2004. A miniature daylight level and activity data recorder for tracking animals over long periods. *Memoirs of the National Institute of Polar Research* 58:227–233.

Åkesson, S. 2003. Avian long-distance navigation: Experiments with migratory birds. Pages 471–492 in *Avian migration* (P. Berthold, E, Gwinner, and E. Sonnenschein, eds.). Springer, Berlin.

Åkesson, S., and A. Hedenström. 2007. How migrants get there: Migratory performance and orientation. *BioScience* 57:123–133.

Åkesson, S., J. Morin, R. Mulheim, and U. Ottosson. 2005. Dramatic orientation shift in White-crowned Sparrows displaced across longitudes in the high Arctic. *Current Biology* 15:1591–1597.

Alerstam, T. 1978. Analysis and theory of visible bird migration. *Oikos* 30:273–349.

Alerstam, T. 1981. The course and timing of bird migration. Pages 9–54 in *Animal migration* (D. J. Aidley, ed.). Cambridge University Press, Cambridge.

Alerstam, T. 1990. *Bird migration*. Cambridge University Press, Cambridge.

Alerstam, T. 1991. Bird flight and optimal migration. *Trends in Ecology and Evolution* 6:210–215.

Alerstam, T. 2001. Detours in bird migration. *Journal of Theoretical Biology* 209:319–331.

Alerstam, T. 2006. Strategies for the transition to breeding in time-selected bird migration. *Ardea* 94:347–357.

Alerstam, T., and A. Hedenström, eds. 1998a. Optimal migration. *Journal of Avian Biology* 29:337–636.

Alerstam, T., and A. Hedenström. 1998b. The development of bird migration theory. *Journal of Avian Biology* 29:343–369.

Alerstam, T., A. Hedenström, and S. Åkesson. 2003. Long-distance migration: Evolution and determinants. *Oikos* 103:247–260.

Alerstam, T., and G. Högstedt. 1982. Bird migration and reproduction in relation to habitats for survival and breeding. *Ornis Scandinavica* 13:25–37.

Alerstam, T., and Å. Lindström. 1990. Optimal bird migration: The relative importance of time, energy, and safety. Pages 331–351 in *Bird migration: Physiology and ecophysiology* (E. Gwinner, ed.). Springer-Verlag, Berlin.

Allan, J. R., and A. P. Orosz. 2001. The costs of birdstrikes to commercial aviation. In *Bird Strike Committee Proceedings, 2001 Bird Strike Committee—USA/Canada*. Third Joint Annual Meeting, Calgary, Alberta. Central Science Laboratory, Birdstrike Avoidance Team, Sand Hutton, York, UK.

Allee, W. C., A. E. Emerson, O. Park, T. Park, and K. P. Schmidt. 1949. *Principles of animal ecology*. Saunders, Philadelphia.

Allen, R. P. 1952. *The whooping crane*. National Audubon Society, New York.

Allen, W. V. 1976. Biochemical aspects of lipid storage and utilization in animals. *American Zoologist* 16:631–647.

Ambuel, B., and S. A. Temple. 1983. Area-dependent changes in the bird communities and vegetation of southern Wisconsin forests. *Ecology* 64:1057–1068.

American Bird Conservancy. 2010. Cats indoors. http://www.abcbirds.org/abcprograms/policy/cats/index.html. Accessed June 7, 2011.

American Ornithologists' Union. 1886. *The code of nomenclature and check-list of North American birds.* American Ornithologists' Union, New York.

American Ornithologists' Union. 1895. *Check-list of North American birds.* American Ornithologists' Union, New York.

American Ornithologists' Union. 1910. *Check-list of North American birds.* American Ornithologists' Union, New York.

American Ornithologists' Union. 1931. *Check-list of North American birds.* American Ornithologists' Union, Lancaster, PA.

American Ornithologists' Union. 1957. *Check-list of North American birds.* American Ornithologists' Union, Baltimore.

American Ornithologists' Union. 1998. *Check-list of North American birds.* 7th ed. American Ornithologists' Union, Baltimore.

American Ornithologists' Union. 2012. *Check-list of North American birds.* American Ornithologists' Union, Washington, DC. http://www.aou.org/checklist/north/index.php. Accessed September 4, 2012.

Anders, A. D., D. C. Dearborn, J. Faaborg, and F. R. Thompson III. 1997. Juvenile survival in a population of Neotropical migrant birds. *Conservation Biology* 11:698–707.

Anders, A. D., J. Faaborg, and F. R. Thompson III. 1998. Postfledging dispersal, habitat use, and home–range size of juvenile Wood Thrushes. *Auk* 115:349–358.

Andrén, H. 1994. Effects of habitat fragmentation on birds and mammals in landscapes with different proportions of suitable habitat—a review. *Oikos* 71:355–366.

Andrewartha, H. G. 1971. *Introduction to the study of animal populations.* 2nd ed. University of Chicago Press, Chicago.

Andrewartha, H. G., and L. C. Birch. 1954. *The distribution and abundance of animals.* University of Chicago Press, Chicago.

Andrle, R. F., and J. R. Carroll. 1988. *The atlas of breeding birds in New York State.* Cornell University Press, Ithaca, NY.

Ankney, C. D. 1984. Nutrient reserve dynamics of breeding and molting Brant. *Auk* 101:361–370.

Aoyama, J. 2009. Life history and evolution of migration in catadromous eels (Genus *Anguilla*). *Aqua-BioScience Monographs* 2(1):1–42.

Arcese, P., M. K. Sogge, A. B. Marr, and M. A. Patten. 2002. Song Sparrow (*Melospiza melodia*). In The birds of North America online (A. Poole, ed.). Cornell Laboratory of Ornithology, Ithaca, NY. http://bna.birds.cornell.edu.bnaproxy.birds.cornell.edu/bna/species/704/articles/introduction. Accessed September 8, 2012.

Armitage, K. 1955. Territorial behavior in fall migrant Rufous Humingbirds. *Condor* 57:239–240.

Arnold, K. E., and I. P. F. Owens. 2002. Extra-pair paternity and egg dumping in birds: Life history, parental care and the risk of retaliation. *Proceedings of the Royal Society B* 363:231–246.

Arnold, T. W., and R. M. Zink. 2011. Collision mortality has no discernible effect on population trends of North American birds. *PLoS One* 6(9):1–6, e24708.

Arrhenius, O. 1921. Species and area. *Ecology* 2:1–57.

Arsenhault, D. P., P. B. Stacey, and G. A. Hoezler. 2005. Mark-recapture and DNA fingerprinting data reveal high breeding-site fidelity, low natal philopatry, and low levels of genetic population differentiation in Flammulated Owls (*Otus flammeolus*). *Auk* 122:329–337.

Askins, R. A. 1993. Population trends in grassland, shrubland, and forest birds in eastern North America. *Current Ornithology* 11:1–34.

Askins, R. A. 2000. *Restoring North America's birds: Lessons from landscape ecology.* Yale University Press, New Haven.

Askins, R. A., F. Chavez–Ramirez, B. C. Dale, C. A. Haas, J. R. Herkert, F. L. Knopf, and P. D. Vickery, eds. 2007. *Conservation of grassland birds in North America: Understanding ecological processes in different regions.* Ornithological Monographs 64. American Ornithologists' Union, Washington, DC.

Askins, R. A., J. F. Lynch, and R. Greenberg. 1990. Population declines in migratory birds in eastern North America. *Current Ornithology* 7:1–57.

Askins, R. A., and M. J. Philbrick. 1987. Effects of changes in regional forest abundance on the decline and recovery of a forest bird community. *Wilson Bulletin* 99:7–21.

Astheimer, L. B., W. A. Buttemer, and J. C. Wingfield. 1992. Interactions of corticosterone with feeding, activity and metabolism in passerine birds. *Ornis Scandinavica* 23:355–365.

Audubon, J. J. 1834. *Ornithological biographies.* Volume 2. Adam and Charles Black, Edinburgh.

Audubon, J. J. 1839. *Ornithological biographies.* Volume 5. Adam and Charles Black, Edinburgh.

Audubon, J. J. 1841. *The birds of America.* Volume 3. J. B. Chevalier, Philadelphia.

Audubon, J. J. 1843. *The birds of America.* Volume 6. J. J. Audubon, New York and Philadelphia.

Austin, J. E., and M. R. Miller. Northern Pintail *Anas acuta.* In The birds of North America online (A. Poole, ed.). Cornell Laboratory of Ornithology, Ithaca, NY. http://bna.birds.cornell.edu.bnaproxy.birds.cornell.edu/bna/species/163/articles/introduction. Accessed September 8, 2012.

Avery, M. L., P. F. Springer, and J. J. Cassell. 1975. The effects of a tall tower on nocturnal bird migration—a portable ceilometer study. *Auk* 93:281–289.

Avery, M. L., E. A. Tillman, and C. C. Laukert. 2001. Evaluation of chemical repellents for reducing crop damage by Dickcissels in Venezuela. *International Journal of Pest Management* 47(4):311–314.

Axelrod, D. 1952. Variables affecting the probabilities of dispersal in geologic time. *Bulletin of the American Museum of Natural History* 99:177–188.

Aymí, R., and G. Gargallo. 2006. Blackcap *Sylvia atricapilla.* Pages 693–694 in *Handbook of the birds of the world.* Volume 11 (J. del Hoyo, A. Elliott, and D. Christie, eds.). Lynx Edicions, Barcelona.

Azara, F. de. 1802–1805. *Apuntamientos para la Historia Natural de los Paxaros del Paraguay y Rio de la Plata.* Volumes 1–3. Imprenta de la Viuda de Ibarra, Madrid.

Bächler, E., S. Hahn, M. Schaub, R. Arlettaz, L. Jenni, J. W. Fox, V. Afanasyev, and F. Liechti. 2010. Year-round tracking of small trans–Saharan migrants using light–level geolocators. *PLoS ONE* 5:e9566.

Bagg, A. M., W. W. H. Gunn, D. S. Miller, J. T. Nichols, W. Smith, and F. P. Wolfarth. 1950. Barometric-pressure patterns and spring migration. *Wilson Bulletin* 62:5–19.

Baird, S. F. 1852. Description of a new species of Sylvicola. *Annals of the Lyceum of Natural History of New York* 5(1):217–218.

Bairlein, F. 1998. The effect of diet composition on migratory fuelling in Garden Warblers *Sylvia borin*. *Journal of Avian Biology* 29:546–551.

Bairlein, F., and E. Gwinner. 1994. Nutritional mechanisms and temporal control of migratory energy accumulation in birds. *Annual Review of Nutrition* 14:187–215.

Baker, A. J. 2002. The deep roots of bird migration: Inferences from the historical record preserved in DNA. *Ardea* 90:503–513.

Baker, A. J., P. M. González, T. Piersma, L. J. Niles, I. de Lima Serrano do Nascimento, P. W. Atkinson, N. A. Clark, C. D. T. Minton, M. K. Peck, and G. Aarts. 2004. Rapid population decline in Red Knots: Fitness consequences of refuelling rates and late arrival in Delaware Bay. *Proceedings of the Royal Society B* 271:875–882.

Baker, C., and M. Baker. 2012. Orchid species culture: daylength for various latitudes. http://www.orchidculture.com/. Accessed September 4, 2012.

Baker, M. C., and E. M. Baker 1973. Niche relationships among six species of shorebirds on their wintering and breeding ranges. *Ecological Monographs* 43:193–212.

Baker, R. R. 1978. *The evolutionary ecology of animal migration.* Hodder and Stougham, London.

Baker, R. R. 1993. The function of post-fledging exploration: A pilot study of three species of passerines ringed in Britain. *Ornis Scandinavica* 24:71–79.

Ball, G. F., and E. D. Ketterson. 2008. Sex differences in the response to environmental cues regulating seasonal reproduction in birds. *Proceedings of the Royal Society B* 363:231–246.

Baltz, M. E., and S. C. Latta. 1998. Cape May Warbler *Setophaga tigrina*. In The birds of North America online (A. Poole, ed.). Cornell Laboratory of Ornithology, Ithaca, NY. http://bna.birds.cornell.edu.bnaproxy.birds.cornell.edu/bna/species/332/articles/introduction. Accessed September 8, 2012.

Baqar, S., C. G. Hayes, J. R. Murphy, and D. M. Watts. 1993. Vertical transmission of West Nile virus by *Culex* and *Aedes* species mosquitoes. *American Journal of Tropical Medicine and Hygiene* 48:757–762.

Barg, J. J., D. M. Aihama, J. Jones, and R. J. Robertson. 2006. Within-territory habitat use and microhabitat selection by male Cerulean Warblers (*Dendroica cerulea*). *Auk* 123:795–806.

Barraclough, R. K., ed. 2006. *Current topics in avian disease research: Understanding endemic and invasive diseases.* Ornithological Monographs 60. American Ornithologists' Union, Washington, DC.

Barrow, W. C., C. C. Chen, R. B. Hamilton, K. Ouchley, and T. J. Spengler. 2000a. Disruption and restoration of en route habitat, a case study: The Chenier Plain. *Studies in Avian Biology* 20:71–87.

Barrow, W. C., R. B. Hamilton, M. A. Powell, and K. Ouchley. 2000b. Contribution of landbird migration to the biological diversity of the Northwest Gulf Coastal Plain. *Texas Journal of Science* 52(4):151–172.

Barry, T. W. 1966. The geese of the Anderson River delta, Northwest Territories. Ph.D. diss., University of Alberta, Edmonton, Canada.

Bartgis, A. C. 1986. A petition to list the Appalachian population of Bewick's Wren as endangered. Maryland Natural Heritage Program, Annapolis, MD.

Bates, H. W. 1863. *The naturalist on the River Amazon*. Murray Press, London.

Bates, J. M. 2000. Allozymic genetic structure and natural habitat fragmentation: Data for five species of Amazonian forest birds. *Condor* 102:770–783.

Batt, B. D. J., A. D. Afton, M. G. Anderson, C. D. Ankney, D. H. Johnson, J. A. Kadlec, and G. L. Krapu, eds. 1992. *Ecology and management of breeding waterfowl*. University of Minnesota Press, Minneapolis.

Bauchinger, U., T. Van't Hof, and H. Biebach. 2007. Testicular development during long distance spring migration. *Hormones and Behavior* 51:295–305.

Bayne, E. M., and K. A. Hobson. 2001. Movement patterns of adult male Ovenbirds during the post-fledging period in fragmented and forested boreal landscapes. *Condor* 103:343–351.

Beadell, J. S., F. Ishtiaq, R. Covas, M. Melo, B. H. Warren, C. T. Atkinson, S. Bensch, G. R. Graves, Y. V. Jhala, M. A. Peirce, A. R. Rahmani, D. M. Fonseca, and R. C. Fleischer. 2006. Global phylogeographic limits of Hawaii's avian malaria. *Proceedings of the Royal Society B* 363:3335–3346.

Bearhop, S., W. Fiedler, R. W. Furness, S. C. Votier, S. Waldron, J. Newton, G. J. Bowen, P. Berthold, and K. Farnsworth. 2005. Assortative mating as a mechanism for rapid evolution of a migratory divide. *Science* 310:502–504.

Beason, R. C. 1978. The influences of weather and topography on water bird migration in the southwestern United States. *Oecologia* 32:153–169.

Bechard, M. J., C. S. Houston, J. H. Sarasola, and A. S. England. 2010. Swainson's Hawk *Buteo swainsoni*. The birds of North America online (A. Poole, ed.). Cornell Laboratory of Ornithology, Ithaca, NY. http://bna.birds.cornell.edu.bnaproxy.birds.cornell.edu/bna/species/265/articles/introduction. Accessed September 8, 2012.

Beddington, J. R. 1974. Age distribution and the stability of simple discrete time population models. *Journal of Theoretical Biology* 47:65–74.

Bednekoff, P. A., and A. I. Houston. 1994. Models of mass-dependent predation, risk-sensitive foraging, and premigratory fattening in birds. *Ecology* 75:1131–1140.

Beebe, W. 1947. Avian migration at Rancho Grande in north-central Venezuela. *Zoologica* 32:153–168.

Beissinger, S. R., J. R. Walters, D. G. Cantanzaro, K. G. Smith, J. B. Dunning, Jr., S. M. Haig, B. R. Noon, and B. M. Stith, eds. 2006. *Modeling approaches in avian conservation and the role of field biologists*. Ornithological Monographs 59. American Ornithologists' Union, Washington, DC.

Bell, C. P. 2000. Process in the evolution of bird migration and pattern in avian ecogeography. *Journal of Avian Biology* 31:258–265.

Bell, C. P. 2005. The origin and development of bird migration: Comments on Rappole and Jones, and an alternative evolutionary model. *Ardea* 93:115–123.

Bellrose, F. C. 1967. Radar in orientation research. *Proceedings of the International Ornithological Congress* 14:281–309.

Bellrose, F. C. 1976. *Ducks, geese, and swans of North America.* Stackpole Books, Harrisburg, PA.

Benkman, C. W. 1993. Adaptation to single resources and the evolution of crossbill (*Loxia*) diversity. *Ecological Monographs* 63:305–325.

Bensch, S., D. Hasselquist, B. Nielsen, and B. Hansson. 1998. Higher fitness for philopatric than for immigrant males in a semi-isolated population of Great Reed Warblers. *Evolution* 53:877–883.

Bent, A. C. 1919–1958. Life histories of North American birds. *U.S. National Museum Bulletin* 107, 113, 121, 126, 130, 135, 142,146, 162, 167, 170, 174, 176, 179, 191, 195–197, 203, 211.

Bent, A. C. 1926. Life histories of North American marsh birds. *U.S. National Museum Bulletin* 135.

Bent, A. C. 1948. Life histories of North American nuthatches, wrens, thrashers, and their allies. *U.S. National Museum Bulletin* 195.

Bent, A. C. 1953. Life histories of North American wood warblers. *U.S. National Museum Bulletin* 203.

Bent, A. C. 1968. Life histories of North American Cardinals, Grosbeaks, Buntings, Towhees, Finches, Sparrows, and their allies. *U.S. National Museum Bulletin* 237.

Berggren, Å, D. P. Armstrong, and R. M. Lewis. 2004. Delayed plumage maturation increases overwinter survival in North Island robins. *Proceedings of the Royal Society B* 271:2123–2130.

Bergman, D., and J. H. Rappole. The ecology of the Northern Pintail (*Anas acuta*) on its wintering grounds in south Texas [unpublished data]. Texas A&I University, Kingsville.

Bermingham, E., S. Rohwer, S. Freeman, and C. Wood. 1992. Vicariance biogeography in the Pleistocene and speciation in North American wood warblers: A test of Mengel's model. *Proceedings of the National Academy of Sciences* 89:6624–6628.

Berthold, P. 1973. Relationships between migratory restlessness and migration distance in six *Sylvia* species. *Ibis* 115:594–599.

Berthold, P. 1988. The control of migration in European warblers. *Proceedings of the International Ornithological Congress* 19:215–249.

Berthold, P. 1993. *Bird migration: A general survey.* Oxford University Press, Oxford.

Berthold, P. 1996. *Control of bird migration.* Chapman and Hall, London.

Berthold, P. 1999. A comprehensive theory for the evolution, control, and adaptability of avian migration. *Ostrich* 70:1–11.

Berthold, P. 2001. *Bird migration.* Oxford University Press, Oxford.

Berthold, P. 1975. Migration: control and metabolic physiology. Pages 77–128 in *Avian Biology.* Volume 5 (D. S. Farner and J. R. King, eds.). Academic Press, New York.

Berthold, P., E. Gwinner, and E. Sonnenschein, eds. 2003. *Avian migration.* Springer, Berlin.

Berthold, P., Kaatz, M., and U. Querner. 2004. Long-term satellite tracking of white stork (*Ciconia ciconia*) migration: Constancy versus variability. *Journal of Ornithology* 145:356–359.

Berthold, P., G. Morh, and U Querner. 1990. Steuerung und potentielle Evolutionsgeschwindigkeit des obligaten Teilzieherverhaltens: Ergebnisse eines Zweiweg-Selektionsexperiments mit der Mönchsgrasmücke (*Sylvia atricapilla*). *Journal of Ornithology* 131:33–45.

Berthold, P., and U. Querner. 1982. Genetic basis of molt, wing length, and body weight in a migratory species, *Sylvia atricapilla. Experientia* 38:801–802.

Berthold, P., and S. B. Terrill. 1988. Migratory behaviour and population growth of Blackcaps wintering in Britain and Ireland: Some hypotheses. *Ringing and Migration* 9:153–159.

Berthold, P., W. van der Bosch, Z. Jacubiek, C. Kaatz, M. Kaatz, and U. Querner. 2002. Long-term satellite tracking sheds light upon variable migration strategies of White Storks (*Ciconia ciconia*). *Journal of Ornithology* 143:489–495.

Beverton, R. J., and S. J. Holt. 1957. On the dynamics of exploited fish populations. *Fishery Investigations, Series II* 19:1–533.

Biebach, H. 1983. Genetic determination of partial migration in the European Robin (*Erithacus rubecula*). *Auk* 100:601–606.

Biebach, H. 1990. Strategies of trans-desert migrants. Pages 352–367 in *Bird migration: Physiology and ecophysiology* (E. Gwinner, ed.). Springer-Verlag, Berlin.

Biebach, H. 1996. Energetics of winter and migratory fattening. Pages 280–323 in *Avian energetics and nutritional ecology* (C. Carey, ed.). Chapman & Hall, New York.

Biebach, H. 1998. Phenotypic organ flexibility in Garden Warblers *Sylvia borin* during long-distance migration. *Journal of Avian Biology* 29:529–535.

Biebach, H., W. Friedrich, and G. Heine. 1986. Interaction of body mass, fat, foraging and stopover period in trans-Sahara migrating passerine birds. *Oecologia* 69:370–379.

Bildstein, K. L. 2006. *Migrating raptors of the world: Their ecology and conservation.* Cornell University Press, Ithaca, NY.

Birdlife International. 2011. Only international action will save migratory birds. http://www.birdlife.org/community/2011/11/only-international-action-will-save-migratory-birds/. Accessed November 29, 2011.

Birkhead, T. R. 2008. *The wisdom of birds: An illustrated history of ornithology.* Bloomsbury, London.

Birkhead, T. R., and A. P. Møller. 1992. *Sperm competition in birds: Evolutionary causes and consequences.* Academic Press, San Diego.

Birkhead, T. R., and A. P. Møller. 1996. Monogamy and sperm competition in birds. Pages 323–343 in *Sperm competition in birds* (J. M. Black, ed.). Oxford University Press, Oxford.

Blacklock, G. W. 1984. *Checklist of birds of the Welder Wildlife Refuge.* Welder Wildlife Foundation, Sinton, TX.

Blake, J. G., G. J. Niemi, and J. M. Hanowski. 1992. Drought and annual variation in bird populations. Pages 419–430 in *Ecology and conservation of Neotropical migrant landbirds* (J. M. Hagan III and D. W. Johnston, eds.). Smithsonian Institution Press, Washington, DC.

Blanchard, B. D. 1941. The White-crowned Sparrows (*Zonotrichia leucophrys*) of the Pacific seaboard: Environment and annual cycle. *University of California Publications in Zoology* 46:1–178.

Bluhm, C. K. 1988. Temporal patterns of pair formation and reproduction in annual cycles and associated endocrinology in waterfowl. *Current Ornithology* 5:123–185.

Blus, L. J., R. G. Heath, C. D. Gish, A. A. Belisle, and R. Prouty. 1971. Eggshell thinning in the brown pelican: implication of DDE. *BioScience* 21:1213–1215.

Böhning-Gaese, K., L. I. González-Guzmán, and J. H. Brown. 1998. Constraints on dispersal and the evolution of the avifauna of the Northern Hemisphere. *Evolutionary Ecology* 12:767–783.

Bosch, I., F. Herrera, J.-C. Navarro, M. Lentino, A. Dupuis, J. Maffei, M. Jones, E. Fernández, N. Pérez, J. Pérez-Emán, A. Erico Guimãres, R. Barrera, N. Valero, J. Ruiz, G. Velásquez, J. Martinez, G. Comach, N. Komar, A. Spielman, and L. Kramer. 2007. West Nile virus, Venezuela. *Emerging Infectious Diseases* 13:651–653.

Boswell, J. 1791. *The life of Samuel Johnson, L.L.D.* Heritage Press Edition, New York [reprinted 1963].

Both, C., and M. E. Visser. 2001. Adjustment of climate change is constrained by arrival date in a long-distance migrant bird. *Nature* 411:296–298.

Both, C., C. A. M. Van Turnhout, R. G. Bilsma, H. Siepel, A. J. Van Strien, and R. P. B. Foppen. 2010. Avian population consequences of climate change are most sever for long-distance migrants in seasonal habitats. *Proceedings of the Royal Society B* 277:1259–1266.

Boucher, J. J., and A. W. Diamond. 2001. *The effects of climate change on migratory birds: An annotated bibliography*. Atlantic Cooperative Wildlife Ecology Research Network, University of New Brunswick, Fredericton, NB, Canada.

Boulet, M., and D. R. Norris. 2006a. Introduction: The past and present of migratory connectivity. Pages 1–13 in *Patterns of migratory connectivity in two nearctic-neotropical songbirds: New insights from intrinsic markers* (M. Boulet and D. R. Norris, eds.). Ornithological Monographs 61. American Ornithologists' Union, Washington, DC.

Boulet, M., and D. R. Norris. 2006b. *Patterns of migratory connectivity in two nearctic-neotropical songbirds: New insights from intrinsic markers*. Ornithological Monographs 61. American Ornithologists' Union, Washington, DC.

Bowlin, M. S., I.-A. Bisson, J. Shamoun-Baranes, J. D. Reichard, N. Sapir, P. P. Marra, T. H. Kunz, D. S. Wilcove, A. Hedenström, C. C. Guglielmo, S. Åkesson, M. Ramenofsky, and M. Wikelski. 2010. Grand challenges in migration biology. *Integrative and Comparative Biology* 50(3):261–279.

Brelsford, A., and D. E. Irwin. 2009. Incipient speciation despite little assortative mating: The Yellow-rumped Warbler hybird zone. *Evolution* 63:3050–3060.

Brewer, R., and K. G. Harrison. 1975. The time of habitat selection by birds. *Ibis* 117:521–522.

Brewster, W. 1886. Bird migration. *Memoirs of the Nuttall Ornithological Club* 1.

Briggs, R. L. 1977. Blue-winged Teal banding project: Panama Canal Zone. *North American Bird Bander* 2:104–105.

British Trust for Ornithology. 2011a. Tracking cuckoos into Africa. http://www.bto.org/science/migration/tracking-studies/cuckoo-tracking. Accessed February 7, 2012.

British Trust for Ornithology. 2011b. Tracking nightingales into Africa. http://www.bto.org/science/migration/tracking-studies/nightingale-tracking. Accessed February 7, 2012.

Brock, V. E., and R. H. Riffenburgh. 1960. Fish schooling: A possible factor in reducing predation. *Journal of Conservation and International Exploration of the Sea* 25:307–317.

Brodeur, S., R. Décarie, D. M. Bird, and M. Fuller. 1996. Complete migration cycle of Golden Eagles breeding in Northern Quebec. *Condor* 98:293–299.

Brodkorb, P. 1971. Origin and evolution of birds. Pages 19–55 in *Avian biology*. Volume 1 (D. S. Farner and J. R. King, eds.). Academic Press, New York.

Bronson, C. L., T. C. Grubb, Jr., G. D. Sattler, and M. J. Braun. 2005. Reproductive success across the Black-capped Chickadee (*Poecile atricapillus*) and Carolina Chickadee (*P. carolinensis*) hybrid zone in Ohio. *Auk* 122:759–772.

Brooks, M. 1938. Bachman's Sparrow in the north-central portion of its range. *Wilson Bulletin* 50:86–109.

Brower, L. P. 1995. Understanding and misunderstanding the migration of the monarch butterfly (Nymphalidae) in North America: 1857–1995. *Journal of the Lepidopterists Society* 49:304–385.

Brower, L. P. 1996. Monarch butterfly orientation: Missing pieces of a magnificent puzzle. *Journal of Experimental Biology* 199:93–103.

Brower, L. P., and S. P. Malcolm. 1991. Animal migrations: Endangered phenomena. *American Zoologist* 31:265–276.

Brown, C. R. 1997. Purple Martin (*Progne subis*). The birds of North America online (A. Poole, ed.). Cornell Laboratory of Ornithology, Ithaca, NY http://bna.birds.cornell.edu.bnaproxy.birds.cornell.edu/bna/species/287/articles/introduction. Accessed September 8, 2012.

Brown, C. R., and M. Bomberger-Brown. 2000. Weather-mediated natural selection on arrival time in Cliff Swallows (*Petrochelidon pyrrhonota*). *Behavioral Ecology and Sociobiology* 47:339–345.

Brown, C. R., and V. A. O'Brien. 2011. *Are wild birds important in the transport of arthropod-borne viruses?* Ornithological Monographs 71. American Ornithologists' Union, Washington, DC.

Brown, J. L. 1964. The evolution of diversity in avian territorial systems. *Wilson Bulletin* 76:160–169.

Brown, J. L. 1969. Territorial behavior and population regulation in birds. *Wilson Bulletin* 81:293–329.

Brown, L. H., E. K. Urban, and K. Newman. 1982. *The birds of Africa*. Volume 1. Academic Press, London.

Bruderer, B., and A. Boldt. 2001. Flight characteristics of birds: Pt. 1. Radar measurements of speeds. *Ibis* 143:178–204.

Bruderer, B., and F. Liechti. 1998. Flight behaviour of nocturnally migratingbirds in coastal areas—crossing or coasting. *Journal of Avian Biology* 29:499–507.

Buechner, H. K., and J. H. Buechner, eds. 1970. A symposium of the Smithsonian Institution on the avifauna of northern Latin America. *Smithsonian Contributions to Zoology* 26.

Buehler, D. M., and T. Piersma. 2008. Travelling on a budget: Prediction and ecological evidence for bottlenecks in the annual cycle of long-distance migrants. *Philosophical Transactions of the Royal Society B* 363:247–266.

Bull, J. 1974. *Birds of New York State*. Doubleday, Garden City, NY.

Burgess, K. 2009. Whooping Crane population declines for first time in almost a decade. *Los Angeles Times*. April 2. http://latimesblogs.latimes.com/outposts/2009/04/whooping-crane-population-declines-for-first-time-in-almost-a-decade.html. Accessed January 25, 2012.

Burleigh, T. D. 1958. *Georgia birds*. University of Oklahoma Press, Norman.

Buskirk, W. H. 1980. Influence of meteorological patterns and trans-Gulf migration on the calendars of migrants to the Neotropics. Pages 485–493 in *Migrant birds in the Neotropics:*

Ecology behavior, distribution, and conservation (A. Keast and E. S. Morton, eds.). Smithsonian Institution Press, Washington, DC.

Buskirk, W. H., G. V. N. Powell, J. F. Wittenberger, R. E. Buskirk, and T. U. Powell. 1972. Interspecific bird flocks in tropical highland Panama. *Auk* 89:612–624.

Butcher, G. S., and S. Rohwer. 1989. The evolution of conspicuous and distinctive coloration for communication in birds. *Current Ornithology* 6:51–108.

Butcher, G. S., W. A. Niering, W. J. Barry, and R. H. Goodwin. 1981. Equilibrium biogeography and the size of nature preserves: an avian case study. *Oecologia* 49:29–37.

Butler, L. K., M. G. Donahue, and S. Rohwer. 2002. Molt-migration in Western Tanagers (*Piranga ludoviciana*): Age effects, aerodynamics, and conservation implications. *Auk* 119:1010–1023.

Butler, P. J. 2000. Stormy seas for some North American songbirds: Are declines related to severe storms during migration? *Auk* 117:518–522.

Butler, P. J., and A. J. Woakes. 1990. The physiology of bird flight. Pages 300–318 in *Bird migration: Physiology and ecophysiology* (E. Gwinner, ed.). Springer-Verlag, Berlin.

Byelich, J. 1976. *Kirtland's Warbler recovery plan.* U.S. Fish and Wildlife Service, Washington, DC.

Cade, T. J., J. H. Enderson, L. F. Kiff, and C. M. White. 1997. Are there enough good data to justify de-listing the American Peregrine Falcon? *Wildlife Society Bulletin* 25:730–738.

Cain, S. A., and G. M. Castro. 1959. *Manual of vegetation analysis.* Harper and Row, New York.

Capparrella, A. P. 1991. Neotropical avian diversity and riverine barriers. *International Ornithologcal Congress* 20:307–316.

Caro, S. P., A. Charmantier, M. M. Lambrechts, J. Blondel, J. Balthazart, and T. D. Williams. 2009. Local adaptation of timing of reproduction: Females are in the driver's seat. *Functional Ecology* 23:172–179.

Castelvecchi, D. 2012. The compass within. *Scientific American* January:48–53.

Caswell, H. 2006. *Matrix population models.* 2nd ed. Wiley, New York.

Catchpole, C. K. 1972. A comparative study of territory in the Reed Warbler (*Acrocephalus scirpaceus*) and the Sedge Warbler (*A. schoenobaenus*). *Journal of Zoology* 166:213–231.

Catesby, M. 1746. Birds of passage. *Philosophical Transactions of the Royal Society* 44:435–444.

Catesby, M. 1748. Bird migration as a fact. *Gentlemen's Magazine.*

Cattermole, P. J. 2000. *Building planet Earth: Five billion years of Earth history.* Cambridge University Press, London.

CBS News. 2012. What's making the 2012 West Nile virus outbreak the worst ever? http://www.cbsnews.com/8301-504763_162-57500089-10391704/whats-making-the-2012-west-nile-virus-outbreak-the-worst-ever/. Accessed September 7, 2012.

Chamberlain, C. P., J. D. Blum, R. T. Holmes, X. Feng, T. W. Sherry, and G. R. Graves. 1997. The use of isotope tracers for identifying populations of migratory birds. *Oecologia* 109:132–141.

Chan, K. 2005. Partial migration in the silvereye (Zosteropidae: Aves): Pattern, synthesis, and theories. *Ethology, Ecology, and Evolution* 17:449–463.

Chandler, C. C. 2011. *Habitat use and survival of neotropical migrant songbirds during the post-fledging period in the White Mountain National Forest.* MS diss., University of Massachusetts, Amherst.

Chandler, C. C., D. I. King, and R. B. Chandler. 2011. Do mature forest birds prefer early-successional habitat during the post-fledging period? *Forest Ecology and Management* 264:1–9.

Chandler, R. B. 2011. *Avian ecology and conservation in tropical agricultural landscapes with emphasis on Vermivora chrysoptera*. Ph.D. diss., University of Massachusetts, Amherst.

Chandler, R. B., and D. I. King. 2012. Dynamic hierarhical models reveal microhabitat specialization across spatial scales for a declining Nearctic migratory bird in the Neotropics. *Journal of Applied Ecology* (in press).

Chandler, W. J. 1986. Migratory bird protection and management. Pages 214–265 in *Audubon wildlife report 1986* (R. L. Di Silvestro, ed.). National Audubon Society, New York.

Chapman, F. M. 1894. Remarks on the origin of bird migration. *Auk* 11:12–17.

Chautauqua County Soil and Water Conservation District. 2010. Chautauqua County Soil and Water Conservation District, Mayville, New York. http://www.soilwater.org/html/about.htm. Accessed 14 November 2011.

Cheke, R. A., J. F. Venn, and P. Jones. 2007. Forecasting suitable breeding conditions for the red-billed quelea *Quelea quelea* in southern Africa. *Journal of Applied Ecology* 44:523–533.

Chen, C. C., W. C. Barrow, Jr., K. Ouchley, and R. B. Hamilton. 2011. Search behavior of arboreal insectivorous migrants at Gulf Coast stopover sites in spring. *Wilson Journal of Ornithology* 123:347–359.

Chen, H., G. J. D. Smith, K. S. Li, J. Wang, X. H. Fan, J. M. Rayner, D. Vijaykrishna, J. X. Zhang, L. J. Zhang, C. T. Guo, C. L. Cheung, K. M. Xu, L. Duan, K. Huang, K. Qin, Y. H. C. Leung, W. L. WU, H. R. Lu, Y. Chen, N. S. Xia, T. S. P. Naipospos, K. Y. Yuen, S. S. Hassan, S. Bahri, T. D. Nguyen, R. G. Webster, J. S. M. Peiris, and Y. Guan. 2006. Establishment of multiple sublineages of H5N1 virus in Asia: Implications for pandemic control. *Proceedings of the National Academy of Sciences* 103:2845–2850.

Chen, H., G. Smith, S. Zhang, K. Qin, J. Wang, K. Li, R. G. Webster, J. S. M. Peiris, and Y. Guan. 2005. H5N1 virus outbreak in migratory waterfowl. *Nature* 436:191–192.

Chernetsov, N. P. 1998. Habitat distribution during the post-breeding and post-fledging period in the Reed Warbler *Acrocephalus scirpaceus* and the Sedge Warbler *A. schoenobaenus* depends on food abundance. *Ornis Svecica* 8:77–82.

Chernetsov, N. P. 2006. Habitat selection by nocturnal passerine migrants en route: Mechanisms and results. *Journal of Ornithology* 147:185–191.

Chernetsov, N. P. 2008. *Migration of passerines: Stopovers and flight*. Habilitation diss. Zoological Institute, St. Petersburg, Russia [in Russian].

Chernetsov, N. P., P. Berthold, and U. Querner. 2004. Migratory orientation of first-year White Storks (*Ciconia ciconia*): Inherited information and social interactions. *Journal of Experimental Biology* 207(6):937–943.

Cherry, J. D. 1985. Early autumn movements and prebasic molt of Swainson's Thrushes. *Wilson Bulletin* 97:368–370.

Chesser, R. T. 1994. Migration in South America: An overview of the austral system. *Bird Conservation International* 4:91–107.

Chesser, R. T. 1997. Patterns of seasonal and geographical distribution of austral migrant flycatchers (Tyrannidae) in Bolivia. Pages 171–204 in *Studies in neotropical ornithology honoring Ted Parker* (J. R. Remsen, Jr., ed.). Ornithological Monographs 48. American Ornithologists' Union, Washington, DC.

Chesser, R. T. 1998. Further perspectives on the breeding distribution of migratory birds: South American austral migrant flycatchers. *Journal of Animal Ecology* 67:69–77.

Chesser, R. T. 2005. Seasonal distribution and ecology of South American austral migrant flycatchers. Pages 168–181 in *Birds of two worlds* (R. Greenberg and P. Marra, eds.). Johns Hopkins University Press, Baltimore.

Chesser, R. T., and D. J. Levey. 1998. Austral migrants and the evolution of migration in New World birds: Diet, habitat, and migration revisited. *American Naturalist* 152:311–319.

Chester-Jones, I., D. Bellamy, D. K. O. Chan, B. K. Follett, I. W. Henderson, J. G. Phillips, and R. S. Snart. 1972. Biological actions of steroid hormones in non-mammalian vertebrates. Pages 414–480 in *Steroids in non-mammalian vertebrates* (D. R. Idler, ed.). Academic Press, New York.

Chiappe, L. M. 2007. *Glorified dinosaurs: The origin and early evolution of birds.* Wiley, New York.

Chilgren, J. D. 1975. *Dynamics and bioenergetics of post-nuptial molt in captive White-crowned Sparrows (Zonotrichia leucophrys gambelii).* Ph.D. diss., Washington State University, Pullman.

Cicero, C., and J. V. Remsen, eds. 2007. *Festschrift for Ned K. Johnson: Geographic variation and evolution in birds.* Ornithological Monographs 63. American Ornithologists' Union, Washington, DC.

Clark, P. 2011. Urban jungle: to save an Ovenbird. *Washington Post,* E2.

Clarke, W. E. 1912. *Studies in bird migration.* Gurney and Jackson, London.

Clegg, S. M., J. F. Kelly, M. Kimura, and T. B. Smith. 2003. Combining genetic markers and stable isotopes to reveal leapfrog migration in a Neotropical migrant, Wilson's Warbler (*Wilsonia pusilla*). *Molecular Ecology* 12:819–830.

Cochran, W. 1987. Orientation and other migratory behaviours of a Swainson's Thrush followed 1500 km. *Animal Behaviour* 35:927–929.

Cochran, W. W., R. R. Graber, and G. G. Montgomery. 1967. Migratory flights of *Hylocichla* thrushes. *Living Bird* 6:213–225.

Cochran, W. W., H. Mouritsen, and M. Wikelski. 2004. Migrating songbirds recalibrate their magnetic compass daily from twilight cues. *Science* 304:405–408.

Cochran, W. W., and M. Wikelski. 2005. Individual migratory tactics of New World *Catharus* thrushes. Pages 274–289 in *Birds of two worlds* (R. Greenberg and P. Marra, eds.). Johns Hopkins University Press, Baltimore.

Cody, M. L. 1966. A general theory of clutch size. *Evolution* 20:174–184.

Cody, M. L. 1984. *Competition and the structure of bird communities.* Princeton University Press, Princeton, NJ.

Coffey, B. B. 1948. Southward migration of herons. *Bird-Banding* 19:1–5.

Cohen, D. 1967. Optimization of seasonal migratory behavior. *American Naturalist* 101:5–18.

Cole, L. C. 1954. The population consequences of life history phenomena. *Quarterly Review of Biology* 29:103–137.

Cole, L. J. 1905. The occurrence of Bewick's Wren, *Thryomanes bewickii* (Aud.), at Grand Rapids. *Bulletin of the Michigan Ornithological Club* 6:8–10.

Coleman, J. S., S. A. Temple, and S. R. Craven. 1997. Cats and wildlife: A conservation dilemma. University of Vermont, Burlington. http://cnre.vt.edu/extension/fiw/wildlife/damage/Cats.pdf. Accessed January 25, 2012.

Colwell, M. A., and J. R. Jehl, Jr. 1994. Wilson's Phalarope (*Phalaropus tricolor*). In The birds of North America online (A. Poole, ed.). Cornell Laboratory of Ornithology, Ithaca, NY. http://bna.birds.cornell.edu.bnaproxy.birds.cornell.edu/bna/species/083/articles/introduction. Accessed September 8, 2012.

Conklin, J. R., P. F. Battley, M. A. Potter, and J. W. Fox. 2010. Breeding latitude drives individual schedules in a trans-hemispheric migrant bird. *Nature Communications* 1, Article 67. doi:10.1038/ncomms 1072.

Cooke, W. W. 1888. *Report of bird migration in the Mississippi Valley for the years 1884 and 1885.* U.S. Department of Agrigulture, Division of Economic Ornithology, Washington, DC.

Cooke, W. W. 1915. *Bird migration.* U.S. Department of Agriculture, Washington, DC.

Coppack, T., and C. Both. 2002. Predicting life-cycle adaptation in migratory birds in response to global climate change. *Ardea* 90(3; Special Issue):369–378.

Coppack, T., and F. Pulido. 2009. Proximate control and adaptive potential of protandrous migration in birds. *Integrative and Comparative Biology* 49:493–506.

Coppack, T., A. P. Tottrup, and C. Spottiswoode. 2006. Degree of protandry reflects level of extrapair paternity in migratory songbirds. *Journal of Ornithology* 147:260–265.

Cottam, C., and E. Higgins. 1946. DDT and its effects on fish and wildlife. *Journal of Economic Entomology* 39:44–52.

Cotton, P. A. 2003. Avian migration phenology and global climate change. *Proceedings of the National Academy of Sciences* 100:12219–12222.

Coues, E. 1872. *Key to North American birds.* Naturalists' Agency, Salem, MA.

Coulson, J. C. and J. Butterfield. 1986. Studies on a colony of colour-ringed Herring Gulls *Larus argentatus*: I. Adult survival rates. *Bird Study* 33:51–54.

Cox, G. W. 1968. The role of competition in the evolution of migration. *Evolution* 22:180–189.

Cox, G. W. 1985. The evolution of avian migration systems between temperate and torpical regions of the New World. *American Naturalist* 126:451–474.

Crawford, R. L. 1981. Bird casualties at a Leon County, Florida TV tower: A 25-year migration study. *Bulletin of Tall Timbers Research Station* 22:1–30.

Cristol, D., M. Baker, and C. Carbone. 1999. Differential migration revisited: Latitudinal segregation by age and sex class. *Current Ornithology* 15:33–38.

Crook, J. H. 1964. The evolution of social organization and visual communication in weaverbirds (Ploceinae). *Behaviour* 10 (Supplemental):1–178.

Crooks, K. R., and M. Sanjayan, eds. 2006. *Connectivity conservation.* Cambridge University Press, Cambridge.

Currie, D., D. B. Thompson, and T. Burke. 2000. Patterns of territory settlement and consequences for breeding success in the Northern Wheatear *Oenanthe oenanthe. Ibis* 142:389–398.

Curry-Lindahl, K. 1958. Internal timer and spring migration in an equatorial migrant, the Yellow Wagtail (*Motacilla flava*). *Ark. Zool.* 11:541–557.

Curry-Lindahl, K. 1963. Molt, body weights, gonadal development and migration in *Motacilla flava. Proceedings of the International Ornithological Congress* 13:960–973.

Curry-Lindahl, K. 1981. *Bird migration in Africa.* Volumes 1 and 2. Academic Press, London.

Dale, S., A. Lunde, and Ø. Steifetten. 2005. Longer breeding dispersal than natal dispersal in the ortolan bunting. *Behavioral Ecology* 16:20–24.

Dalley, K. L., P. D. Taylor, and D. Shutler. 2009. Success of migratory songbirds breeding in harvested boreal forests of northwestern Newfoundland. *Condor* 111:314–325.

Dallimer, M., and P. J. Jones. 2002. Migration orientation behaviour of the red-billed quelea *Quelea quelea*. *Journal of Avian Biology* 33:89–94.

Darwin, C. 1859. *On the origin of species*. J. Murray, London.

Darwin, C. 1871. *The descent of man*. Modern Library Edition. Random House, New York.

Davidson, N. C., and P. R. Evans 1989. Prebreeding accumulation of fat and muscle protein by arctic-breeding birds. *Proceedings of the International Ornithological Congress* 19:342–352.

Davidson, N. C., and T. Piersma. 1992. The migration of knots: Conservation needs and implications. *Wader Study Group Bulletin* 64, Supplement:198–209.

Davis, J. 1973. Habitat preferences and competition of wintering juncos and Golden-crowned Sparrows. *Ecology* 54:174–180.

Davis, L. 1988. Coordination of incubation routines and mate choice in Adelie Penguins (*Pygoscelis adeliae*). *Auk* 105:428–432.

Davis, T. H. 1968. Winter recoveries of Snowy Egrets banded on Long Island. *Bird-Banding* 39:317.

Dawkins, R. 2012. The descent of Edward Wilson. *Prospect* 24 May 2012. http://www.prospectmagazine.co.uk/magazine/edward-wilson-social-conquest-earth-evolutionary-errors-origin-species/. Accessed September 4, 2012.

De Jong, M. J. 1996. Northern Rough-winged Swallow *Stelgidopteryx serripennis*. In The birds of North America online (A. Poole, ed.). Cornell Laboratory of Ornithology, Ithaca, NY. http://bna.birds.cornell.edu.bnaproxy.birds.cornell.edu/bna/species/234/articles/introduction. Accessed September 8, 2012.

Dean, W. R. J. 2004. *Nomadic desert birds*. Springer, London.

DeAngelis, D. L. 1975. Global asymptotic stability criteria for models of density-dependent population growth. *Journal of Theoretical Biology* 50:35–43.

DeAngelis, D. L., L. J. Svoboda, S. W. Christensen, and D. S. Vaughan. 1980. Stability and return times of Leslie matrices with density-dependent survival: Applications to fish populations. *Ecological Modeling* 8:149–163.

DeGraaf, R. M., and J. H. Rappole. 1995. *Neotropical migratory birds: Natural history, distribution, and population change*. Cornell University Press, Ithaca, NY.

del Hoyo, J., A. Elliot, and D. Christie, eds. 1992–2011. *Handbook of the birds of the world*. Volumes 1–16. Lynx Edicions, Barcelona, Spain.

Derrickson, S. R. 1977. *Aspects of breeding behavior in the pintail (Anas acuta)*. Ph. D. diss., University of Minnesota, Minneapolis.

DeSante, D. F. 1983. Annual variability in the abundance of migrant landbirds on southeast Farallon Island, California. *Auk* 100:826–852.

DeSante, D. F., K. Burton, J. F. Saracco, and B. L. Walker. 1995. Productivity indices and survival rate estimates from MAPS, a continent-wide programme of constant-effort mist netting in North America. *Journal of Applied Statistics* 22:935–947.

DeSante, D. F., K. M. Burton, P. Velez, D. Froehlich, and D. Kaschube. 2008. Monitoring avian productivity and survivorship (MAPS) 2008 protocol. http://www.birdpop.org/DownloadDocuments/manual/08/MAPSManual08.pdf. Accessed September 8, 2012.

Di Giacomo, A. S., A. G. Di Giacomo, and J. R. Contreras. 2005. Status and conservation of the Bobolink (*Dolichonyx oryzivorus*) in Argentina. U.S. Forest Service, Washington, DC. http://gis.fs.fed.us/psw/publications/documents/psw_gtr191/Asilomar/pdfs/519–524.pdf. Accessed January 25, 2012.

Dickerman, R. W., W. F. Scherer, A. S. Moorhouse, E. Toaz, M. E. Essex, and R. E. Steele. 1972. Ecologic studies of Venezuelan encephalitis virus in southeastern Mexico: VI. Infection of wild birds. *American Journal of Tropical Medicine and Hygiene* 21:66–78.

Dickey, D. R., and J. van Rossem. 1938. *The birds of El Salvador.* Field Museum of Natural History, Zoological Series 23.

Dingle, H. 1980. Ecology and evolution of migration. Pages 1–101 in *Animal migration, orientation, and navigation* (S. A. Gauthreaux, Jr., ed.). Academic Press, New York.

Dingle, H. 1996. *Migration: The biology of life on the move.* Oxford University Press, New York.

Dingle, H. 2004. The Australo-Papuan bird migration system: Another consequence of Wallace's line. *Emu* 104(2):95–108.

Dinnerstein, E., D. M. Olson, D. J. Graham, A. L. Webster, S. A. Primm, M. P. Bookbinder, and G. Ledec. 1995. *A conservation assessment of the terrestrial ecoregions of Latin America and the Caribbean.* World Bank, Washington, DC.

Dittman, A. H., and T. P. Quinn. 1996. Homing in Pacific salmon: Mechanisms and ecological basis. *Journal of Experimental Biology* 199:83–91.

Dolbeer, R. A. 2006. Height distribution of birds reported by collisions with aircraft. *Journal of Wildlife Managment* 70(5):1345–1350.

Doligez, B., L. Gustafsson, and T. Pärt. 2009. Heritability of dispersal propensity in apatchy population. *Proceedings of the Royal Society B* 276:2829–2836.

Doligez, B., and T. Pärt. 2008. Estimating fitness consequences of dispersal: a road to know-where? Non-random dispersal and the underestimation of dispersers' fitness. *Journal of Animal Ecology* 77:1199–1211.

Dolnik, V. R., and T. I. Blyumental. 1964. Bioenergetika migratsii ptits. *Usp. Sovrem. Biol.* 58:280–301.

Dolnik, V. R., and T. I. Blyumental. 1967. Autumnal premigratory and migratory periods in the Chaffinch an some other temperate–zone passerine birds. *Condor* 69:435–468.

Dolphign, J. 2000–2009. Avibirds European birdguide online. http://www.avibirds.com/ Accessed February 9 2012.

Donovan, T. M., and F. R. Thompson III. 2001. Modeling the ecological trap hypothesis: A habitat and demographic analysis for migrant songbirds. *Ecological Applications* 11:871–882.

Dorst, J. 1962. *The migrations of birds.* Houghton Mifflin, Boston.

Drent, R., C. Both, M. Green, J. Madsen, and T. Piersma. 2003. Pay-offs and penalties of competing migratory schedules. *Oikos* 103:274–292.

Drent, R., and T. Piersma. 1990. An exploration of the energetics of leap-frog migration in arctic breeding waders. Pages 399–412 in *Bird migration: physiology and ecophysiology* (E. Gwinner, ed.). Springer-Verlag, Berlin.

Drilling, N., R. Titman, and F. McKinney. 2002. Mallard (*Anas platyrhynchos*). In The birds of North America online (A. Poole, ed.). Cornell Laboratory of Ornithology, Ithaca, NY. http://bna.birds.cornell.edu.bnaproxy.birds.cornell.edu/bna/species/658/articles/introduction. Accessed September 8, 2012.

Drury, W. H., and J. A. Keith. 1962. Radar studies of songbird migration in coastal New England. *Ibis* 104:449–489.

Ducatez, M. F., Z. Tarnagda, M. C. Tahita, A. Sow, S. de Landtsheer, B. Z. Londt, I. H. Brown, A. D. M. E. Osterhaus, R. A. M. Fouchier, J.-B. B. Quedraogo, and C. P. Mull. 2007. Genetic characterization of HPAI (H5N1) viruses from poultry and wild vultures, Burkina Faso. *Emerging Infectious Diseases* 13:611–614.

Duffy, D., N. Atkins, and D. Schneider. 1981. Do shorebirds compete on their wintering grounds? *Auk* 98:215–219.

Dugger, B. D. 1997. Factors influencing the onset of spring migration in Mallards. *Journal of Field Ornithology* 68:331–337.

Dunning, J. B., Jr. 1993. Bachman's Sparrow *Peucaea aestivalis*. In The birds of North America online (A. Poole, ed.). Cornell Laboratory of Ornithology, Ithaca, NY. http://bna.birds.cornell.edu.bnaproxy.birds.cornell.edu/bna/species/038/articles/introduction. Accessed September 8, 2012.

Dunning, J. B., Jr., and B. D. Watts 1990. Regional differences in habitat occupancy by Bachman's Sparrow. *Auk* 107:463–472.

Dwight, J., Jr. 1900. The sequence of plumages and moults of the passerine birds of New York. *Annals of the New York Academy of Sciences* 13(1):73–160.

Dyke, G., and D. M. Waterhouse. 2001. A mousebird (Aves: Coliiformes) from the Eocene of England. *Journal for Ornithology* 142:7–15.

Dyson, F. 2003. A new Newton. *New York Review of Books* 50(11):4–6.

Eaton, E. H. 1910. *Birds of New York*. Memoir 12, New York State Museum. State University of New York, Albany.

Eaton, S. W. 1953. Wood warblers wintering in Cuba. *Wilson Bulletin* 65:169–174.

Egevang, C., I. J. Stenhouse, R. A. Phillips, A. Petersen, J. W. Fox, and J. R. D. Silk. 2010. Tracking of Arctic terns *Sterna paradisaea* reveals longest animal migration. *Proceedings of the National Academy of Sciences* 107:2078–2081.

Elgood, J. H., C. H. Fry, and R. J. Dorsett. 1973. African migrants in Nigeria. *Ibis* 115:1–45.

Elgood, J. H., J. B. Heigham, A. M. Moore, A. M. Nason, R. E. Sharland, and N. J. Skinner. 1994. *The birds of Nigeria*. 2nd ed. British Ornithologists' Union Checklist No. 4. British Ornithologists' Union, Tring, UK.

Ellis, T., R. Bousfield, L. Bisset, K. Dyrting, G. Luk, S. Tsim, K. Sturm-Ramirez, R. G. Webster, Y. Guan, and J. S. Malik Peiris. 2004. Investigation of outbreaks of highly pathogenic H5N1 avian influenza in waterfowl and wild birds in Hong Kong in late 2002. *Avian Pathology* 33:492–505.

Ely, C. A. 1973. Returns of North American birds to their wintering grounds in southern Mexico. *Bird-Banding* 44:228–229.

Ely, C. A., P. J. Latas, and R. R. Lohefener. 1977. Additional returns and recoveries of North American birds banded in southern Mexico. *Bird-Banding* 48:275–276.

Emlen, J. M. 1973. *Ecology: An evolutionary approach*. Addison Welsey, Reading, MA.

Emlen, J. T. 1973. Territorial aggression in wintering warblers at Bahama agave blossoms. *Wilson Bulletin* 85:71–74.

Emlen, S. T. 1975. Migration: Orientation and navigation. Pages 129–219 in *Avian biology*. Volume 5 (D. S. Farner and J. R. King, eds.). Academic Press, New York.

Emlen, S. T., and J. T. Emlen. 1966. A technique for recording migratory orientation in birds. *Auk* 83:361–367.

Emlen, S. T., and L. R. Oring. 1977. Ecology, sexual selection, and the evolution of mating systems. *Science* 197:215–223.

Emlen, S. T., J. D. Rising, and W. L. Thompson. 1975. A behavioral and morphological study of sympatry in the Indigo and Lazuli buntings of the Great Plains. *Wilson Bulletin* 87:145–179.

England, A. S. 1998. Swainson's Hawk *Buteo swainsoni*. Bureau of Land Management, U.S. Department of the Interior. http://www.blm.gov/ca/pdfs/cdd_ pdfs/swainhawk1.PDF. Accessed January 25, 2012.

Epstein, H. 2011. Flu warning: Beware the drug companies. *New York Review of Books* May 12. http://www.nybooks.com/articles/archives/2011/may/12/flu-warning-beware-drug-companies/. Accessed June 7, 2011.

Erickson, W. P., G. D. Johnson, and D. P. Young, Jr. 2005. *A summary and comparison of bird mortality from anthropogenic causes with an emphasis on collisions*. U.S. Department of Agriculture, Forest Service, Washington, DC.

Ericson, P. G., C. L. Anderson, T. Britton, A. Elzanowski, U. S. Johansson, M. Kallersjo, J. I. Ohlson, T. J. Parsons, D. Zuccon, and G. Mayr. 2006. Diversification of Neoaves: Integration of molecular sequence data and fossils. *Biological Letters* 2:543–547.

Estrada-Franco, J. G., R. Navarro-Lopez, D. W. C. Beasley, L. Coffey, A.-S. Carrara, A. Travassos da Rosa, T. Clements, E. Wang, G. V. Ludwig, A. Campomanes Cortes, P. Paz Ramirez, R. B. Tesh, A. D. T. Barrett, and S. C. Weaver. 2003. West Nile virus in Mexico: Evidence of widespread circulation since July 2002. *Emerging Infectious Diseases* 9:1604–1607.

Ewald, P. W., and S. Rohwer. 1980. Age, coloration and dominance in nonbreeding hummingbirds: A test of the asymmetry hypothesis. *Behavioral Ecology and Sociobiology* 7:273–279.

Ezaki, Y. 1990. Female choice and the causes and adaptiveness of polygyny in great reed warblers. *Journal of Animal Ecology* 59:103–119.

Faaborg, J., R. T. Holmes, A. D. Anders, K. L. Bildstein, K. M. Dugger, S. A. Gauthreaux, Jr., P. Heglund, K. A. Hobson, A. E. Jahn, D. H. Johnson, S. C. Latta, D. J. Levey, P. P. Marra, C. L. Merkord, E. Nol, S. I. Rothstein, T. W. Sherry, T. S. Sillet, F. R. Thompson III, and N. Warnock. 2010a. Conserving migratory landbirds in the New World: Do we know enough? *Ecological Applications* 20:398–418.

Faaborg, J., R. T. Holmes, A. D. Anders, K. L. Bildstein, K. M. Dugger, S. A. Gauthreaux, Jr., P. Heglund, K. A. Hobson, A. E. Jahn, D. H. Johnson, S. C. Latta, D. J. Levey, P. P. Marra, C. L. Merkord, E. Nol, S. I. Rothstein, T. W. Sherry, T. S. Sillet, F. R. Thompson III, and N. Warnock. 2010b. Recent advances in understanding migration systems of New World landbirds. *Ecological Monographs* 80:3–48.

Fahrig, L. 1997. Relative effects of habitat loss and fragmentation on population extinction. *Journal of Wildlife Management* 61:603–610.

Farmer, A. H., and J. A. Wiens. 1998. Optimal migration schedules depend on the landscape and the physical environment: A dynamic modeling view. *Journal of Avian Biology* 29:405–415.

Farner, D. S. 1955. The annual stimulus for migration: Experimental and physiologic aspects. Pages 198–237 in *Recent studies in avian biology* (A. Wolfson, ed.). University of Illinois Press, Urbana.

Farner, D. S. 1958. Photoperiodism in animals with special reference to avian testicular cycles. Pages 17–29 in *19th Annual Biological Colloquium*. Oregon State College, Corvalis.

Feare, C. J., and M. Yasué. 2006. Asymptomatic infection with highly pathogenic avian H5N1 in wild birds: How sound is the evidence? *Virology Journal* 3:96–103.

Fergus, R., M. Fry, W. B. Karesh, P. P. Marra, S. Newman, and E. Paul. 2006. Migratory birds and avian flu. *Science* 312:845.

Fernandez, L. 2002. The paleoclimate record and climate models. University of Michigan Global Change Project, Ann Arbor. http://www.sprl.umich.edu/GCL/Notes-1998-Fall/climate_rec.html. Accessed June 11, 2010.

Ficken, M. S., and R. W. Ficken. 1962. The comparative ethology of wood warblers: A review. *Living Bird* 1:103–118.

Ficken, M. S., and R. W. Ficken. 1967. Differences in the breeding behavior and ecology of the American Redstart. *Wilson Bulletin* 79:188–199.

Fine, A., M. Layton, J. Miller, D. Cimini, M. C. Vargas, and A. Inglesby. 1999. Update: West Nile-like viral encephalitis—New York, 1999. *Morbidity and Mortality Weekly Report* 48(39):890–892.

Finlayson, C. 2011. *Avian survivors: The history and biogeography of Palearctic birds*. T. & A. D. Poyser, London.

Finlayson, C., and J. S. Carrión. 2007. Rapid ecological turnover and its impact on Neanderthal and other human populations. *Trends in Ecology and Evolution* 22:213–222.

Fischer, J. H., U. Munro, and J. B. Phillips. 2003. Magnetic navigation by an avian migrant? Pages 423–432 in *Avian migration* (P. Berthold, E, Gwinner, and E. Sonnenschein, eds.). Springer, Berlin.

Fisher, R. A. 1930. *The genetical theory of natural selection*. Dover, New York.

Flint, V. E., and P. S. Tomkovich. 1982. *Ecological and ethological isolation in the Pectoral Sandpiper and the Sharp-tailed Sandpiper*. Pages 241–261. Ornithological Studies in the USSR. Volume 2. Nauka, Moscow.

Fodor, J., and M. Piattelli-Palmarini. 2009. *What Darwin got wrong*. Farrar, Strauss and Giroux, New York.

Fogden, M. P. 1972. The seasonality and population dynamics of equatorial birds in Sarawak. *Ibis* 114:307–343.

Forbush, E. H., and J. R. May. 1939. *Natural history of the birds of eastern and central North America*. Houghton Mifflin, New York.

Foster, M. S. 1967. Molt cycles of the Orange-crowned Warbler. *Condor* 69:169–200.

Francis, C. M., and F. Cooke. 1986. Differential timing of spring migration in wood warblers (Parulinae). *Auk* 103:548–556.

Fransson, T. 1998. Patterns of fuelling in Whitethroats *Sylvia communis* in relation to departure. *Journal of Avian Biology* 29:569–573.

Fransson, T., and B.-O. Stolt. 1993. Is there an autumn migration of continental Blackcaps (*Sylvia atricapilla*) into northern Europe. *Die Vogelwarte* 37:89–95.

Freemark, K. E., and B. Collins. 1992. Landscape ecology of birds breeding in temperate forest fragments. Pages 443–454 in *Ecology and conservation of Neotropical migrant landbirds* (J. M. Hagan III and D. W. Johnston, eds.). Smithsonian Institution Press, Washington, DC.

Fretwell, S. D. 1972. *Populations in a seasonal environment*. Princeton University Press, Princeton, NJ.

Frith, H. J. 1969. *Birds in the Australian high country*. A. H. & A. W. Reed, Sydney.

Fromme-Bouman, H. 1962. Jahresperiodische Untersuchungen und der Nebennieren-rinde der Amsel (*Turdus merula*). *Die Vogelwarte* 21:188–198.

Fry, C. H., I. J. Ferguson-Lees, and R. J. Dowsett. 1972. Flight muscle hypertrophy and ecophysiological variation of Yellow Wagtail *Motacilla flava* races at Lake Chad. *Journal of Zoology* 167:293–306.

Fry, C. H., S. Keith, and E. K. Urban. 1988. *The birds of Africa*. Volume 3. Academic Press, New York.

Fry, C. H., S. Keith, and E. K. Urban. 2000. *The birds of Africa*. Volume 6. Princeton University Press, Princeton, NJ.

Fry, C. H., S. Keith, L. Brown, and E. K. Urban. 2004. *The birds of Africa*. Volume 7. Princeton University Press, Princeton, NJ.

Fuller, M., D. Holt, and L. S. Schueck. 2003. Snowy Owl movements: Variation on the migration theme. Pages 359–363 in *Avian migration* (P. Berthold, E. Gwinner, and E. Sonnenschein, eds.). Springer-Verlag, Berlin.

Gaidet, N., T. Dodman, A. Caron, G. Balança, S. Desvaux, F. Goutard, G. Cattoli, F. Lamarque, W. Hagemeijer, and F. Monicat. 2007. Avian influenza viruses in water birds, Africa. *Emerging Infectious Diseases* 13:626–629.

Galli, A. E., C. F. Leck, and R. T. T. Forman. 1976. Avian distribution patterns in forest islands of different sizes in central New Jersey. *Auk* 93:356–364.

Gätke, H. 1891. *Die Vogelwarte. Braunschweig*. Meier Verlag, Helgoland.

Gauthier-Clerc, M., C. Lebarbenchon, and F. Thomas. 2007. Recent expansion of highly pathogenic avian influenza H5N1: A critical review. *Ibis* 149:202–214.

Gauthreaux, S. A., Jr. 1971. A radar and direct visual study of passerine spring migration in southern Louisiana. *Auk* 88:343–365.

Gauthreaux, S. A., Jr. 1978a. The influence of global climatological factors on the evolution of bird migratory pathways. *Proceedings of the International Ornithological Congress* 17:517–525.

Gauthreaux, S. A., Jr. 1978b. The ecological significance of behavioral dominance. Pages 17–54 in *Perspectives in ethology* (P. P. G. Bateson and P. H. Klopfer, eds.). Plenum, New York.

Gauthreaux, S. A., Jr. ed. 1980. *Animal migration, orientation, and navigation*. Academic Press, New York.

Gauthreaux, S. A., Jr. 1982. The ecology and evolution of avian migration systems. Pages 93–168 in *Avian biology*, Volume 6 (D. S. Farner, and J. R. King, eds.). Academic Press, New York.

Gauthreaux, S. A., Jr., and C. G. Belser. 1999. Bird migration in the region of the Gulf of Mexico. *Proceedings of the International Ornithological Congress* 22:1931–1947.

Gauthreaux, S. A., Jr., C. G. Belser, and C. M. Welch. 2006. Atmospheric trajectories and spring bird migration across the Gulf of Mexico. *Journal of Ornithology* 147:317–325.

Gavrilov, V. M. 1974. Metabolism of molting birds. *Zool. Zh.* 53:1363–1375.

George, J. C., and A. J. Berger. 1966. *Avian myology*. Academic Press, New York.

George, R. R., R. E. Tomlinson, R. W. Engel-Wilson, G. L. Waggerman, and A. G. Spratt. 1994. White-winged dove. Pages 29–50 in *Migratory shore and upland game bird management in North America* (T. C. Tacha and C. E. Braun, eds.). Allen Press, Lawrence, KS.

George, R. R. 2000. *Migration, harvest, and population dynamics of white-winged doves banded in Texas and northeastern Mexico, 1950–1978.* Texas Parks and Wildlife Department, Austin.

Gilbert, M., X. Xiao, J. Domenech, J. Lubroth, V. Martin, and J. Singenbergh. 2006. Anatidae migration in the western Palearctic and spread of highly pathogenic avian influenza H5N1 virus. *Emerging Infectious Diseases* 12:1650–1656.

Gill, F. B. 2007. *Ornithology.* Freeman, New York.

Gill, F. B., and M. Wright. 2006. *Birds of the world.* Princeton University Press, Princeton, NJ.

Gill, R. E., Jr., C. A. Babcock, C. M. Handel, W. R. Butler, Jr., and D. G. Raveling. 1996. Migration, fidelity, and use of autumn staging grounds in Alaska by Cackling Canada Geese *Branta canadensis* minima. *Wildfowl* 47:42–61.

Gill, R. E., Jr., P. Canevari, and E. H. Iversen. 1998. Eskimo Curlew *Numenius borealis.* In The birds of North America online (A. Poole, ed.). Cornell Laboratory of Ornithology, Ithaca, NY. http://bna.birds.cornell.edu.bnaproxy.birds.cornell.edu/bna/species/347/articles/introduction. Accessed September 8, 2012.

Gill, R. E., Jr., T. L. Tibbitts, D. C. Douglas, C. M. Handel, D. M. Mulcahy, J. C. Gottschalck, N. Warnock, B. J. McCaffery, P. F. Battley, and T. Piersma. 2009. Extreme endurance flights by landbirds crossing the Pacific Ocean: ecological corridor rather than barrier? *Proceedings of the Royal Society B* 276:447–457.

Gillis, E. A., D. J. Green, H. A. Middleton, and C. A. Morrissey. 2008. Life history correlates of alternative migratory strategies in American dippers. *Ecology* 89:1687–1695.

Gilpin, M. E., and F. J. Ayala. 1973. Global models of growth and competition. *Proceedings of the National Academy of Sciences* 70:3590–3593.

Ginn, H. B., and D. S. Melville. 1983. *Moult in birds.* British Trust for Ornithology, Hertfordshire, UK.

Gleason, H. A. 1922. On the relationship between species and area. *Ecology* 3:158–162.

Glenn, T. C., W. Stephan, and M. J. Braun. 1999. Effects of a population bottleneck on Whooping Crane mitochondrial DNA variation. *Conservation Biology* 13:1097–1107.

Gluck, E. 1984. Habitat selection in birds and the role of early experience. *Z. tierpsychol.* 66:45–54.

Gordo, O. 2007. Why are bird migration dates shifting? A review of weather and climate effects on avian migratory phenology. *Climate Research* 35:37–58.

Gorman, M. L., and H. Milne. 1971. Seasonal changes in the adrenal steroid tissue of the Common Eider *Somateria mollissima* and its relation to organic metabolism in normal and oil-polluted birds. *Ibis* 113:218–228.

Gosler, A. G. 1996. Environmental and social determinants of winter fat storage in the Great Tit. *Journal of Animal Ecology* 65:1–17.

Goss-Custard, J. D., R. G. Caldow, R. T. Clarke, S. E. A. le V. dit Durell, and W. J. Sutherland. 1995a. Deriving population parameters from individual variations in foraging behaviour. Empirical game theory distribution model of oystercatchers *Haematopus ostralegus* feeding on mussels *Mytilus edulis. Journal of Animal Ecology* 64:265–276.

Goss-Custard, J. D., R. T. Clarke, S. E. A. le V. dit Durell, R. G. Caldow, and B. J. Ens. 1995b. Population consequences of winter habitat loss in a migratory shorebird. I. Estimating model parameters. *Journal of Applied Ecology* 32:337–351.

Gould, S. J., and E. S. Vrba. 1982. Exaptation—a missing term in the science of form. *Paleobiology* 8:4–15.

Gourley, R. S., and C. E. Lawrence. 1977. Stable population analysis in periodic environments. *Theoretical Population Biology* 11:49–59.

Gowaty, P. A. 1993. Differential dispersal, local resource competition, and sex ratio variation in birds. *American Naturalist* 141:263–280.

Grant, G. S., and T. L. Quay. 1970. Sex and age criteria in the Slate-colored Junco. *Bird-Banding* 41:274–278.

Graves, G. R., C. S. Romanek, and A. Rodriguez Navarro. 2002. Stable isotope signature of philopatry and dispersal in a migratory songbird. *Proceedings of the National Academy of Sciences* 99:8096–8100.

Greenberg, R. 1981. Dissimilar bill shapes in New World tropical versus temperate forest foliage-gleaning birds. *Oecologia* 49:143–147.

Greenberg, R. 1984. *The winter exploitation systems of Bay-breasted and Chestnut-sided warblers in Panama.* University of California Publications in Zoology 116.

Greenberg, R. 1986. Competition in migrant birds in the nonbreeding season. *Current Ornithology* 3:281–307.

Greenberg, R., P. Bichier, A. C. Angon, and R. Reitsma. 1997. Bird populations in shade and sun coffee in Guatemala. *Conservation Biology* 11:448–459.

Greenberg, R., and J. Gradwohl. 1980. Observations of paired Canada Warblers *Wilsonia canadensis* during migration in Panama. *Ibis* 122:509–512.

Greenberg, R., and P. Marra, eds. 2005. *Birds of two worlds.* Johns Hopkins University Press, Baltimore, MD.

Greenewalt, C. H. 1962. Dimensional relationships for flying animals. *Smithsonian Miscellaneous Collections* 144:1–46.

Greenwood, P. J. 1980. Mating systes, philopatry, and dispersal in birds and mammals. *Animal Behaviour* 28:1140–1162.

Greenwood, P. J., and P. H. Harvey. 1982. The natal and breeding dispersal of birds. *Annual Review of Ecology and Systematics* 13:1–21.

Gregory, R. D., D. W. Gibbons, and P. F. Donald. 2004. Bird census and survey techniques. Pages 17–55 in *Bird ecology and conservation: A handbook of techniques* (W. J. Sutherland, I. Newton, and R. E. Green, eds.). Oxford University Press, Oxford.

Gregory, R. D., N. I. Wilkinson, D. G. Noble, J. A. Robinson, A. F. Brown, J. Hughes, D. Procter, D. W. Gibbons, and C. A. Galbraith. 2002. The population status of birds in the United Kingdom, Channel Islands, and Isle of Man. *British Birds* 95:410–448.

Grimes, L. G. 1987. *The birds of Ghana.* British Ornithologists' Union Checklist No. 9. British Ornithologists' Union, Tring, UK.

Groebbels, F. 1928. Zur Physiologie der Vogelzuges. *Verh. Ornithol. Ges. Bayern* 18:44–74.

Groebbels, F. 1937. *Der Vogel. II. Geschlecht und Fortpflanzung.* Gebruder Borntraeger, Berlin.

Grzybowski, J. A. 1991. *Black-capped Vireo recovery plan.* Region 2 Office, U.S. Fish and Wildlife Service, Albuquerque, NM.

Gudmundsson, G., and T. Alerstam. 1998a. Why is there no transpolar bird migration? *Journal of Avian Biology* 29:93–96.

Gudmundsson, G., and T. Alerstam 1998b. Optimal map projection for analysing long-distance migration routes. *Journal of Avian Biology* 29:597–605.

Guilford, T., J. Meade, J. Willis, R. A. Phillips, D. Boyle, S. Roberts, M. Collett, R. Freeman, and C. M. Perrins. 2009. Migration and stopover in a small pelagic seabird, the Manx shearwater *Puffinus puffinus*: insights from machine learning. *Proceedings of the Royal Society B* 276:1215–1223.

Gunnarsson, T. G., J. A. Gill, T. Sigurbjörnsson, and W. J. Sutherland. 2004. Arrival synchrony in migratory birds. *Nature* 431:646.

Gwinner, E. 1968. Circannuale Periodik al Grundlage des jahreszeitlichen Funktionswandels bei Zugvöfeln. Untersuchungen am Fitis (*Phylloscopus trochilus*) un am Waldlaubsänger (*P. sibilatrix*). *Journal of Ornithology* 109:70–95.

Gwinner, E. 1971. Orientierung. Pages 299–348 in *Grundriss der Vogelzugkunde* (E. Schüz, ed.). Springer, Berlin.

Gwinner, E. 1972. Endogenous factors in bird migration. *NASA Special Publication* 262:321–338.

Gwinner, E. 1975. Circadian and circannual rhythms in birds. Pages 221–285 in *Avian biology*. Volume 5 (D. S. Farner and J. R. King, eds.). Academic Press, New York.

Gwinner, E. 1986. *Circannual rhythms. Endogenous annual clocks in the organization of seasonal processes.* Springer, Berlin.

Gwinner, E., ed. 1990. *Bird migration: physiology and ecophysiology.* Springer-Verlag, Berlin.

Gwinner, E., and D. Czeschlik. 1978. On the significance of spring migratory restlessness in caged birds. *Oikos* 30:364–372.

Gwinner, E., and B. Helm. 2003. Circannual and ciradian contributions to the timing of avian migration. Pages 81–95 in *Avian migration* (P. Berthold, E. Gwinner, and E. Sonnenschein, eds.). Springer, Heidelberg.

Gwinner, E., S. König, and C. S. Haley. 1995. Genetic and environmental factors influencing clutch size in equatorial and temperate zone Stonechats (*Saxicola torquata axillaris* and *S. t. rubicola*): An experimental study. *Auk* 112:748–755.

Hagan, J. M., III, and D. W. Johnston, eds. 1992. *Ecology and conservation of Neotropical migrant landbirds.* Smithsonian Institution Press, Washington, DC.

Hagar, C., and F. M. Packard. 1952. *Checklist of the birds of the Central Coast of Texas.* Privately printed, Corpus Christi, TX [archives of the Welder Wildlife Foundation, Sinton, TX].

Haig, S. M., C. L. Gatto-Trevor, D. T. Mullins, and M. A. Colwell. 1997. Population identification of Western Hemisphere shorebirds throughout the annual cycle. *Molecular Ecology* 6(5):413–427.

Hamel, P. B. 1995. Bachman's Warbler *Vermivora bachmanii*. In The birds of North America online (A. Poole, ed.). Cornell Laboratory of Ornithology, Ithaca, NY. http://bna.birds.cornell.edu.bnaproxy.birds.cornell.edu/bna/species/150/articles/introduction. Accessed September 8, 2012.

Hamilton, T. H. 1961. On the functions and causes of sexual plumage in breeding plumage characters of North American species of warblers and orioles. *American Naturalist* 95:121–123.

Hamilton, W. J. I. 1962. Evidence concerning the function of nocturnal call notes of migratory birds. *Condor* 64:390–401.

Hamilton, W. J. I. 1966. Social aspects of bird orientation mechanisms. Pages 57–72 in *Animal orientation and navigation* (R. M. Storm, ed.). Oregon State University Press, Corvalis.

Hamilton, W. J. I., and W. M. Gilbert. 1969. Starling dispersal from a winter roost. *Ecology* 50:886–898.

Hamilton, W. J. I., W. M. Gilbert, F. H. Heppner, and R. J. Planck. 1967. Starling roost dispersal and a hypothetical mechanism regulating rhythmical animal movement to and from dispersal centers. *Ecology* 48:825–833.

Haney, J. C., D. S. Lee, and M. Walsh-McGeHee. 1998. A quantitative analysis of winter distribution and habitats of Kirtland's Warblers in the Bahamas. *Condor* 100:201–217.

Hann, H. W. 1937. Life history of the Oven-bird in southern Michigan. *Wilson Bulletin* 49:145–237.

Harari, A. R., D. Ben-Yakir, and D. Rosen. 2000. Male pioneering as a mating strategy: The case of the beetle *Maladera matrida*. *Ecological Entomology* 25:387–394.

Harrington, B. A. 1996. *The flight of the Red Knot*. W. W. Norton, New York.

Harrison, J. L. 1962. The distribution of feeding habits among animals in a tropical rain forest. *Journal of Animal Ecology* 31:53–63.

Hayes, F. E., P. A. Scharf, and R. S. Ridgely. 1994. Austral bird migrants in Paraguay. *Condor* 96:83–97.

Hayslette, S. E., T. C. Tacha, and G. L. Waggerman. 1996. Changes in white-winged dove reproduction in southern Texas, 1954-1993. *Journal of Wildlife Management* 50:298–301.

Hayslette, S. E., T. C. Tacha, and G. L. Waggerman. 2000. Factors affecting white-winged, white-tipped, and mourning dove reproduction in lower Rio Grande Valley. *Journal of Wildlife Management* 54:286–295.

Haywood, S. 1993a. Role of extrinsic factors in the control of clutch size in the Blue Tit *Parus caeruleus*. *Ibis* 135:79–84.

Haywood, S. 1993b. Sensory and hormonal control of clutch size in birds. *Quarterly Review of Biology* 68:33–60.

Healy, S., and W. A. Calder. 2006. Rufous Hummingbird (*Selasphorus rufus*). In The birds of North America online (A. Poole, ed.). Cornell Laboratory of Ornithology, Ithaca, NY. http://bna.birds.cornell.edu.bnaproxy.birds.cornell.edu/bna/species/053/articles/introduction. Accessed September 8, 2012.

Hedenström, A., and T. Alerstam. 1998. How fast can birds migrate? *Journal of Avian Biology* 29:424–432.

Hegemann, A., H. P. van der Jeugd, M. de Graaf, L. L. Oostebrink, and B. I. Tieleman. 2010. Are Dutch Skylarks partial migrants? Ring recovery data and radio-telemetry suggest local coexistence of contrasting migration strategies. *Ardea* 98:135–145.

Heinroth, O. 1911. Beitrage zur Biologie, namentlich Ethologie und Psychologie der Anatiden. *Proceedings of the International Ornithological Congress* 5:589–702.

Heise, C. D., and C. C. Rimmer. 2000. Definitive Prebasic molt of Gray Catbirds at two sites in New England. *Condor* 102:894–904.

Heitmayr, M. E. 1987. The Prebasic moult and basic plumage of female Mallards (*Anas platyrhynchos*). *Canadian Journal of Zoology* 65:2248–2261.

Helbig, A. J. 1991. Inheritance of migratory direction in a bird species: a cross-breeding experiment with SE- and SW-migrating blackcaps (*Sylvia atricapilla*). *Behavioral Ecology and Sociobiology* 28:9–12.

Helbig, A. J. 1996. Genetic basis, mode of inheritance and evolutionary changes of migratory directions in Palaearctic warblers (Aves: Sylviidae). *Journal of Experimental Biology* 199:49–55.

Helbig, A. J. 2003. Evolution of bird migration: A phylogenetic and biogeographic perspective. Pages 3–21 in *Avian migration* (P. Berthold, E. Gwinner, and E. Sonnenschein, eds.). Springer, Heidelberg.

Heldbjerg, H., and A. D. Fox. 2008. Long-term population declines in Danish trans-Saharan migrant birds. *Bird Study* 55:267–279.

Helm, B. 2003. Seasonal timing in different environments: Comparative studies in stonechats. Ph.D. diss., University of Munich, Munich.

Helm, B. 2006. Zugunruhe of migratory and non-migratory birds in circannual context. *Journal of Avian Biology* 37:533–540.

Helm, B., and E. Gwinner 2005. Carry-over effects of day length during spring migration. *Journal of Ornithology* 146:348–354.

Helm, B., and E. Gwinner. 2006a. Migratory restlessness in an equatorial nonmigratory bird. *PLoS Biology* 4:611–614.

Helm, B., and E. Gwinner. 2006b. Migratory restlessness in an equatorial nonmigratory bird. *PLoS Biology* 4:611–614.

Helm, B., E. Gwinner, and L. Trost. 2005. Flexible seasonal timing and migratory behavior: results from Stonechat breeding programs. *Annals of the New York Academy of Sciences* 1046:216–227.

Helm, B., T. Piersma, and H. Van der Jeugd. 2006. Sociable schedules: Interplay between avian social and seasonal behavior. *Animal Behaviour* 72:245–262.

Helms, C. W. 1963. The annual cycle and Zugunruhe in birds. *Proceedings of the International Ornithological Congress* 13:925–939.

Helms, C. W., and W. H. Drury. 1960. Winter and migratory fat field studies on some North American buntings. *Bird-Banding* 31:1–40.

Heneberg, P. 2006. Migration behaviour of mute swans (*Cygnus olor*) wintering in ýeské BudČjovice, Czech Republic. *Linzer Biologische Beitrage* 38(2):1–10.

Henny, C. J., and G. B. Herron. 1989. DDE selenium mercury and White-faced Ibis reproduction at Carson Lake, Nevada, USA. *Journal of Wildlife Management* 48:1–13.

Hensley, M. M., and J. B. Cope. 1951. Further data on removal and repopulation of the breeding birds in a spruce-fir community. *Auk* 68:483–493.

Herder, F., J. Pfaender, and U. Schliewen. 2008. Adaptive sympatric speciation of polychromatic roundfin sailfin silverside fish in Lake Matano (Sulawesi). *Evolution* 62:2178–2195.

Herrera, C. M. 1978. Ecological correlates of residence and non-residence in a Mediterranean passerine bird community. *Journal of Animal Ecology* 47:871–890.

Herzog, P., and D. M. Keppie. 1980. Migration in a local population of Spruce Grouse. *Condor* 82:366–372.

Hesse, R. 1922. Die Bedeutung der Tagesdauer für die Vögel. *Sitzber. Nathist. Ver. Bonn* 1922–1923:A13–A17.

Hill, R. D. 1994. Theory of geolocation by light levels. Pages 227–236 in *Elephant seals: population ecology, behavior and physiology* (B. J. Le Boeuf and R. M. Laws, eds.). University of California Press, Berkeley.

Hinde, R. A. 1956. The biological significance of the territories of birds. *Ibis* 98:340–369.

Hobson, K. A. 1995. Stable isotopes and the determination of avian migratory connectivity and seasonal interactions *Auk* 122:1037–1048.

Hobson, K. A. 2005. Flying fingerprints: making connections with stable isotopes and trace elements. Pages 235–246 in *Birds of two worlds* (R. Greenberg and P. Marra, eds.), Johns Hopkins University Press, Baltimore, MD.

Hobson, K. A. 2008. Applying isotopic methods to tracking animal movements. Pages 45–78 in *Tracking animal migration with stable isotopes* (K. A. Hobson and L. I. Wassenaar, eds.). Elsevier, Amsterdam.

Hobson, K. A., and L. I. Wassenaar. 1997. Linking breeding and wintering grounds of neotropical migrant songbirds using stable hydrogen isotopic analysis of feathers. *Oecologia* 109:142–148.

Hobson, K. A., and L. I. Wassenaar, eds. 2008. *Tracking animal migration with stable isotopes*. Academic Press, London.

Hochachka, P. W., J. R. Neely, and W. R. Driedzic. 1977. Integration of lipid utilization with Krebs cycle activity in muscle. *Proceedings of the Federation of the American Societies for Experimental Biology* 36(7):2009–2014.

Holberton, R., and A. M. Dufty, Jr. 2005. Hormones and variation in life history strategies of migratory and nonmigratory birds. Pages 290–302 in *Birds of two worlds* (R. Greenberg and P. Marra, eds.). Johns Hopkins University Press, Baltimore.

Holberton, R., and A. M. Dufty, Jr. 2005. Hormones and variation in life history strategies of migratory and nonmigratory birds. Pages 290–302 in *Birds of two worlds* (R. Greenberg and P. Marra, eds.). Johns Hopkins University Press, Baltimore.

Holberton, R., J. D. Parrish, and J. C. Wingfield. 1996. Modulation of the adrenocortical stress response in neotropical migrants during autumn migration. *Auk* 113:558–564.

Holland, R. A., M. Wikelski, F. Kümmeth, and C. Bosque. 2009. The secret life of Oilbirds: New insights into the movement ecology of a unique avian frugivore. *PLoS ONE* 4:1–11.

Holmes, R. T., and F. A. Pitelka 1998. Pectoral Sandpiper (*Calidris melanotos*). In The birds of North America online (A. Poole, ed.). Cornell Laboratory of Ornithology, Ithaca, NY. http://bna.birds.cornell.edu.bnaproxy.birds.cornell.edu/bna/species/348/articles/introduction. Accessed September 8, 2012.

Holmes, R. T., N. L. Rodenhouse, and T. S. Sillett. 2005. Black-throated Blue Warbler (*Setophaga caerulescens*). In The birds of North America online (A. Poole, ed.). Cornell Laboratory of Ornithology, Ithaca, NY. http://bna.birds.cornell.edu.bnaproxy.birds.cornell.edu/bna/species/087/articles/introduction. Accessed September 8, 2012.

Holmes, R. T., J. C. Schultz, and P. Nothnagle. 1979. Bird predation on forest insects: An exclosure experiment. *Science* 206:462–463.

Holmes, R. T., and T. W. Sherry. 1992. Site fidelity of migratory warblers in temperate breeding and Neotropical wintering areas: Implications for population dynamics, habitat selection, and conservation. Pages 563–575 in *Ecology and conservation of Neotropical migrant landbirds* (J. M. Hagan III and D. W. Johnston, eds.). Smithsonian Institution Press, Washington, DC.

Holmes, R. T., T. W. Sherry, and L. Reitsma. 1989. Population structure, territoriality, and overwinter survival of two migrant warbler species in Jamaica. *Condor* 91:545–561.

Holmes, R. T., T. W. Sherry, and F. W. Sturges. 1986. Bird community dynamics in a temperate deciduous forest: Long-term trends at Hubbard Brook. *Ecological Monographs* 56:201–220.

Houston, A. I. 1998. Models of optimal avian migration: State, time and predation. *Journal of Avian Biology* 29:395–404.

Hubálek, Z. 2003. Spring migration of birds in relation to North Atlantic Oscillation. *Folia Zoologica* 52:287–298.

Hubálek, Z. 2004. An annotated checklist of pathogenic microorganisms associated with migratory birds. *Journal of Wildlife Diseases* 40:639–659.

Hubálek, Z., and J. Halouzka. 1999. West Nile fever—a reemerging mosquito-borne viral disease in Europe. *Emerging Infectious Diseases* 5:643–650.

Hughes, J. M. 2009. *The migration of birds: seasons on the wing.* Firefly, Buffalo, NY.

Humphrey, P. S., and K. C. Parkes. 1959. An approach to the study of molts and plumages. *Auk* 76:1–31.

Hüppop, O., and K. Hüppop. 2003. North Atlantic oscillation and timing of spring migration in birds. *Proceedings of the Royal Society B* 270:233–240.

Hustings, F. 1988. *European monitoring studies of breeding birds.* SOVON, Dutch Center for Field Ornithology, Beek, The Netherlands.

Hutchinson, G. E. 1957. Concluding remarks. *Cold Spring Harbor Symposium on Quantitative Biology* 22:415–427.

Hutto, R. L. 1988. Is tropical deforestation responsible for the reported declines in Neotropical migrant populations? *American Birds* 42:375–379.

Hutto, R. L. 1994. The composition and social organization of mixed-species flocks in a tropical deciduous forest in western Mexico. *Condor* 96:105–118.

Hutto, R. L. 2000. On the importance of *en route* periods to the conservation of migratory landbirds. *Studies in Avian Biology* 20:109–114.

Igual, J. M., M. G. Forero, G. Tavecchia, J. Gonzalez-Solis, A Martinez-Abrain, K. A. Hobson, X. Ruiz, and D. Oro. 2005. Short-term effects of data-loggers on Cory's Shearwater (*Calonectris diomedea*). *Marine Biology* 146:619–624.

International Union for Conservation of Nature. 2011. http://www.iucn.org/. Accessed June 7, 2011.

Irwin, D. E., and J. H. Irwin. 2005. Siberian migratory divides—the role of seasonal migration in speciation. Pages 27–40 in *Birds of two worlds* (R. Greenberg and P. Marra, eds.). Johns Hopkins University Press, Baltimore.

Jacobsen, E. M. 1991. *Ynglefuglerapport 1990.* Dansk Ornitologisk Forening, Copenhagen.

Jaeger, M. M., R. L. Bruggers, B. E. Johns, and W. A. Erickson. 1986. Evidence of itinerant breeding of the red-billed quelea *Quelea quelea* in the Ethiopian Rift Valley. *Ibis* 128:469–482.

Jallageas, M., and I. Assenmacher. 1979. Further evidence for reciprocal interactions between the annual and sexual thyroid cycles in male Peking ducks. *General Comparative Endocrinology* 37:44–51.

James, F. C., and H. H. Shugart, Jr. 1974. The phenology of the nesting season of the American Robin (*Turdus migratorius*) in the United States. *Condor* 76:159–168.

James, F. C., D. A. Wiedenfeld, and C. E. McCulloch. 1992. Trends in breeding populations of warblers: Declines in the southern highlands and increases in the lowlands. Pages 43–56 in *Ecology and conservation of Neotropical migrant landbirds* (J. M. Hagan III and D. W. Johnston, eds.). Smithsonian Institution Press, Washington, DC.

James, P. 1956. Destruction of warblers on Padre Island, Texas, in May 1951. *Wilson Bulletin* 68:224–227.

Janda, J., K. Štastný, and R. Fuchs. 1990. Czechoslovak breeding bird counts: New indices for 1981–88. Pages 427–428 in *Bird census and atlas studies. Proceedings of the 11th International Conference on Bird Census and Atlas Work* (K. Štastný and V. Bejček, eds.). Czech Ornithological Society, Prague.

Jehl, J. R., Jr. 1990. Aspects of the molt migration. Pages 102–113 in *Bird migration: Physiology and ecophysiology* (E. Gwinner, ed.). Springer-Verlag, Berlin.

Jehl, J. R., Jr. 1997. Cyclical changes in body composition in the annual cycle and migration of the Eared Grebe Podiceps nigricollis. *Journal of Avian Biology* 28:132–142.

Jenni, L., and S. Jenni-Eiermann. 1992. High plasma triglyceride levels in small birds during migratory flight: A new pathway for fuel supply during endurance locomotion at very high mass-specific metabolic rates? *Physiological Zoology* 65:112–123.

Jenni, L., and S. Jenni-Eiermann. 1998. Fuel supply and metabolic constraints in migrating birds. *Journal of Avian Biology* 29:521–528.

Jenni, L., and M. Schaub. 2003. Behavioural and physiological reactions to environmental variation in bird migration: A review. Pages 115–171 in *Avian migration* (P. Berthold, E, Gwinner, and E. Sonnenschein, eds.). Springer, Berlin.

Jenni, L., and R. Winkler. 1994. *Moult and ageing of European passerines*. Academic Press, San Diego.

Jettj, L. A., T. J. Hayden, and J. Cornelius. 1998. *Demographics of the Golden-cheeked Warbler (Dendroica chrysoparia) on Fort Hood, Texas*. U.S. Army Corps of Engineers, Champaign, IL.

Jetz, W., D. S. Wilcove, and A. P. Dobson. 2007. Projected impacts of climate and land-use change on the global diversity of birds. *PLoS Biology Online* 5(6):e157.

Johnsen, S. 1929. Draktskiftet hos lirypen (*Lagopus lagopus* Linnaeus) i Norge. *Bergen Museum Aarb*. 1.

Johnson, M. D., T. W. Sherry, R. T. Holmes, and P. P. Marra. 2006. Assessing habitat quality for a migratory songbird wintering in natural and agricultural habitats. *Conservation Biology* 20:1433–1444.

Johnson, N. K. 1963. Comparative molt cycles in the Tyrannid genus *Empidonax*. *Proceedings of the International Ornithological Congress* 13:870–884.

Jones, P. 1985. The migration strategies of Palaearctic passerines in West Africa. Pages 9–21 in *Migratory birds: problems and prospects in Africa* (A. MacDonald and P. Goriup, eds.). International Council for Bird Preservation, Cambridge, UK.

Jones, P. 1995. Migration strategies of Palaearctic passerines in Africa: An overview. *Israel Journal of Zoology* 41:393–406.

Jones, P. 1996. Community dynamics of arboreal insectivorous birds in African savannas in relation to seasonal rainfall patterns and habitat change. Pages 421–447 in *Dynamics of tropical communities* (D. M. Newberry, H. H. T. Prins, and N. D. Brown, eds.). Blackwell, Oxford.

Jones, P. 1999. Community dynamics of arboreal insectivorous birds in African savannas in relation to seasonal rainfall patterns and habitat change. Pages 421–447 in *Dynamics of tropical communities* (D. M. Newberry, H. H. T. Prins, and N. D. Brown, eds.). Blackwell, Oxford.

Jones, P., J. Vickery, S. Holt, and W. Cresswell. 1996. A preliminary assessment of some factors influencing the density and distribution of Palaearctic passerine migrants in the Sahel zone of West Africa. *Bird Study* 43:73–84.

Jourdain, E., M. Gauthier-Clerc, D. J. Bicout, and P. Sabatier. 2007. Bird migration routes and risk for pathogen dispersion into western Mediterranean wetlands. *Emerging Infectious Diseases* 13:365–372.

Kaitala, A. V. Kaitala, and P. Lundberg. 1993. A theory of partial migration. *American Naturalist* 142:59–81.

Kale, H. W., II. 1967. Aggressive behavior by a migrating Cape May Warbler. *Auk* 84:120–121.

Kalela, O. 1954. Populationsökologische Gesichtspunkte zur Entstehung des Vogelzuges. *Annales Zoologici Societatis Zoologicae Botanicae Fennicae Vanamo* 16:1–30.

Karlsson, H., J. Bäckman, C. Nilsson, and T. Alerstam. 2010. Exaggerated orientation scatter of nocturnal migrants close to breeding grounds: Comparisons between seasons and latitudes. *Behavioral Ecology and Sociobiology* 64:2021–2031.

Karr, J. R. 1971. Wintering Kentucky Warblers. *Bird-Banding* 42:299.

Karr, J. R. 1976. On the relative abundance of migrants from the North Temperate Zone in tropical habitats. *Wilson Bulletin* 88:433–458.

Katti, M., and T. Price 1996. Effects of climate on Palaearctic warblers over-wintering in India. *Journal of the Bombay Natural History Society* 93:411–427.

Katti, M., and T. Price. 1999. Annual variation in fat storage by a migrant warbler overwintering in the Indian tropics. *Journal of Animal Ecology* 68:815–823.

Kaufmann, J. H. 1983. On the definitions and functions of dominance and territoriality. *Biological Review* 58:1–20.

Kawaoka Y. T. C., W. Sladen, and R. G. Webster. 1988. Is the gene pool of influenza viruses in shorebirds and gulls different from that in wild ducks? *Virology Journal* 163:247–250.

Keast, A., and E. S. Morton, eds. 1980. *Migrant birds in the Neotropics: Ecology, behavior, distribution, and conservation.* Smithsonian Institution Press, Washington, DC.

Keddy-Hector, D. P., and C. Beardmore. 1992. *Golden-cheeked Warbler recovery plan.* Region 2 Office, U.S. Fish and Wildlife Service, Albuquerque, NM.

Keeton, W. T. 1980. Avian orientation and navigation: new developments in an old mystery. *Proceedings of the International Ornithological Congress* 17:137–157.

Keeton, W. T., T. S. Larkin, and D. M. Windsor. 1974. Normal fluctuations in the earth's magnetic field influence pigeon orientation. *Journal of Comparative Physiology A* 952:95–103.

Keith, S., E. K. Urban, and C. H. Fry. 1992. *The birds of Africa.* Academic Press, New York.

Keller, G. S., and R. H. Yahner. 2006. Declines of migratory songbirds: Evidence for wintering-ground causes. *Northeastern Naturalist* 13:83–92.

Kelly, J. F. 2006. Stable isotope evidence links breeding geography and migration timing in wood warblers (Parulidae). *Auk* 123:431–437.

Kemper, C. 1996. A study of bird mortality at a west central Wisconsin TV tower from 1957–1995. *Passenger Pigeon* 58:235.

Kennedy, E. D., and D. W. White. 1997. Bewick's Wren (*Thryomanes bewickii*). In The birds of North America online (A. Poole, ed.). Cornell Laboratory of Ornithology, Ithaca, NY. http://bna.birds.cornell.edu.bnaproxy.birds.cornell.edu/bna/species/315/articles/introduction. Accessed September 8, 2012.

Kerlinger, P. 1982. The migration of Common Loons through eastern New York. *Condor* 84:97–100.

Kerlinger, P. 1989. *Flight strategies of migrating hawks*. University of Chicago Press, Chicago.

Kerlinger, P. 1995. *How birds migrate*. Stackpole, Books, Mechanicsburg, PA.

Kerlinger, P., M. R. Lein, and B. J. Sevick. 1985. Distribution and population fluctuations of wintering Snowy Owls (*Nyctea scandiaca*) in North America. *Canadian Journal of Zoology* 63:1829–1834.

Kerlinger, P., and F. R. Moore. 1989. Atmospheric structure and avian migration. *Current Ornithology* 6:109–142.

Kershner, E. L., J. W. Walk, and R. E. Warner. 2004. Postfledging movements of juvenile Eastern Meadowlarks (*Sturnella magna*) in Illinois. *Auk* 121:1146–1154.

Ketterson, E. D., and V. Nolan, Jr. 1976. Geographic variation and its climatic correlates in the sex ratio of eastern-wintering Dark-eyed Juncos (*Junco hyemalis hyemalis*). *Ecology* 57:679–693.

Ketterson, E. D., V. Nolan, Jr. 1983. The evolution of differential bird migration. *Current Ornithology* 1:357–402.

Kimura, M., S. M. Clegg, I. J. Lovette, K. R. Holder, D. J. Girman, B. Milá, P. Wade, and T. B. Smith. 2002. Phylogenetic approaches to asessing demographic connectivity between breeding and overwintering regions in a Nearctic-Neotropical warbler (*Wilsonia pusilla*). *Molecular Ecology* 11:1605–1616.

King, B., M. Woodcock, and E. C. Dickinson. 1987. *A field guide to the birds of south-east Asia.* 2nd ed. Houghton, London.

King, D. I., C. C. Chandler, J. H. Rappole, R. B. Chandler, and D. W. Mehlman. 2012. Establishing quantitative habitat targets for an endangered Neotropical migrant (*Setophaga chrysoparia*) during the non-breeding season. *Bird Conservation International* 22 (in press).

King, D. I., R. M. DeGraaf, M.-L. Smith, and J. P. Buonaccorsi. 2006a. Habitat selection and habitat-specific survival of fledgling ovenbirds (*Seiurus aurocapilla*). *Journal of Zoology* 269:414–421.

King, D. I., and J. H. Rappole. 2000. Mixed-species foraging flocks in montane pine forests of Middle America. *Condor* 102:664–672.

King, D. I., and J. H. Rappole. 2001a. Mixed-species bird flocks in dipterocarp forest of north-central Burma (Myanmar). *Ibis* 143:380–390.

King, D. I., and J.H. Rappole. 2001b. Kleptoparasitism of laughing thrushes *Garrulax* by Greater Racket-tailed Drongos *Dicrurus paradiseus* in Myanmar. *Forktail* 117:121–122.

King, D. I., and J. H. Rappole. 2002. Commensal foraging relationships of the White-browed Fantail (*Rhipidura aureola*) in Myanmar (Burma). *Journal of the Bombay Natural History Society* 99:308–312.

King, D. I., J. H. Rappole, and J. P. Bounaccoursi. 2006b. Long-term population trends of forest-dwelling Nearctic-Neotropical migrant birds: A question of temporal scale. *Bird Populations* 7:1–9.

King, D. T. 2005. Interactions between the American White Pelican and aquaculture in the southeastern United States: An overview. *Waterbirds* 28:83–86.

King, J. R. 1968. Cycles of fat deposition and molt in White-crowned Sparrows in contant environmental conditions. *Comparative Biochemical Physiology* 24:827–837.

King, J. R. 1972. Adaptive periodic fat storage by birds. *Proceedings of the International Ornithological Congress* 15:200–217.

King, J. R. 1974. Seasonal allocation of time and energy reserves in birds. Pages 4–70 in *Avian energetics* (R. A. Payne, Jr., ed.). Nuttall Ornithologial Club, Cambridge, MA.

King, J. R., and D. S. Farner. 1959. Premigratory changes in body weight and fat in wild and captive male White-crowned Sparrows. *Condor* 61:315–324.

King, J. R., and D. S. Farner. 1965. Studies of fat deposition in migratory birds. *Annals of the New York Academy of Sciences* 131:422–440.

King, J. R., D. S. Farner, and M. L. Morton 1965. The lipid reserves of White-crowned Sparrows on the breeding ground in central Alaska. *Auk* 82:236–252.

King, J. R., S. Barker, and D. Farner. 1963. A comparison of energy reserves during the autumnal and vernal migratory periods in the White-crowned Sparrow, *Zonotrichia leucophrys gambelii*. *Ecology* 44:513–521.

Kipling, R. 1912. *Just so stories*. Doubleday, New York.

Kjellén, N. 1994. Moult in relation to migration—a review. *Ornis Svecica* 21:1–24.

Klaassen, M., and H. Biebach. 2000. Flight altitude of trans-Sahara migrants in autumn: A comparison of radar observations with predictions from meteorological conditions and water and energy balance models. *Journal of Avian Biology* 31:47–55.

Knopf, F. L. 1994. Avian assemblages on altered grasslands. *Studies in Avian Biology* 15:247–257.

Kociolek, A. V., A. P. Clevenger, C. C. St. Clair, and D. S. Proppe. 2010. Effects of road networks on bird populations. *Conservation Biology* 25:241–249.

Kodric-Brown, A., and J. H. Brown. 1978. Influence of economics, interspecific competition, and sexual dimorphism on territoriality of migrant Rufous Hummingbirds. *Ecology* 59:285–296.

Koenig, W., and J. Dickinson. 2004. *Ecology and evolution of cooperative breeding in birds*. Cambridge University Press, Cambridge.

Kokko, H., T. G. Gunnarsson, L. J. Morrell, and J. A. Gill. 2006. Why do female migratory birds arrive later than males? *Journal of Animal Ecology* 75:1293–1303.

Komar, N., S. Langevin, S. Hinten, N. Nemeth, E. Edwards, D. Hettler, B. Davis, R. Bowen, and M. Bunning. 2003. Experimental infection of North American birds with the New York 1999 strain of West Nile virus. *Emerging Infectious Diseases* 9:311–322.

Kot, M. 2001. *Elements of mathematical ecology*. Cambridge University Press, Cambridge.

Kramer, G. 1952. Experiments on bird orientation. *Ibis* 94:265–285.

Kramer, G. 1957. Experiments on bird orientation and their interpretation. *Ibis* 99:196–227.

Krebs, C. J. 1994. *Ecology, the experimental analysis of distribution and abundance*. Harper Collins, New York.

Krebs, J. R. 1971. Territory and breeding density in the Great Tit, *Parus major*. *Ecology* 52:2–22.

Kreithen, M. L. 1978. Sensory mechanisms for animal orientation—can any new ones be discovered? Pages 25–34 in *Animal migration, navigation, and homing* (K. Schmidt-Koenig, and W. T. Keeton, eds.). Springer-Verlag, Berlin.

Kreithen, M. L., and W. T. Keeton. 1974. Detection of polarized light by the homing pigeon *Columba livia. Journal of Comparative Physiology A* 891:83–92.

Kren, J., and A. C. Zoerb. 1997. Northern Wheatear *Oenanthe oenanthe*. In The birds of North America online (A. Poole, ed.). Cornell Laboratory of Ornithology, Ithaca, NY. http://bna.birds.cornell.edu.bnaproxy.birds.cornell.edu/bna/species/316/articles/introduction. Accessed September 8, 2012.

Kristoffersen, A. V. 2002. An early Paleogene trogon (Trogoniformes) from the Fur Formation, Denmark. *Journal of Vertebrate Paleontology* 22(3):661–666.

Krulwich, R., and M. Block. 2011. Flamingos drop from the Siberian sky: locals mystified, National Public Radio. http://www.npr.org/blogs/krulwich/2011/03/07/134229725/flamingos-drop-from-siberian-sky-locals-mystified. Accessed March 15, 2011.

Kung, N. Y., R. S. Morris, N. R. Perkins, L. D. Sims, T. M. Ellis, L. Bissett, M. Chow, K. F. Shortridge, Y. Guan, and M. J. S. Peiris. 2007. Risk for infection with highly pathogenic influenza A virus in chickens, Hong Kong, 2002. *Emerging Infectious Diseases* 13:412–418.

Kuno, E. 1991. Some strange properties of the logistic equation defined with r and K: inherent defects or artifacts? *Research in Population Ecology* 33:33–39.

Kunzmann, M. R., L. S. Hall, and R. R. Johnson. 1998. Elegant Trogon *Trogon elegans*. In The birds of North America online (A. Poole, ed.). Cornell Laboratory of Ornithology, Ithaca, NY. http://bna.birds.cornell.edu.bnaproxy.birds.cornell.edu/bna/species/357/articles/introduction. Accessed September 8, 2012.

Kushlan, J. A. 1976. Wading bird predation in a seasonally fluctuating pond. *Auk* 93:464–476.

Lack, D. L. 1943. *The life of the robin*. Witherby, London.

Lack, D. L. 1944a. The problem of partial migration. *British Birds* 37:122–130.

Lack, D. L. 1944b. Ecological aspects of species formation in birds. *Ibis* 86:260–286.

Lack, D. L. 1947–1948. The significance of clutch size. *Ibis* 89:302–352; 90:25–45.

Lack, D. L. 1954. *The natural regulation of animal numbers*. Oxford University Press, Oxford.

Lack, D. L. 1960. The influence of weather on passerine migration: A review. *Auk* 77:171–209.

Lack, D. L. 1966. *Population studies of birds*. Oxford University Press, Oxford.

Lack, D. L. 1968a. *Ecological adaptations for breeding in birds*. Methuen, London.

Lack, D. L. 1968b. Bird migration and natural selection. *Oikos* 19:1–9.

Lack, D. L., and R. Moreau. 1965. Clutch-size in tropical passerine birds of forest and savanna. *Oiseau* 35(Special):76–89.

LaDeau, S. L., P. P. Marra, A. M. Kilpatrick, and C. A. Calder. 2008. West Nile virus revisited: Consequences for North American ecology. *BioScience* 58(10):937–946.

Lajtha, K., and R. H. Michener, eds. 1994. *Stable isotopes in ecology and environmental studies*. Blackwell Scientific Publications, Oxford.

Lamb, R. A. 1989. Genes and proteins of the influenza viruses. Pages 1–87 in *The influenza viruses* (R. M. Krug, H. Fraenkel-Conrat, and R. R. Wagner, eds.). Plenum Press, New York.

Lampe, H. M., and Y. O. Espmark. 2003. Mate choice in Pied Flycatchers *Ficedula hypoleuca*: Can females use song to find high-quality mates and territories? *Ibis* 145(1):E24–E33.

Lanciotti, R. S., J. T. Roehrig, V. Deubel, J. Smith, M. Parker, K. Steele, B. Crise, K. E. Volpe, M. B. Crabtree, J. H. Scherret, R. A. Hall, J. S. MacKenzie, C. B. Cropp, B. Panirahy, E. Ostlund, B. Schmitt, M. Malkinson, C. Banet, J. Weissman, N. Komar, H. M. Savage, W. Stone, T. McNamara, and D. J. Gubler. 1999. Origin of the West Nile virus responsible for an outbreak of encephalitis in the northeastern United States. *Science* 286:2333–2337.

Lang, J. D., L. A. Powell, D. G. Krementz, and M. J. Conroy. 2002. Wood Thrush movements and habitat use: Effects of forest management for Red-cockaded Woodpeckers. *Auk* 119:109–124.

Langin, K. M., P. P. Marra, Z. Németh, F. R. Moore, T. K. Kyser, and L. M. Ratcliffe. 2009. Breeding latitude and timing of spring migration in songbirds crossing the Gulf of Mexico. *Journal of Avian Biology* 40:309–316.

Lanyon, S. M., and C. F. Thompson. 1986. Site fidelity and habitat quality as determinants of settlement pattern in male Painted Buntings. *Condor* 88:206–210.

Larkin, R. P., and R. E. Szafoni. 2008. Evidence for widely dispersed birds migrating together at night. *Integrative and Comparative Biology* 48:40–49.

Latta, S. C., and M. E. Baltz. 1997. Population limitation in Neotropical migratory birds: Comments on Rappole and McDonald 1994. *Auk* 114:754–762.

Least Heat Moon, W. 1982. *Blue highways: A journey into America*. Fawcett Crest, New York.

Leberg, P. L., T. J. Spengler, and W. C. Barrow, Jr. 1996. Lipid and water depletion in migrating passerines following passage over the Gulf of Mexico. *Oecologia* 106:1–7.

Leck, C. F. 1972. The impact of some North American migrants at fruiting trees in Panama. *Auk* 89:842–850.

Lee, D. S. 1995. Status and seasonal distributions of Bicknell's and Gray-cheeked thrushes in North Carolina. *Chat* 59:1–8.

Leimgruber, P. 1998. Landscape ecology of small mammals and songbirds in a managed forest mosaic of the Appalachian Mountains. Ph.D. diss., University of Oklahoma, Norman.

Leisler, B. 1990. Selection and use of habitat of wintering migrants. Pages 156–174 In *Bird migration: Physiology and ecophysiology* (E. Gwinner, ed.). Springer-Verlag, Berlin.

Leisler, B., and K. Schulze-Hagen. 2012. *Diversity in a uniform bird family—the reed warblers and their allies*. Zeist, The Netherlands (in press).

Leisler, B., H. Winkler, and M. Wink. 2002. Evolution of breeding systems in Acrocephaline warblers. *Auk* 119:379–390.

Lekagul, B., and P. D. Round. 1991. *A guide to the birds of Thailand*. Saha Karn Bhaet, Bangkok.

Leopold, M. F., C. J. G. van Damme, and H. W. van der Veer. 1998. Diet of cormorants and the impact of cormorant predation on juvenile flatfish in the Dutch Wadden Sea. *Journal of Sea Research* 40:93–107.

Leslie, P. H. 1945. The use of matrices in certain population mathematics. *Biometrika* 33:183–221.

Leslie, P. H. 1948. Some further notes on the use of matrices in population mathematics. *Biometrika* 35:213–245.

Leu, M., and C. W. Thompson. 2002. The potential importance of migratory stopover sites as flight feather molt staging areas: a review for neotropical migrants. *Biological Conservation* 106:45–56.

Levey, D. J., and F. G. Stiles. 1992. Evolutionary precursors to long-distance migration: Resource availability and movement patterns in neotropical landbirds. *American Naturalist* 140:447–476.

Lewis, J. C. 1995. Whooping Crane *Grus americana*. In The birds of North America online (A. Poole, ed.). Cornell Laboratory of Ornithology, Ithaca, NY. http://bna.birds.cornell.edu.bnaproxy.birds.cornell.edu/bna/species/153/articles/introduction. Accessed September 8, 2012.

Lewontin, R. 2010. Not so natural selection. *New York Review of Books* 57(9):34–36.

Li, K., Y. Guan, J. Wang, G. Smith, K. Xu, L. Duan, A. P. Rahardjo, P. Puthavathana, C. Buranathai, T. D. Nguyen, A. T. S. Estoepangestie, A. Chaisingh, P. Auewarakui, H. T. Long, N. T. H. Hanh, R. J. Webby, L. L. M. Poon, H. Chen, K. F. Shortridge, K. Y. Yuen, R. G. Webster, and J. S. M. Peiris. 2004. Genesis of a highly pathogenic and potentially pandemic H5N1 influenza virus in eastern Asia. *Nature* 430:209–213.

Lichtenstein, M. H. C. 1823. *Verzeichnis der Doubletten des Zoologischen Museums der Koeniglichen Universität zu Berlin nebst beschreibung vieler bisher unbekannter Arten von Säugethieren, Voegeln, Amphibien und Fischen*. Koeniglichen Universität zu Berlin, Berlin.

Liechti, F., and B. Bruderer. 1998. The relevance of wind for optimal migration theory. *Journal of Avian Biology* 29:561–568.

Liechti, F., M. Klaassen, and B. Bruderer. 2000. Predicting migratory flight altitudes by physiological models. *Auk* 117:205–214.

Liedvogel, M., S. Åkesson, and S. Bensch. 2011. The genetics of migration on the move. *Trends in Ecology and Evolution* 26:561–569.

Lincoln, F. C. 1952. *Migration of birds*. Doubleday, Garden City, NJ.

Lindström, Å. 1989. Finch flock size and risk of hawk predation at a migratory stopover site. *Auk* 106:225–232.

Lindström, Å. 1990. The role of predation risk in stopover habitat selection in migrating Bramblings *Fringilla montifringilla*. *Behavioral Ecology* 1:102–106.

Lindström, Å. 2005. Overview. Pages 249–250 In *Birds of two worlds* (R. Greenberg and P. Marra, eds.). Johns Hopkins University Press, Baltimore.

Lindström, Å., S. Daan, and G. H. Visser. 1994. The conflict between moult and migratory fat deposition: A photoperiodic experiment with bluethroats. *Animal Behaviour* 48:1173–1181.

Lindström, Å., M. Klaassen, and A. Kvist. 1999. Variation in energy intake and basal metabolic rate of a bird migrating in a wind tunnel. *Functional Ecology* 13:352–359.

Lindström, Å., A. Kvist, T. Piersma, A. Dekinga, and M. W. Dietz. 2000. Avian pectoral muscle size rapidly tracks body mass changes during flight, fasting, and fuelling. *Journal of Experimental Biology* 203:913–919.

Lindström, Å., and T. Piersma. 1993. Mass changes in migrating birds: the evidence for fat and protein storage re-examined. *Ibis* 135:70–78.

Lindström, Å., G. H. Visser, and S. Daan. 1993. The energetic cost of feather synthesis is proportional to basal metabolic rate. *Physiological Zoology* 66:490–510.

Linnaeus, C. 1757. *Migrationes Avium Sistems*. L. M. Hojer, Uppsala, Sweden.

Lockwood, M. W., and B. Freeman. 2004. *Handbook of Texas birds*. Texas A&M University Press, College Station.

Lockwood, R., J. P. Swaddle, and J. M. V. Rayner. 1998. Avian wingtip shape reconsidered: wingtip shape indices and morphological adaptations to migration. *Journal of Avian Biology* 29:273–292.

Lodé, T., M.-J. Holveck, and D. Lesbarrères. 2005. Asynchronous arrival pattern, operational sex ratio, and occurrence of multiple paternities in a territorial anuran, *Rana damatina*. *Biological Journal of the Linnean Society* 86:191–200.

Lofts, B., A. J. Marshall, and A. Wolfson 1963. The experimental demonstrations of pre-migration activity in the abscence of fat deposition in birds. *Ibis* 105:99–105.

Löhrl, H. 1959. Zur Frage des Zeitpunkts einer Prägung auf die Heimatregion beim Halsbandschnäpper (*Ficedula albicollis*). *Journal of Ornithology* 100:132–140.

Lopez Ornat, A., and R. Greenberg. 1990. Sexual segregation by habitat in migratory warblers in Quintana Roo, Mexico. *Auk* 107:539–543.

Lorenzen, L. C., and D. S. Farner. 1964. An annual cycle in the interrenal tissue of the adrenal gland of the White-crowned Sparrow, *Zonotrichia leucophrys gambellii*. *General and Comparative Endocrinology* 4:253–263.

Lotka, A. J. 1922. The stability of the normal age distribution. *Proceedings of the National Academy of Sciences* 8:339–345.

Lotka, A. J. 1956. *Elements of mathematical biology*. Dover Publications, New York.

Louchart, A. 2008. Emergence of long distance bird migrations: a new model integrating global climate changes. *Naturwissenschaften* 95:1109–1119.

Lowery, G. H. 1945. Trans-Gulf spring migration of birds and the coastal hiatus. *Wilson Bulletin* 57:92–121.

Lowery, G. H. 1946. Evidence of trans-Gulf migration. *Auk* 63:175–211.

Lowery, G. H. 1951. A quantitative study of the nocturnal migration of birds. *University of Kansas Publications of the Museum of Natural History* 3:361–472.

Lowery, G. H. 1955. *Louisiana birds*. Louisiana State University Press, Baton Rouge.

Lowery, G. H., and R. J. Newman. 1966. A continentwide view of bird migration on four nights in October. *Auk* 83:547–586.

Lowther, P. E. 1975. Geographic and geological variation in the family Icteridae. *Wilson Bulletin* 87:481–495.

Lowther, P. E. 1999. Alder Flycatcher *Empidonax alnorum*. In The birds of North America online (A. Poole, ed.). Cornell Laboratory of Ornithology, Ithaca, NY. http://bna.birds.cornell.edu.bnaproxy.birds.cornell.edu/bna/species/446/articles/introduction. Accessed September 8, 2012.

Lowther, P. E. 2000. Pacific-slope Flycatcher (*Empidonax difficilis*) and Cordilleran Flycatcher (*Empidonax occidentalis*). In The birds of North America online (A. Poole, ed.). Cornell Laboratory of Ornithology, Ithaca, NY. http://bna.birds.cornell.edu.bnaproxy.birds.cornell.edu/bna/species/556a/articles/introduction. Accessed September 8, 2012.

Lundberg, A., R. Alatalo, A. Carlson, and S. Ulfstrand. 1981. Biometry, habitat distribution and breeding success in the Pied Flycatcher *Ficedula hypoleuca*. *Ornis Scandinavica* 12:68–79.

Lundgren, B. O., and K.-H. Kiessling. 1988. Comparative aspects of fibre types, areas, and capillary supply in the pectoralis muscle of some passerine birds with differing migratory behavior. *Journal of Comparative Physiology. B. Biochemistry, Systematics, and Environmental Physiology* 158:165–173.

Lustick, S. 1970. Energy requirements of molt in cowbirds. *Auk* 87:742–746.

L'vov, D. K., M. Shchelkanov, P. G. Deriabin, I. T. Fediakina, E. I. Burtseva, A. G. Prilipov, D. E. Kireev, E. V. Usachev, T. I Aliper, A. D. Zaberezhnyi, T. V. Grebennikova, I. V. Galkina, A. A. Slavskii, K. E. Litvin, A. M. Dongur-ool, B. A. Medvedev, M. D. Dokper-ool, A. A. Mongush, M. S. Arapchor, A. O. Kenden, N. A. Vlasov, E. A. Nepoklonov, and D. Suarez. 2006. [Isolation of highly pathogenic avian influenza (HPAI) A/5H5N1 strains from wild birds in the epizootic outbreak on the Ubsu-Nur Lake (June 2006) and their incorporation to the Russian Federation State collection of viruses (July 3, 2006)]. *Voprosy Virusologii* 51(6):14–18.

Lynch, J. F., E. S. Morton, and M. E. Van der Voort. 1985. Habitat segregation between the sexes of wintering Hooded Warblers (*Wilsonia citrina*). *Auk* 102:714–721.

Lynch, J. R., and D. F. Whigham. 1984. Effects of forest fragmentation on breeding bird communities in Maryland, USA. *Biological Conservation* 28:287–324.

Lynch, J. F., and R. F. Whitcomb. 1978. Effects of the insularization of the eastern deciduous forest on avifaunal diversity and turnover. Pages 461–489 in *Classification, inventory, and analysis of fish and wildlife habitat* (A. Marmelstein, ed.). U.S. Fish and Wildlife Service, Washington, DC.

MacArthur, R. H. 1957. On the relative abundance of bird species. *Proceedings of the National Academy of Sciences* 43:293–295.

MacArthur, R. H. 1972. *Geographical ecology*. Harper and Row, New York.

MacArthur, R. H., and E. O. Wilson. 1967. *The theory of island biogeography*. Princeton University Press, Princeton, NJ.

Mackenzie, S. A., and C. A. Friis. 2006. Long Point Bird Observatory 2005 field operations report. Bird Studies Canada. http://www.bsc–eoc.org/download/LPBO%20Annual%20Report%202005.pdf. Accessed March 15, 2011.

Mackworth-Praed, C. W., and C. H. B. Grant. 1973. *African handbook of birds: Series three*. Longman, London.

Macwhirter, B., P. Austin-Smith, Jr., and D. Kroodsma. 2002. Sanderling (*Calidris alba*). In The birds of North America online (A. Poole, ed.). Cornell Laboratory of Ornithology, Ithaca, NY. http://bna.birds.cornell.edu.bnaproxy.birds.cornell.edu/bna/species/653/articles/introduction. Accessed September 8, 2012.

Marchant, J. H. 1992. Recent trends in breeding bird population of some common trans-Saharan migrant birds in northern Europe. *Ibis* 134, Supplement 1:113–119.

Marchant, J. H., R. Hudson, S. P. Carter, and P. Whittington. 1990. *Population trends in British breeding birds*. British Trust for Ornithology, Tring, UK.

Marra, P. P., K. A. Hobson, and R. T. Holmes. 1998. Linking winter and summer events in a migratory bird by using stable-carbon isotopes. *Science* 282:1884–1886.

Marra, P. P., and R. L. Holberton. 1998. Corticosterone levels as indicators of habitat quality: Effects of habitat segregation in a migratory bird during the non-breeding season. *Oecologia* 116:284–292.

Marra, P. P., and R. T. Holmes. 2001. Consequences of dominance-mediated habitat segregation in American Redstarts during the nonbreeding season. *Auk* 118:92–104.

Marra, P. P., D. R. Norris, S. M. Haig, M. Webster, and J. A. Royle. 2006. Migratory connectivity. Pages 157–183 in *Connectivity conservation* (K. R. Crooks and M. Sanjayan, eds.). Cambridge University Press, Cambridge.

Marshall, M. R., J. A. DeCecco, A. B. Williams, G. A. Gale, and R. J. Cooper. 2003. Use of regenerating clearcuts by late-successional bird species and their young during the post-fledging period. *Forest Ecology and Management* 183:27–135.

Martin, P. S. 1973. The discovery of America. *Science* 179:969–974.

Martin, T. E. 1995. Avian life history evolution in relation to nest predation, and food. *Ecological Monographs* 65:101–127.

Mathot, K. J., B. D. Smith, and R. W. Elner, 2007. Latitudinal clines in food distribution correlate with differential migration in the Western Sandpiper. *Ecology* 88:781–791.

Matthews, S. N., R. J. O'Connor, L. R. Iverson, and A. M. Prasad. 2004. *Atlas of climate change effects in 150 bird species of the eastern United States*. U.S. Department of Agriculture, Forest Service, Delaware, OH.

May, R. M. 1973. On relationships among various types of population models. *American Naturalist* 107:46–57.

May, R. M. 1976. Simple mathematical models with very complicated dynamics. *Nature* 261:459–467.

May, R. M., and G. F. Oster. 1976. Bifurcations and dynamic complexity in simple ecological models. *American Naturalist* 110:573–599.

Mayfield, H. F. 1992. Kirtland's Warbler (*Setophaga kirtlandii*). In The birds of North America online (A. Poole, ed.). Cornell Laboratory of Ornithology, Ithaca, NY. http://bna.birds. cornell.edu.bnaproxy.birds.cornell.edu/bna/species/019/articles/introduction. Accessed September 8, 2012.

Maynard Smith, J. 1968. *Mathematical ideas in biology*. Cambridge University Press, Cambridge.

Maynard Smith, J. 1972. *Models in ecology*. Cambridge University Press, Cambridge.

Maynard Smith, J. 1974. *Models in ecology*. Cambridge University Press, Cambridge.

Maynard Smith, J. 1977. Parental investment: a prospective analysis. *Animal Behaviour* 25:1–9.

Mayr, E. 1946. History of the North American bird fauna. *Wilson Bulletin* 58:2–41.

Mayr, E. 1982. *The growth of biological thought: diversity, evolution, and inheritance*. Belknap Press, Cambridge, MA.

Mayr, E., and W. Meise. 1930. Theories on the history of migrants [in German]. *Vogelzug* 1:149–172.

Mayr, E., and L. L. Short. 1970. Species taxa of North American birds. *Publications of the Nuttall Ornithological Club* 9.

Mayr, G. 1999. A new trogon from the Middle Oligocene of Céreste, France. *Auk* 116:427–434.

Mayr, G. 2001. Comments on the osteology of *Masillapodargus longipes* Mayr 1999 and *Paraprefica major* Mayr 1999, caprimulgiform birds from the Middle Eocene of Messel (Hessen, Germany). *Neues Jahrbuch für Geologie und Paleontologie, Monatshefte* 2001(2):65–76.

Mayr, G., and M. Daniels. 1998. Eocene parrots from Messel (Hessen, Germany) and the London Clay of Walton-on-the-Naze (Essex, England). *Senckenbergiana Lethaea* 78:157–177.

McCarty, J. P. 1996. Eastern Wood-Pewee *Contopus virens*. In The birds of North America online (A. Poole, ed.). Cornell Laboratory of Ornithology, Ithaca, NY. http://bna.birds.cornell.edu.bnaproxy.birds.cornell.edu/bna/species/245/articles/introduction. Accessed September 8, 2012.

McClure, E. 1974. *Migration and survival of the birds of Asia*. U.S. Army Component, SEATO Medical Research Laboratory, Bangkok, Thailand.

McGowan, K. J., and K. Corwin. 2008. *The second atlas of breeding birds in New York State.* Cornell University Press, Ithaca, NY.

McGrath, R. D. 1991. Nestling weight and juvenile survival in the blackbird, *Turdus merula*. *Journal of Animal Ecology* 60:335–351.

McKinney, F., S. R. Derrickson, and P. Mineau. 1983. Forced copulation in waterfowl. *Behaviour* 86:250–294.

McLaren, I. A., A. C. Lees, C. Field, and K. J. Collins. 2006. Origins and characteristics of Nearctic landbirds in Britain and Ireland in autumn: A statistical analysis. *Ibis* 148:707–726.

McLean, R. G. 2006. West Nile virus in North American birds. Pages 44–64 in *Current topics in avian disease research: understanding endemic and invasive diseases* (R. K. Barraclough, ed.). Ornithological Monographs 60. American Ornithologists' Union, Washington, DC.

McNamara, J. M., R. K. Welham, and A. I. Houston. 1998. The timing of migration within the contex of an annual routine. *Journal of Avian Biology* 29:416–423.

McNeil, R. 1969. La determination du contenue lipidique et la capacité de vol chez qulques espèces d'oiseau de rivage (Charadriidae et Scolopacidae). *Canadian Journal of Zoology* 47:525–536.

McNeil, R. 1971. Lean-season fat in a South American population of Black-necked Stilts. *Condor* 73:472–475.

McNeil, R., and M. Carrera de Itriago. 1968. Fat deposition in the Scissor-tailed Flycatcher (*Muscivora t. tyrannus*) and the Small-billed Elaenia (*Elaenia parvirostris*) during the austral migratory period in northern Venezuela. *Canadian Journal of Zoology* 46:123–128.

McNeil, R., M. Tulio Diaz, and A. Villeneuve. 1994. The mystery of shorebird over-summering: A new hypothesis. *Ardea* 82:143–151.

McShea, W. J., and J. H. Rappole 1997. Variable song rates in three species of passerines and implications for estimating bird populations. *Journal of Field Ornithology* 68:367–375.

Meier, A. H., D. S. Farner, and J. R. King. 1965. A possible endocrine basis for migratory behaviour in the White-crowned Sparrow, *Zonotrichia leucophrys gambellii*. *Animal Behaviour* 13:453–465.

Meier, A. H., and B. R. Ferrell. 1978. Avian endocrinology. *Chemical Zoology* 10:214–260.

Meier, A. H., B. R. Ferrell, and L. J. Miller. 1980. Circadian components of the circannual mechanism in the White-throated Sparrow. *Proceedings of the International Ornithological Congress* 17:458–462.

Mengel, R. M. 1964. The probable history of species formation in some northern wood warblers. *Living Bird* 3:9–43.

Mengel, R. M. 1970. The North American Great Plains as an isolating agent in bird speciation. *University of Kansas Department of Geology Special Publications* 3:279–340.

Menzbier, M. A. 1886. Die Zugstrassen der Vögel im Europäischen Russland. *Bull. de la Societe imp. des natur. de Moskou.* 61:291–369.

Merkel, F. 1963. Long-term effects of constant photoperiods on European robins and white-throats. *Proceedings of the International Ornithological Congress* 13:950–959.

Merkel, F. W. 1966. The sequence of events leading to migratory restlessness. *Ostrich* Supplement 6:239–248.

Mettke-Hofmann, C., and E. Gwinner. 2003. Long-term memory for a life on the move. *Proceedings of the National Academy of Sciences* 100:5863–5866.

Mewaldt, L. R. 1964. Effects of bird removal on winter populations of sparrows. *Bird-Banding* 35:184–195.

Mewaldt, L. R., S. S. Kibby, and M. L. Morton. 1968. Comparative biology of Pacific coastal White-crowned Sparrows. *Condor* 70:14–30.

Middendorf, A. 1855. Die Isepiptesen Russlands. *Mem. de l'Akad. des Sciences de St. Petersbourg Sc. naturelles* 8:1–137.

Mills, L. S. 2002. False samples are not the same as blind controls. *Nature* 415:471.

Mittelbach, G. G., D. W. Schemske, H. V. Cornell, A. P. Allen, J. M. Brown, M. B. Bush, S. P. Harrison, A. H. Hurlbert, N. Knowlton, H. A. Lessios, C. M. McCain, A. R. McCune, L. A. McDade, M. A. McPeek, T. J. Near, T. D. Price, R. E. Ricklefs, K. Roy, D. F. Sax, J. M. Sobel, and M. Turelli, 2007. Evolution and the latitudinal diversity gradient: Speciation, extinction, and biogeography. *Ecology Letters* 10:315–331.

Mithen, S. 2012. How fit is E. O. Wilson's evolution? *New York Review of Books.* http://www.nybooks.com/articles/archives/2012/jun/21/how-fit-eo-wilson-evolution/. Accessed September 4, 2012.

Moffat, C. B. 1903. The spring rivalry of birds: Some views on the limit to multiplication. *Irish Naturalist* 12:152–166.

Moldenhauer, R. R., and D. J. Regelski. 1996. Northern Parula *Setophaga americana.* In The birds of North America online (A. Poole, ed.). Cornell Laboratory of Ornithology, Ithaca, NY. http://bna.birds.cornell.edu.bnaproxy.birds.cornell.edu/bna/species/215/articles/introduction. Accessed September 8, 2012.

Molnar, P., and M. A. Cane. 2002. El Niño's tropical climate and teleconnections as a blueprint for pre-Ice Age climates. *Paleoceanography* 17(2):11–13.

Mönkkönen, M., P. Helle, and D. Welsh. 1992. Perspectives on Palearctic and Nearctic bird migration: Comparisons and overview of life-history and ecology of migrant passerines. *Ibis* 134 (Supplement 1):7–13.

Monroe, B. L., Jr. 1968. *A distributional survey of the birds of Honduras.* Ornithological Monographs 7. American Ornithologists' Union, Washington, DC.

Montgomerie, R. D., and B. E. Lyon. 1986. Does longevity influence the evolution of delayed plumage maturation in passerine birds. *American Naturalist* 128:930–936.

Moore, F. R. 1982. Sunset and the orientation of a nocturnal bird migrant—a mirror experiment. *Behavioral Ecology and Sociobiology* 10:153–155.

Moore, F. R. 1987. Sunset and the orientation behaviour of migrants. *Biological Reviews* 62:65–86.

Moore, F. R. 2000. Stopover ecology of Nearctic-Neotropical migrants: Habitat relations and conservation implications: Preface. *Studies in Avian Biology* 20:1–3.

Moore, F. R., S. A. Gauthreaux, P. Kerlinger, and T. R. Simons. 1995. Habitat requirements during migration: Important link in conservation. Pages 121–144 in *Ecology and management of neotropical birds* (T. E. Martin and D. M. Finch, eds.). Oxford University Press, Oxford.

Moore, F. R., P. Kerlinger, and T. R. Simons. 1990. Stopover on a Gulf coast barrier island by spring trans-Gulf migrants. *Wilson Bulletin* 102:487–500.

Moore, F. R., and T. R. Simmons. 1992. Habitat suitability and stopover ecology of Neotropical landbird migrants. Pages 345–355 in *Ecology and conservation of Neotropical migrant landbirds* (J. M. Hagan III and D. W. Johnston, eds.). Smithsonian Institution Press, Washington, DC.

Moore, F. R., R. J. Smith, and R. Sandberg. 2005. Stopover ecology of intercontinental migrants. Pages 251–261 in *Birds of two worlds* (R. Greenberg and P. Marra, eds.). Johns Hopkins University Press, Baltimore.

Moore, J., and R. Ali. 1984. Are dispersal and inbreeding avoidance related? *Animal Behaviour* 32:94–112.

Morales, M. A., M. Barrandeguy, C. Fabbri, J. B. Garcia, A. Vissani, K. Trono, G. Gutierrez, S. Pigretti, H. Menchaca, N. Garrido, N. Taylor, F. Fernandez, S. Levis, and D. Enría. 2006. West Nile virus isolation from equines in Argentina. *Emerging Infectious Diseases* 12:1559–1561.

Morbey, Y. E., and R. C. Ydenberg. 2001. Protandrous arrival times to breeding areas: A review. *Ecology Letters* 4(6):663–673.

Moreau, R. E. 1950. The breeding seasons of African birds. 1. Land birds. *Ibis* 92:223–267.

Moreau, R. E. 1952. The place of Africa in the Palaearctic migration system. *Journal of Animal Ecology* 21:250–271.

Moreau, R. E. 1972. *The Palaearctic-African bird migration systems.* Academic Press, New York.

Morel, G., and F. Bourlière. 1962. Ecological relations of the sedentary and migratory avifauna in a Sahel savannah of lower Senegal [in French]. *Terre et Vie* 4:371–393.

Morris, R., and R. Jackson. 2005. *Epidemiology of H5N1 avian influenza in Asia and implications for regional control.* United Nations Food and Agricultural Organization, Rome, Italy.

Morse, D. H. 1970. Ecological aspects of some mixed-species foraging flocks of birds. *Ecological Monographs* 40:119–168.

Morse, D. H. 1971. The insectivorous bird as an adaptive strategy. *Annual Review of Ecology and Systematics* 2:177–200.

Morse, D. H. 1980. Population limitation: Breeding or wintering grounds? Pages 506–516 in *Migrant birds in the neotropics: Ecology, behavior, distribution, and conservation* (A. Keast and E. S. Morton, eds.). Smithsonian Institution Press, Washington, DC.

Morton, E. S. 1971. Food and migration habits of the Eastern Kingbird in Panama. *Auk* 88:925–926.

Morton, E. S. 1980a. Adaptation to seasonal change by migrant land birds in the Panama Canal Zone. Pages 437–453 in *Migrant birds in the neotropics: Ecology, behavior, distribution, and conservation* (A. Keast and E. S. Morton, eds.). Smithsonian Institution Press, Washington, DC.

Morton, E. S. 1980b. The importance of migrant birds to the advancement of evolutionary theory. Pages 555–557 in *Migrant birds in the neotropics: Ecology, behavior, distribution, and conservation* (A. Keast and E. S. Morton, eds.). Smithsonian Institution Press, Washington, DC.

Morton, M. L. 1991. Postfledging dispersal of Green-tailed Towhees to a subalpine meadow. *Condor* 93:466–468.

Morton, M. L. 1992. Effects of sex and birth date on premigration biology, migration schedules, return rates and natal dispersal in the mountain white-crowned sparrows. *Condor* 94:117–133.

Morton, M. L., M. M. Wakamatsu, M. E. Pereyra, and G. A. Morton. 1991. Postfledging dispersal, habitat imprinting, and philopatry in a montane migratory sparrow. *Ornis Scandinavica* 22:98–106.

Moskoff, W. 1995. Solitary Sandpiper *Tringa solitaria*. In The birds of North America online (A. Poole, ed.). Cornell Laboratory of Ornithology, Ithaca, NY. http://bna.birds.cornell.edu.bnaproxy.birds.cornell.edu/bna/species/156/articles/introduction. Accessed September 8, 2012.

Mouritsen, H. 1960. Yes, the clock-and-compass strategy can explain the distribution of ringing recoveries: reply to Thorup et al. *Animal Behaviour* 60:F9–F14.

Mouritsen, H. 2003. Spatiotemporal orientation strategies of long-distance migrants. Pages 493–514 in *Avian migration* (P. Berthold, E, Gwinner, and E. Sonnenschein, eds.). Springer, Berlin.

Mouritsen, H., and O. N. Larsen. 2001. Migrating songbirds tested in computer-controlled Emlen funnels use stellar cues for a time-independent compass. *Journal of Experimental Biology* 204:3855–3865.

Mowbray, T. B., C. R. Ely, J. S. Sedinger, and R. E. Trost. 2002. Canada Goose *Branta canadensis*. In The birds of North America online (A. Poole, ed.). Cornell Laboratory of Ornithology, Ithaca, NY. http://bna.birds.cornell.edu.bnaproxy.birds.cornell.edu/bna/species/682/articles/introduction. Accessed September 8, 2012.

Moynihan, M. 1962. The organization and probable evolution of mixed-species flocks of neotropical birds. *Smithsonian Miscellaneous Collection* 143:1–140.

Mueller, H. C., and D. D. Berger. 1966. Analyses of weight and fat variations in transient Swainson's Thrushes. *Bird-Banding* 37:83–112.

Muheim, R., J. B. Phillips, and S. Akesson. 2006. Polarized light cues underlie compass calibration in migratory songbirds. *Science* 313:837–839.

Mukhin, A. L., N. S. Chernetson, and D. A. Kishkinev. 2005. Reed Warbler, *Acrocephalus scirpaceus* (Aves, Sylviidae), song as an acoustic marker of wetland biotope during migration (in Russian). *Zool. zhurnal* 84:995–1002.

Mukhin, A., V. Grinkevich, and B. Helm. 2009. Under cover of darkness: nocturnal life of diurnal birds. *Journal of Biological Rhythms* 24:225–231.

Munn, C. A. 1985. Permanent canopy and understory flocks in Amazonia: species composition and population density. Pages 683–712 in *Neotropical ornithology* (P. A. Buckley, M. S. Foster, E. S. Morton, R. S. Ridgely, and F. C. Buckley, eds.). Ornithological Monographs 36. American Ornithologists' Union, Washington, DC.

Munster, V., A. Wallensten, C. Baas, G. Rimmelzwann, M. Schutten, B. Olsen, A. D. Osterhaus, and R. A. M. Fouchier. 2005. Mallards and highly pathogenic avian influenza ancestral viruses, northern Europe. *Emerging Infectious Diseases* 11:1545–1551.

Murphy, M. E. 1996. The energetics and nutrition of molt. Pages 158–198 in *Avian energetics and nutritional ecology* (C. Carey, ed.). Chapman and Hall, New York.

Murray, B. G., Jr. 1966. Migration of age and sex classes of passerines on the Atlantic coast in autumn. *Auk* 83:352–360.

Myers, J. M. 2011. Acrimony in Athens: A Georgia town's contentious choice to promote cat colonies. http://joomla.wildlife.org/index.php?option=com_content&task=view&id=842&Itemid=183. Accessed June 7, 2011.

Myers, J. P. 1981. A test of three hypotheses for latitudinal segregation of the sexes in wintering birds. *Canadian Journal of Zoology* 59:1527–1534.

Nager, R. G. 2006. The challenges of making eggs. *Ardea* 94:323–346.

Nasci, R. S., H. M. Savage, D. J. White, J. R. Miller, B. C. Cropp, M. S. Godsey, A. J. Kerst, P. Bennett, K. Gottfried, and R. S. Lanciotti. 2001. West Nile virus in overwintering *Culex* mosquitoes, New York City, 2000. *Emerging Infectious Diseases* 7:742–747.

Nathan, R., W. M. Getz, E. Revilla, M. Holyoak, R. Kadmon, D. Saltz, and P. E. Smouse. 2008. A movement ecology paradigm for unifying organismal movement research. *Proceedings of the National Academy of Sciences* 105:19052–19059.

Natural Resources Conservation Service. 2011. Whooping Crane *Grus americana*. U.S. Department of Agriculture. ftp://ftp-fc.sc.egov.usda.gov/MT/www/news/factsheets/Whooping_Crane.pdf. Accessed January 25, 2012.

Nebel, S. 2007. Differential migration of shorebirds in the East Asian-Australasian Flyway. *Emu* 107:14–18.

Neill, W. E. 1990. Induced vertical migration in copepods as a defence against invertebrate predation. *Nature* 345:524–526.

Nelson, J. B. 1978. *The Sulidae*. Oxford University Press, London.

Newton, I. 1995. Relationship between breeding and wintering ranges in Palaearctic-African migrants. *Ibis* 134:241–249.

Newton, I. 1998. *Population limitation in birds*. Academic Press, London.

Newton, I. 2003. Geographical patterns in bird migration. Pages 211–224 in *Avian migration* (P. Berthold, E. Gwinner, and E. Sonnenschein, eds.). Springer, Heidelberg.

Newton, I. 2006a. Movement patterns of Common Crossbills *Loxia curvirostra* in Europe. *Ibis* 148:782–788.

Newton, I. 2006b. Can conditions experienced during migration limit the population levels of birds? *Journal of Ornithology* 148:782–788.

Newton, I. 2008. *The migration ecology of birds*. Academic Press, London.

Newton, I. 2012. Obligate and facultative migration in birds: Ecological aspects. *Journal of Ornithology* 153 (Supplement 1):171–180.

Nice, M. M. 1937. Studies in the life history of the Song Sparrow. Volume 1. *Transactions of the Linnean Society of New York* 37.

Nice, M. M. 1941. The role of territory in bird life. *American Midland Naturalist* 26:441–487.

Nichols, J. D., F. A. Johnson, and B. K. Williams. 1995. Managing North American waterfowl in the face of uncertainty. *Annual Review of Ecology and Systematics* 26:177–199.

Nichols, J. D., R. E. Tomlinson, and G. L. Waggerman. 1986. Estimating nest detection probabilities for White-winged Dove nest transects in Tamaulipas, Mexico. *Auk* 103:825–828.

Nickell, W. P. 1968. Return of northern migrants to tropical winter quarters and banded birds recovered in the United States. *Bird-Banding* 39:107–116.

Nilsson, S. G., and I. N. Nilsson. 1976. Numbers, food consumption, and fish predation by birds in Lake Möckeln, southern Sweden. *Ornis Scandinavica* 7:61–70.

Nisbet, I. C. T. 1955. Atmospheric turbulence and bird flight. *British Birds* 48:557–559.

Nisbet, I. C. T., and W. H. Drury, Jr. 1968. Short-term effects of weather on bird migration: A field study using multivariate statistics. *Animal Behaviour* 16:496–530.

Nisbet, I. C. T., W. H. Drury, Jr., and J. Baird. 1963. Weight loss during migration. Part 1. Deposition and consumption of fat by the Blackpoll Warbler (*Dendroica striata*). *Bird-Banding* 34:107–138.

Nisbet, I. C. T., and L. Medway. 1972. Dispersion, population ecology, and migration of eastern great reed warblers *Acrocephalus orientalis* wintering in Malaysia. *Ibis* 114:451–494.

Nolan, V., Jr. 1978. *The ecology and behavior of the Prairie Warbler Dendroica discolor.* Ornithological Monographs 26. American Ornithologists' Union, Washington, DC.

Nolan, V., Jr., E. D. Ketterson, and C. A. Buerkle. 1999. Prairie Warbler *Setophaga discolor.* In The birds of North America online (A. Poole, ed.). Cornell Laboratory of Ornithology, Ithaca, NY. http://bna.birds.cornell.edu.bnaproxy.birds.cornell.edu/bna/species/455/articles/introduction. Accessed September 8, 2012.

Nolan, V., Jr., E. D. Ketterson, D. A. Cristol, C. M. Rogers, E. D. Clotfelter, R. C. Titus, S. J. Schoech, and E. Snajdr. 2002. Dark-eyed Junco *Junco hyemalis.* In The birds of North America online (A. Poole, ed.). Cornell Laboratory of Ornithology, Ithaca, NY. http://bna.birds.cornell.edu.bnaproxy.birds.cornell.edu/bna/species/716/articles/introduction. Accessed September 8, 2012.

Nolan, V., Jr., and R. E. Mumford. 1965. An analysis of Prairie Warblers killed in Florida during nocturnal migration. *Condor* 67:322–338.

Normile, D. 2006. Avian influenza: evidence points to migratory birds in H5N1 spread. *Science* 311:1225.

Norris, D. R., M. B. Wunder, and M. Boulet. 2006. Perspectives on migratory connectivity. Pages 79–88 in *Patterns of migratory connectivity in two nearctic-neotropical songbirds: New insights from intrinsic markers* (M. Boulet and D. R. Norris, eds.). Ornithological Monographs 61. American Ornithologists' Union, Washington, DC.

North American Bird Conservation Initiative. 2009. The state of the birds: United States of America. http://www.stateofthebirds.org/2009/pdf_files/State_of_the_Birds_2009.pdf. Accessed January 25, 2012.

Norwine, J., and K. John. 2007. Welcome to the Anthropocene. Pages 1–4 in *South Texas climate 2100. The changing climate of south Texas: 1900–2100* (J. Norwine and K. John, eds.). Texas A&M University, Kingsville.

Noss, R. F., M. A. O'Connell, and D. D. Murphy. 1997. *The science of conservation planning.* Island Press, Washington, DC.

Oberholser, H. C. 1898. Description of a new North American thrush. *Auk* 15:303–306.

Oberholser, H. C. 1974. *The bird life of Texas.* University of Texas Press, Austin.

Odum, E. P. 1971. *Fundamentals of ecology.* W. B. Saunders Company, Philadelphia.

Odum, E. P., C. E. Connell, and H. L. Stoddard. 1961. Flight energy and estimated flight ranges of some migratory birds. *Auk* 78:515–527.

Ogden, J. C. 1978. *Population trends of colonial wading birds on the Atlantic and Gulf coasts.* Pages 137–154 In *Wading birds: Research Report 7* (A. Sprunt, IV, J. C. Ogden, and S. Winckler, eds.) National Audubon Society, New York. Wading birds: Research Report 7. A. Sprunt, IV, J. C. Ogden, and S. Winckler. National Audubon Society, New York.

Ohman, M. D., B. W. Frost, and E. B. Cohen. 1983. Reverse diel vertical migration: An escape from invertebrate predators. *Science* 220:1404–1407.

Olsen, B., V. J. Munster, A. Wallensten, J. Waldenström, A. D. M. E. Osterhaus, and R. A. M. Fouchier. 2006. Global patterns of influenza A virus in wild birds. *Science* 312:384–388.

Orians, G. H. 1980. *Some adaptations of marsh-nesting blackbirds.* Princeton University Press, Princeton, NJ.

Oring, L. W. 1982. Avian mating systems. Pages 1–92 in *Avian biology.* Volume 6 (D. S. Farner, and J. R. King, eds.). Academic Press, New York.

Oring, L. W., and D. B. Lank. 1982. Sexual selection, arrival times, philopatry and site fidelity in the polyandrous Spotted Sandpiper. *Behavioral Ecology and Sociobiology* 10:185–191.

Ostrom, J. H. 1974. *Archaeopteryx* and the origin of flight. *Quarterly Review of Biology* 49:27–47.

Otter, K., L. Ratcliffe, and P. T. Boag. 1994. Extra-pair paternity in the Black-capped Chickadee. *Condor* 96:218–222.

Owen, M., and J. M. Black. 1991. A note on migration mortality and its significance in goos population dynamics. *Ardea* 79:195–196.

Packard, F. M. 1951. *Birds of the Central Coast of Texas.* Welder Wildlife Foundation (archives), Sinton, TX.

Pagen, R. W., F. R. Thompson, and D. E. Burhans. 2000. Breeding and post-breeding habitat use by forest migrant songbirds in the Missouri Ozarks. *Condor* 102:738–747.

Palmén, I. A. 1876. *über die Zugstraßen der Vögel.* Th. Griens, Leipzig.

Palmer, R. S. 1972. Patterns of molting. Pages 65–102 in *Avian biology.* Volume 2 (D. S. Farner, and J. R. King, eds.). Academic Press, New York.

Palmgren, P. 1949. On the diurnal rhythm of activity and rest in birds. *Ibis* 91:567–576.

Papi, F. 1989. Pigeons use olfactory cues to navigate. *Ethology, Ecology, and Evolution* 1:219–231.

Paradis, E. 1998. Interactions between spatial and temporal scales in the evolution of dispersal rates. *Evolutionary Ecology* 12:235–244.

Parkes, K. C. 1963. The contribution of museum collections to knowledge of the living bird. *Living Bird* 2:121–130.

Parmelee, D. F. 1992. Snowy Owl *Bubo scandiacus.* In The birds of North America online (A. Poole, ed.). Cornell Laboratory of Ornithology, Ithaca, NY. http://bna.birds.cornell. edu.bnaproxy.birds.cornell.edu/bna/species/010/articles/introduction. Accessed September 8, 2012.

Parmesan, C. 2006. Ecological and evolutionary responses to recent climate change. *Annual Review of Ecology and Systematics* 37:637–669.

Parnell, J. F. 1969. Habitat relations of the Parulidae during spring migration. *Auk* 86:505–521.

Parrish, J. D., and T. W. Sherry. 1994. Sexual habitat segregation by American Redstarts wintering in Jamaica: Importance of resource seasonality. *Auk* 111:38–49.

Parsons, K. C., and T. L. Master. Snowy Egret *Egretta thula.* In The birds of North America online (A. Poole, ed.). Cornell Laboratory of Ornithology, Ithaca, NY. http://bna.birds.cornell.edu.

bnaproxy.birds.cornell.edu/bna/species/489/articles/introduction. Accessed September 8, 2012.

Pärt, T. 1990. Natal dispersal in the Collared Flycatcher: possible causes and reproductive consequences. *Ornis Scandinavica* 21:83–88.

Pärt, T., and B. Doligez. 2003. Gathering public information for habitat selection: prospecting birds cue on parental activity. *Proceedings of the Royal Society B* 270:1809–1813.

Patten, M. A., and C. A. Marantz. 1996. Implications of vagrant southeastern vireos and warblers in California. *Auk* 113:911–923.

Payne, R. B. 1972. Mechanisms and control of molt. Pages 103–155 in *Avian biology*. Volume 2 (D. S. Farner, and J. R. King, eds.). Academic Press, New York.

Payne, R. B. 1984. *Sexual selection, lek and arena behavior, and sexual size dimorphism in birds.* Ornithological Monographs 33. American Ornithologists' Union, Washington, DC.

Pearl, R. 1927. The growth of populations. *Quarterly Review of Biology* 2:532–548.

Pearl, R., L. J. Reed, and J. F. Kish. 1940. The logistic curve and the census count of 1940. *Science* 92:486–488.

Pearson, D. J. 1973. Moult of some Palaearctic warblers wintering in Uganda. *Bird Study* 20:24–36.

Pearson, D. J. 1975. The timing of complete moult in the Great Reed Warbler *Acrocephalus arundinaceus. Ibis* 117:506–509.

Pearson, D. J. 1990. Palearctic passerine migrants in Kenya and Uganda: Temporal and spatial patterns of their movements. Pages 44–59 in *Bird migration: Physiology and ecophysiology* (E. Gwinner, ed.). Springer-Verlag, Berlin.

Pearson, D. J., and G. C. Backhouse. 1976. The southward migration of Palaearctic birds over Ngulia, Kenya. *Ibis* 118:78–105.

Pearson, D. J., and P. C. Lack. 1992. Migration patterns and habitat use by passerine and near-passerine migrant birds in eastern Africa. *Ibis* 134(Supplement 1):89–98.

Pearson, D. J., G. Nokolans, and J. S. Ash. 1980. The southward migration of Palaearctic passerines through northeast and east tropical Africa: A review. *Proceedings of the Pan-African Ornithological Congress* 4:243–262.

Peer, B. D., H. J. Homan, G. M. Linz, and W. J. Bleier. 2003. Impact of blackbird damage to sunflower: Bioenergetic and economic models. *Ecological Applications* 13:248–256.

Peiris J. S. M., and F. P. Amerasinghe. 1994. West Nile fever. Pages 139–148 in *Handbook of zoonoses. Section B: Viral.* 2nd ed. (G. W. Beran, and J. H. Steele, eds.). CRC Press, Boca Raton, FL.

Pennycuick, C. J. 1969. The mechanics of bird migration. *Ibis* 111:525–556.

Pennycuick, C. J. 1975. Mechanics of flight. Pages 1–75 in *Avian biology*. Volume 5 (D. S. Farner, and J. R. King, eds.). Academic Press, New York.

Pennycuick, C. J. 1998. Field observations of thermals and thermal streets, and the theory of cross-country soaring flight. *Journal of Avian Biology* 29:33–43.

Pennycuick, C. J., R. M. Compton and L. Beckingham. 1968. A computer model for simulating the growth of a population. *Journal of Theoretical Biology* 18:316–329.

Perdeck, A. C. 1958. Two types of orientation of migrating Starlings, *Sturnus vulgaris* L. and Chaffinches, *Fringilla coelebs* as revealed by displacement experiments. *Ardea* 46:1–37.

Perez, E. S., E., Secaira, C. Macias, S. Morales, and I. Amezcua. 2008. *Plan de conservación de los bosques de pino-encino de Centroamérica y el ave migratoria Dendroica chrysoparia*. Fundación Defensores de la Naturaleza y The Nature Conservancy, Guatemala City, Guatemala.

Petit, D. R., J. F. Lynch, R. L. Hutto, J. G. Blake, and R. B. Waide. 1995. Habitat use and conservaton in the neotropics. Pages 145–197 in *Ecology and management of neotropical migratory birds* (T. E. Martin, and D. M. Finch, eds.). Oxford University Press, Oxford.

Pfaender, J. 2011. Phenotypic traits meet patterns of resource use: The radiation of sharpfin sailfin silverside fish in Lake Matano. In *Tropical vertebrates in a changing world* (K.–L. Schuchmann, ed.). Alexander Koening Museum, Bonn, Germany.

Pfaender, J., U. K. Schliewen, and F. Herder. 2010. Phenotypic traits meet patterns of resource use in the radiation of sharpfin sailfin silverside fish in Lake Matano. *Evolutionary Ecology* 24:957–974.

Phillips, A. R. 1975. The migrations of Allen's and other hummingbirds. *Condor* 77:196–205.

Phillips, A. R. 1986. *The known birds of North and Middle America. Part I.* Allan R. Phillips, Denver.

Phillips, A. R. 1991. *The known birds of North and Middle America. Part II.* Alan R. Phillips, Denver.

Phillips, A. R., J. Marshall, and G. Monson. 1964. *The birds of Arizona*. University of Arizona Press, Tucson.

Phillips, R. A., P. Catry, J. R. D. Silk, S. Bearhop, R. McGill, V. Afanasyev, and I. J. Strange. 2007. Movements, winter distribution and activity patterns of Flakland and brown skuas: Insights from loggers and isotopes. *Marine Ecology Progress Series* 345:281–291.

Phillips, R. A., and K. C. Hamer. 1999. Lipid reserves, fasting capability and the evolution of nestling obesity in procellariiform seabirds. *Proceedings of the Royal Society B* 266:1329–1334.

Phillips, R. A., J. R. D. Silk, J. P. Croxall, V. Afansyev, and D. R. Briggs. 2004. Accuracy of geolocation estimates for flying seabirds. *Marine Ecology Progress Series* 266:265–272.

Phillips, R. A., J. C. Xavier, and J. P. Croxall. 2003. Effects of satellite transmitters on albatrosses and petrels. *Auk* 120:1082–1109.

Philpott, S. M., and T. Dietsch. 2003. Coffee and conservation: A global context and the value of farmer involvement. *Conservation Biology* 17:1844–1846.

Pianka, E. R. 1970. On r- and K-selection. *American Naturalist* 104:592–597.

Pielou, E. C. 1969. *An introduction to mathematical ecology*. Wiley-Interscience, New York.

Pierotti, R. J., and T. P. Good. 1994. Herring Gull *Larus argentatus*. In The birds of North America online (A. Poole, ed.). Cornell Laboratory of Ornithology, Ithaca, NY. http://bna.birds.cornell.edu.bnaproxy.birds.cornell.edu/bna/species/124/articles/introduction. Accessed September 8, 2012.

Piersma, T., M. Klaassen, J. H. Bruggemann, A.-M. Blomert, A. Gueye, Y. Ntiamoa-Baidu, and N. E. Van Brederode. 1990. Seasonal timing of the spring departure of waders from the Banc D'Arguin, Mauritania. *Ardea* 78:123–134.

Piersma, T., and Å. Lindström. 1997. Rapid reversible changes in organ size as a component of adaptive behaviour. *Trends in Ecology and Evolution* 12:134–138.

Piersma, T., J. Pérez-Tris, H. Mouritsen, U. Bauchinger, and F. Bairlein. 2005a. Is there a migratory syndrome common to all migrant birds? *Annals of the New York Academy of Sciences* 1046:282–293.

Piersma, T., D. I. Rogers, P. M. González, L. Zwarts, L. J. Niles, I. DeLima Serrano do Nascimento, C. D. T. Minton, and A. J. Baker. 2005b. Fuel storage rates before northward flights in Red Knots worldwide. Pages 262–273 in *Birds of two worlds* (R. Greenberg and P. Marra, eds.). Smithsonian Institution Press, Washington, DC.

Piersma T, and J. A. van Gils. 2011. *The flexible phenotype: A body-centred integration of ecology, physiology and behaviour.* Oxford University Press, Oxford.

Pigliucci, M. 2001. *Phenotypic plasticiy: Beyond nature and nurture.* Johns Hopkins University Press, Baltimore.

Poole, A. F., R. O. Bierregaard, and M. S. Martell. 2002. Osprey *Pandion haliaetus.* In The birds of North America online (A. Poole, ed.). Cornell Laboratory of Ornithology, Ithaca, NY. http://bna.birds.cornell.edu.bnaproxy.birds.cornell.edu/bna/species/683/articles/introduction. Accessed September 8, 2012.

Poole, A., ed. 2010. The birds of North America online. Cornell Laboratory of Ornithology Ithaca, NY. http://bna.birds.cornell.edu.bnaproxy.birds.cornell.edu/bna. Accessed September 8, 2012.

Powell, G. V. N. 1980. Migrant participation in Neotropical mixed-species flocks. Pages 477–483 in *Migrant birds in the Neotropics: Ecology, behavior, distribution, and conservation* (A. Keast and E. S. Morton, eds.). Smithsonian Institution Press, Washington, DC.

Powell, G. V. N. 1985. Sociobiology and adaptive significance of interspecific foraging flocks in the neotropics. Pages 713–732 in *Neotropical ornithology* (P. A. Buckley, M. S. Foster, E. S. Morton, R. S. Ridgely, and F. C. Buckley, eds.). Ornithological Monographs 36. American Ornithologists' Union, Washington, DC.

Powell, G. V. N., and J. H. Rappole. 1986. The Hooded Warbler. Pages 827–853 in *Audubon wildlife report 1986* (R. L. D. Silvestro, ed.). National Audubon Society, New York.

Powell, G. V. N., J. H. Rappole, and S. A. Sader. 1992. Nearctic migrant use of lowland Atlantic habitats in Costa Rica: A test of remote sensing for identification of habitat. Pages 287–298 in *Ecology and conservation of Neotropical migrant landbirds* (J. M. Hagan III and D. W. Johnston, eds.). Smithsonian Institution Press, Washington, DC.

Power, D. M. 1971. Warbler ecology: Diversity, similarity, and seasonal differences in habitat segregation. *Ecology* 52:434–443.

Press, W. H., S.A. Teukolsky, W. T. Vetterling and B. P. Flannery. 1992. *Numerical recipes in FORTRAN.* Cambridge University Press, Cambridge.

Price, T. 2008. *Speciation in birds.* Roberts, Greenwood Village, CT.

Prum, R. O., and A. H. Brush. 2002. The evolutionary origin and the diversification of feathers. *Quarterly Review of Biology* 77:261–295.

Przybylo, R., B. C. Sheldon, and J. Merilä. 2000. Climatic effects on breeding and morphology; evidence phenotypic plasticity. *Journal of Animal Ecology* 69:395–403.

Pulido, F. 2007. The genetics and evolution of avian migration. *BioScience* 57:165–174.

Pulido, F., and P. Berthold. 1998. The microevolution of migratory behaviour in the blackcap: Effects of genetic covariances on evolutionary trajectories. *Biological Conservation of Fauna* 102:206–211.

Pulido, F., and P. Berthold. 2003. Quantitative genetic analysis of migratory behavior. Pages 53–77 in *Avian migration* (P. Berthold, E. Gwinner, and E. Sonnenschein, eds.). Springer, Heidelberg.

Pulido, F., and P. Berthold. 2010. Current selection for lower migratory activity will drive the evolution of residency in a migratory bird population. *Proceedings of the National Academy of Sciences of the United States of America* 107(16):7341–7346.

Pulido, F., P. Berthold, P. Mohr, and U. Querner. 2001. Heritability of the timing of autumn migration in a natural bird population. *Proceedings of the Royal Society B* 268:953–959.

Pulido, F., P. Berthold, and A. J. van Noordwijk. 1996. Frequency of migrants and migratory activity are genetically controlled in a bird population: Evolutionary implications. *Proceedings of the National Academy of Sciences* 93:14642–14647.

Pulliam, H. R. 1988. Sources, sinks, and population regulation. *American Naturalist* 132:652–661.

Pulliam, H. R., and T. Caraco. 1984. Living in groups: Is there an optimal group size? Pages 122–147 in *Behavioural ecology: An evolutionary approach* (J. R. Krebs and N. B. Davies, eds.). Sinauer, Sunderland, MA.

Pyle, P. 1997. *Identification guide to North American birds.* Slate Creek Publishing, Bolinas, CA.

Qian, H., and R. E. Ricklefs. 2008. Global concordance in diversity patterns of vascular plants and terrestrial vertebrates. *Ecology Letters* 11:547–553.

Quay, W. B. 1985. Sperm release in migrating wood-warblers (Parulinae) nesting at higher latitudes. *Wilson Bulletin* 97:283–295.

Quay, W. B. 1986. Timing and location of spring sperm release in northern thrushes. *Wilson Bulletin* 98:526–534.

Rabøl, J. 1969a. Reversed migration as the cause of westward vagrancy by four *Phylloscopus* warblers. *British Birds* 62:89–92.

Rabøl, J. 1969b. Orientation of autumn migrating garden warblers (*Sylvia borin*) after displacement from western Denmark (Blåvand) to eastern Sweden (Ottenby). A preliminary experiment. *Dansk Orn Foren Tidsskr* 63:93–104.

Rabøl, J. 1970. Displacement and phaseshift experiments with night-migrating passerines. *Ornis Scandinavica* 1:27–43.

Rabøl, J. 1978. One-direction orientation versus goal area navigation in migratory birds. *Oikos* 30:216–223.

Rabøl, J. 1985. The moving goal area and the orientation system of migrant birds. *Dansk Orn Foren Tidsskr* 79:29–42.

Radabaugh, B. E. 1974. Kirtland's Warbler and its Bahama wintering grounds. *Wilson Bulletin* 86:374–383.

Raess, M. 2008. Continental efforts: migration speed in spring and autumn in an inner-Asian migrant. *Journal of Avian Biology* 39:13–18.

Raikow, R. J. 1985. Locomotor system. Pages 37–147 in *Form and function in birds*. Volume 3 (A. S. King and J. McLelland, eds.). Academic Press, London.

Raitasuo, K. 1964. Social behaviour of the mallard (*Anas platyrhynchos*) in the course of the annual cycle. *Papers on Game Research, Helsinki* 24:1–72.

Ralph, C. J. 1978. Disorientation and possible fate of young passerine coastal migrants. *Bird-Banding* 49:237–247.

Ramenofsky, M. 2010. Behavioral endocrinology of migration. Pages 191–199 in *Encyclopedia of animal behavior*. Volume 1 (M. Breed and J. Moore, eds.). Oxford Academic Press, Oxford.

Ramenofsky, M., and J. C. Wingfield. 2007. Regulation of migration. *Bioscience* 57:135–143.

Ramos, M. A. 1983. *Seasonal movements of bird populations at a Neotropical study site in southern Veracruz, Mexico*. Ph.D. diss., University of Minnesota, Minneapolis.

Ramos, M. A. 1988. Eco-evolutionary aspects of bird movements in the northern neotropical region. *Proceedings of the International Ornithological Congress* 19:251–293.

Ramos, M. A. 1991. Geographic variation in the Swainson's Thrush *Catharus ustulatus*. Pages 89–93 in *Known birds of North and Middle America. Part II* (A. Phillips, ed.). Privately published, Denver.

Ramos, M. A., and J. H. Rappole. 1994. Relative homing abilities of migrants and residents in tropcial rain forest of southern Veracruz, Mexico. *Bird Conservation International* 4:175–180.

Ramsar. 2011. The Ramsar Convention on Wetlands. http://www.ramsar.org/cda/en/ramsar-home/main/ramsar/1_4000_0__. Accessed September 8, 2012.

Randler, C. 2002. Avian hybridization, mixed pairing and female choice. *Animal Behaviour* 63:103–119.

Rappole, J. H. 1978. Seasonal distribution notes on birds from the Welder Refuge, San Patricio County, Texas. *Bulletin of the Texas Ornithological Society* 11:30–34.

Rappole, J. H. 1983. Analysis of plumage variation in the Canada Warbler. *Journal of Field Ornithology* 54:152–159.

Rappole, J. H. 1988. Intra- and intersexual competition in migratory passerine birds during the nonbreeding season. *Proceedings of the International Ornithological Congress* 17:2308–2317.

Rappole, J. H. 1995. *The ecology of migrant birds: A Neotropical perspective*. Smithsonian Institution Press, Washington, DC.

Rappole, J. H. 1996. The importance of forest for the world's migratory bird species. Pages 389–406 in *Conservation of faunal diversity in forested landscapes* (R. M. DeGraaf and R. I. Miller, eds.). Chapman and Hall, London.

Rappole, J. H. 2005a. Evolution of old and new world migration systems: A response to Bell. *Ardea* 93:125–131.

Rappole, J. H. 2005b. Not all deaths are equal. *Audubon Naturalist Society Newsletter*, June–July.

Rappole, J. H. 2006. Origins and timing of avian migrant evolution in the New World. *Acta Zoologica Sinica* 52: Supplement [Proceedings of the 23rd International Ornithological Congress].

Rappole, J. H. 2007. *Wildlife of the Mid-Atlantic region*. University of Pennsylvania Press, Philadelphia.

Rappole, J. H. 2011. Did big pharma create the flu panic? *New York Review of Books* 58(14):100. http://www.nybooks.com/articles/archives/2011/sep/29/did-big-pharma-create-flu-panic/?pagination=false.

Rappole, J. H., T. Aung, P. C. Rasmussen, and S. C. Renner. 2011b. Ornithological exploration in the southeastern sub-Himalayan region of Myanmar. Pages 10–29 in *Avifauna of the eastern Himalayas and southeastern sub-Himalayan Mountains: Center of endemism or many species in marginal habitats?* (S. C. Renner and J. H. Rappole, eds.). Ornithological Monographs 70. American Ornithologists' Union, Washington, DC.

Rappole, J. H., and K. Ballard. 1987. Passerine post-breeding movements in a Georgia old field community. *Wilson Bulletin* 99:475–480.

Rappole, J. H., and G. W. Blacklock. 1983. *Birds of the Texas coastal bend*. Texas A&M University Press, College Station.

Rappole, J. H., G. W. Blacklock, and J. Norwine. 2007. Apparent rapid range change in South Texas birds: Response to climate change? Pages 131–142 in *Texas Climate 2100* (J. Norwine and K. John, eds.). Texas A&M University, Kingsville, TX.

Rappole, J. H., B. W. Compton, P. Leimgruber, J. Robertson, D. I. King, and S. C. Renner. 2006. Modeling movement of West Nile virus in the Western Hemisphere. *Vector-Borne Zoonotic Diseases* 6:128–139.

Rappole, J. H., and R. M. DeGraaf. 1996. Research and effective management of Neotropical migrant birds. *Transactions of the North American Wildlife and Natural Resources Conference* 61:450–462.

Rappole, J. H., S. D. Derrickson, and Z. Hubálek. 2000b. Migratory birds and spread of West Nile virus in the Western Hemisphere. *Emerging Infectious Diseases* 6:319–328.

Rappole, J. H., T. Fulbright, J. Norwine, and R. Bingham. 1994. The forest-grassland boundary in Texas: Population differentiation without geographic isolation. Pages 167–177 in *Proceedings of the Third International Rangelands Congress* (P. Singh, ed.). Range Management Society of India, Jhansi, India.

Rappole, J. H., S. Glasscock, K. Goldberg, D. Song, and S. Faridani. 2011c. Range change among New World tropical and subtropical birds. Pages 151–167 in *Tropical vertebrates in a changing world* (K.-L. Schuchmann, ed.). Bonner Zoologische Monographen 57. Bonner Zoologische, Bonn, Germany.

Rappole, J. H., B. Helm, and M. A. Ramos. 2003a. An integrative framework for understanding the origin and evolution of avian migration. *Journal of Avian Biology* 34:124–128.

Rappole, J. H., and Z. Hubálek. 2003. Migratory birds and West Nile virus. *Journal of Applied Microbiology* 94:47S–58S.

Rappole, J. H., and Z. Hubálek. 2006. Birds and influenza H5N1 virus movement to and within North America. *Emerging Infectious Diseases* 12:1486–1492.

Rappole, J. H., and P. Jones. 2002. Evolution of Old and New World migration systems. *Ardea* 90:525–537.

Rappole, J. H., A. H. Kane, R. H. Flores, and A. R. Tipton. 1989a. Seasonal variation in habitat use by great-tailed grackles in the lower Rio Grande Valley of Texas. Animal Damage Control Symposium, Ft. Collins, CO, U.S. Department of Agriculture.

Rappole, J. H., A. H. Kane, R. H. Flores, and A. R. Tipton. 1989b. Seasonal variation in habitat use by great-tailed grackles in the lower Rio Grande Valley of Texas. Animal Damage Control Symposium, Ft. Collins, CO, U.S. Department of Agriculture.

Rappole, J. H., D. I. King, and W. Barrow. 1999. Winter ecology of the endangered Golden-cheeked Warbler. *Condor* 101:762–770.

Rappole, J. H., D. I. King, and J. Diez. 2003b. Winter versus breeding habitat limitation for an endangered avian migrant. *Ecological Applications* 13:735–742.

Rappole, J. H., D. I. King, and P. Leimgruber. 2000a. Winter habitat and distribution of the endangered Golden-cheeked Warbler. *Animal Conservation* 2:45–59.

Rappole, J. H., D. I. King, J. Diez, and J. H. Vega Rivera. 2005. Factors affecting population size in Texas' Golden-cheeked Warbler. *Endangered Species Update* 22(3):95–103.

Rappole, J. H., D. I. King, and J. H. Vega Rivera. 2003c. Coffee and conservation. *Conservation Biology* 17:334–336.

Rappole, J. H., D. I. King, and J. H. Vega Rivera. 2003d. Coffee and conservation III: Reply to Phipott and Dietsch. *Conservation Biology* 17:1847–1849.

Rappole, J. H., and M. V. McDonald. 1994. Cause and effect in population declines of migratory birds. *Auk* 111:652–660.

Rappole, J. H., and M. V. McDonald. 1998. Response to Latta and Baltz (1997). *Auk* 115:246–251.

Rappole, J. H., W. J. McShea, and J. Vega-Rivera. 1993. Evaluation of two survey methods in upland avian breeding communities. *Journal of Field Ornithology* 64:55–70.

Rappole, J. H., E. S. Morton, T. E. Lovejoy, and J. S. Ruos. 1983. *Nearctic avian migrants in the Neotropics.* U.S. Fish and Wildlife Service, Washington, D.C.

Rappole, J. H., A. Pine, D. Swanson, and G. Waggerman. 2007. Conservation and management for migratory birds: Insights from population data and theory for the White-winged Dove. Pages 4–20 in *Wildlife science: Linking ecological theory and management applications* (T. Fulbright and D. Hewitt, eds.). CRC Press, Gainesville.

Rappole, J. H., and M. A. Ramos. 1994. Factors affecting migratory routes over the Gulf of Mexico. *Bird Conservation International* 4:251–262.

Rappole, J. H., and M. A. Ramos. 1995. Determination of habitat requirements for migratory birds. Pages 235–241 in *Conservation of neotropical migratory birds in Mexico* (M. H. Wilson, and S. A. Sader, eds.). University of Maine, Orono.

Rappole, J. H., M. A. Ramos, R. J. Oehlenschlager, D. W. Warner, and C. P. Barkan. 1979. Timing of migration and route selection in North American songbirds. Pages 199–214 in *Proceedings of the First Welder Wildlife Foundation Symposium* (L. Drawe, ed.). Welder Wildlife Foundation, Sinton, TX.

Rappole, J. H., M. A. Ramos, and K. Winker. 1989a. Wintering Wood Thrush movements and mortality in southern Veracruz. *Auk* 106:402–410.

Rappole, J. H., and K.-L. Schuchmann. 2003. The ecology and evolution of hummingbird population movements: A review. Pages 39–51 in *Avian migration* (P. Berthold, E. Gwinner, and E. Sonnenschein, eds.). Springer, Heidelberg.

Rappole, J. H., N. M. Shwe, and W. J. McShea. 2011a. Seasonality in avian communities of a dipterocarp monsoon forest and related habitats in Myanmar's Central Dry Zone. Pages 131–146 in *The ecology of dry tropical forest* (W. J. McShea, ed.). Smithsonian Institution Press, Washington, DC.

Rappole, J. H., and A. R. Tipton. 1992. The evolution of avian migration in the Neotropics. *Ornitología Neotropical* 3:45–55.

Rappole, J. H, and G. Waggerman. 1986. Calling males as an index of density for breeding White-winged doves. *Wildlife Society Bulletin* 14:151–155.

Rappole, J. H., and D. W. Warner. 1976. Relationships between behavior, physiology, and weather in avian transients at a migration stopover point. *Oecologia* 26:193–212.

Rappole, J. H., and D. W. Warner. 1980. Ecological aspects of avian migrant behavior in Veracruz, Mexico. Pages 353–394 in *Migrant birds in the Neotropics: Ecology, behavior, conservation, and distribution* (A. Keast and E. S. Morton, eds.). Smithsonian Institution Press, Washington, DC.

Rappole, J. H., D. W. Warner, and M. A. Ramos. 1977. Territoriality and population structure in a small passerine community. *American Midland Naturalist* 97:110–119.

Rasmussen, P. C., and J. C. Anderton. 2005a. *Birds of South Asia: The Ripley guide.* Volume 1. Smithsonian Institution and Lynx Edicions, Washington, DC, and Barcelona.

Rasmussen, P., and J. C. Anderton. 2005b. *Birds of South Asia: The Ripley guide,* Volume 2. Smithsonian Institution and Lynx Edicions, Washington, DC, and Barcelona.

Rau, G. H., A. J. Mearns, D. R. Young, R. J. Olson, H. A. Schafer, and I. R. Kaplan. 1983. Animal $^{13}C/^{12}C$ correlates with trophic level in pelagic food webs. *Ecology* 64:1314–1318.

Raveling, D. G. 1979. The annual cycle of body composition of Canada Geese with special reference to control of reproduction. *Auk* 96:234–252.

Raynor, G. S. 1956. Meterorlogical variables and the northward movement of nocturnal land bird migrants. *Auk* 73:153–175.

Recher, H. F., and J. T. Recher. 1969. Some aspects of the ecology of migrant shorebirds. 2. Aggression. *Wilson Bulletin* 81:140–154.

Reed, J. M., T. Boulinier, E. Danchin, and L. W. Oring. 1999. Informed dispersal: Prospecting by birds for breeding sites. *Current Ornithology* 15:189–259.

Rees, E. C. 1987. Conflict of choice within pairs of Bewick's Swans regarding their migratory movement to and from the wintering grounds. *Animal Behaviour* 35:1685–1693.

Rees, E. C. 1989. Consistency in the timing of migration for individual Bewick's Swans. *Animal Behaviour* 38:384–393.

Reese, M. 2012. Letters: Universities under attack. *London Review of Books* 34(5).

Reeves, W. C. 1974. Overwintering of arboviruses. *Progress in Medical Virology* 17:193–220.

Regelski, D. J., and R. R. Moldenhauer. 1997. Tropical Parula *Setophaga pitiayumi.* In The birds of North America online (A. Poole, ed.). Cornell Laboratory of Ornithology, Ithaca, NY.

Renfrew, R., and A. M. Saavedra. 2007. Ecology and conservation of Bobolinks (*Dolichonyx oryzivorus*) in rice production regions of Bolivia. *Ornitologia Neotropical* 18:61–73.

Reudink, M. W., P. P. Marra, T. K. Kyser, P. T. Boag, K. M. Langin, and L. M. Ratcliffe. 2009. Non-breeding season events influence sexual selection in a long-distance migratory bird. *Proceedings of the Royal Society B* 276:1619–1626.

Reynolds, R. E., T. L. Shaffer, R. W. Renner, W. E. Newton, and B. D. J. Batt. 2001. Impact of the conservation reserve program on duck recruitment in the U.S. prairie pothole region. *Journal of Wildlife Managment* 65:765–780.

Rich, T. D., C. J. Beardmore, H. Berlanga, P. J. Blancher, M. S. W. Bradstreet, G. S. Butcher, D. W. Demarest, E. H. Dunn, W. C. Hunter, E. E. Iñigo-Elias, J. A. Kennedy, A. M. Martell, A. O. Panjabi, D. N. Pashley, K. V. Rosenberg, C. M. Rustay, J. S. Wendt, and T. C. Will. 2004. *Partners in Flight North American landbird conservation plan.* Cornell Laboratory of Ornithology, Ithaca, NY.

Richards, F. J. 1959. A flexible growth function for empirical use. *Journal of Experimental Biology* 10:290–300.

Richardson, W. J. 1976. Autumn migration over Puerto Rico and the western Atlantic: A radar study. *Ibis* 118:309–332.

Richardson, W. J. 1978. Timing and amount of bird migration in relation to weather: A review. *Oikos* 30:224–272.

Richardson, W. J. 1990. Timing of bird migration in relation to weather. Pages 78–101 in *Bird migration: Physiology and ecophysiology* (E. Gwinner, ed.). Springer-Verlag, Berlin.

Ricker, W. E. 1954. Stock and recruitment. *Journal of Fisheries Research Board of Canada* 11:559–623.

Ricklefs, R. E. 1972. Latitudinal variation in breeding productivity in the Rough-winged Swallow. *Auk* 89:826–836.

Ricklefs, R. E. 1973. *Ecology*. Chiron Press, Newton, MA.

Ricklefs, R. E., and M. Wikelski. 2002. The physiology/life history nexus. *Trends in Ecology and Evolution* 17(10):462–468.

Ridgway, R. 1889. *Thryothorus bewickii* (Aud.). Pages 92–93 in *Ornithology of Illinois*. Volume 1. Natural History Survey of Illinois, Springfield, IL.

Rimmer, C. C. 1988. Timing of the definitive Prebasic molt in Yellow Warblers at James Bay, Ontario. *Condor* 90:141–156.

Robbins, C. S. 1979. Effect of forest fragmentation on bird populations. Pages 198–212 in *Proceedings of the workshop on management of north central and northeastern forests for nongame birds* (R. M. DeGraaf, and K. E. Evans, eds.). Report Number GTR NC–51. U.S. Forest Service, Washington, DC.

Robbins, C. S., D. Bridge, and R. Feller. 1959. Relative abundance of adult male redstarts at an inland and a coastal locality during fall migration. *Maryland Birdlife* 15:23–25.

Robbins, C. S., D. K. Dawson, and B. A. Dowell. 1989a. Habitat area requirements of breeding forest birds of the Middle Atlantic states. *Wildlife Monographs* 103.

Robbins, C. S., J. R. Sauer, R. Greenberg, and S. Droege. 1989b. Population declines in North American birds that migrate to the Neotropics. *Proceedings of the National Academy of Sciences* 86:7658–7662.

Robbins, C. S., J. W. Fitzpatrick, and P. B. Hamel. 1992. A warbler in trouble: *Dendroica cerulea*. Pages 165–184 in *Ecology and conservation of Neotropical migrant landbirds* (J. M. Hagan III and D. W. Johnston, eds.). Smithsonian Institution Press, Washington, DC.

Roberts, J. O. L. 1971. Survival among North American wood warblers. *Bird-Banding* 42:165–184.

Roberts, T. S. 1936. *The birds of Minnesota*. Volume 1. University of Minnesota Press, Minneapolis.

Robinson, D. W., M. S. Bowlin, I. Bisson, J. Shamoun-Baranes, K. Thorup, R. Diehl, T. Kunz, S. Mabey, and D. W. Winkler. 2010. Integrating concepts and technologies to advance the study of bird migration. *Frontiers in Ecology and the Environment* 8(7):354–361.

Robinson, R. A., J. A. Learmonth, A. M. Hutson, C. D. Macleod, T. H. Sparks, D. I. Leech, G. J. Pierce, M. M. Rehfisch, and H. Q. P. Crick. 2005. *Climate change and migratory species*. British Trust for Ornithology Report No. 414. British Trust for Ornithology, Thetford, England.

Robinson, S. K., F. R. Thompson III, T. M. Donovan, D. R. Whitehead, and J. Faaborg. 1995. Regional forest fragmentation and the nesting success of migratory birds. *Science* 267:1987–1990.

Robinson, T. R., R. R. Sargent, and M. B. Sargent. 1996. Ruby-throated Hummingbird (*Archilochus colubris*). In The birds of North America online (A. Poole, ed.). Cornell Laboratory of Ornithology, Ithaca, NY. http://bna.birds.cornell.edu.bnaproxy.birds.cornell.edu/bna/species/204/articles/introduction. Accessed September 8, 2012.

Robl, J. 1972. Weight gains and losses by recaptures. *Vogelwarte* 27:50–65.

Robson, C. 2000. *A guide to the birds of southeast Asia.* Princeton University Press, Princeton, NJ.

Rodenhouse, N. L., T. S. Sillett, P. J. Doran, and R. T. Holmes. 2003. Multiple density-dependent mechanisms regulate a migratory bird population during the breeding season. *Proceedings of the Royals Society of London B* 270:2105–2110.

Roe, N. A., and W. E. Rees. 1979. Notes on the puna avifauna of Azángaro Province, Department of Puno, southern Peru. *Auk* 96:475–482.

Rogriguez, A., J. J. Negro, J. W. Fox, and V. Afanasyev. 2009. Effects of geolocator attachments on breeding parameters of Lesser Kestrels. *Journal of Field Ornithology* 80:399–407.

Rohwer, S. 1975. The social significance of avian winter plumage variability. *Evolution* 29:593–610.

Rohwer, S. 1986. A previously unknown plumage of first-year Indigo Buntings and theories of delayed plumage maturation. *Auk* 103:281–292.

Rohwer, F. C., and M. G. Anderson. 1988. Female-biased philopatry, monogamy, and the timing of pair-formation in migratory waterfowl. *Current Ornithology* 5:187–221.

Rohwer, S., E. Bermingham, and C. Wood. 2001. Plumage and mitochondrial DNA haplotype variation across a moving hybrid zone. *Evolution* 55:405–422.

Rohwer, S., and G. S. Butcher. 1988. Winter versus summer explanations of delayed plumage maturation in temperate passerine birds. *American Naturalist* 131:556–572.

Rohwer, S., K. A. Hobson, and V. G. Rohwer. 2009. Migratory double breeding in neotropical migrant birds. *Proceedings of the National Academy of Sciences* 106:19050–19055.

Rohwer, S., and J. Manning. 1990. Differences in timing and number of molts for Baltimore and Bullock's Orioles: Implications to hybrid fitness and theories of delayed plumage maturation. *Condor* 92:125–140.

Rohwer, S., D. M. Niles, and S. D. Fretwell. 1980. Delayed maturation and the deceptive acquisition of resources. *American Naturalist* 115:400–437.

Rohwer, V. G., S. Rohwer, and J. H. Barry. 2008. Molt scheduling of western Neotropical migrants and up-slope movement of Cassin's Vireo. *Condor* 110:365–370.

Rosengren, R., and W. Fortelius. 1986. Ortstreue in foraging ants of the *Formica rufa* group—hierarchy of orienting cues and long-term memory. *Insectes Sociaux* 33:306–337.

Roth, R. R., M. S. Johnson, and T. J. Underwood. 1996. Wood Thrush *Hylocichla mustelina*. In The birds of North America online (A. Poole, ed.). Cornell Laboratory of Ornithology, Ithaca, NY. http://bna.birds.cornell.edu.bnaproxy.birds.cornell.edu/bna/species/246/articles/introduction. Accessed September 8, 2012.

Roth, T., P. Sprau, R. Schmidt, M. Naguib, and V. Amrhein. 2009. Sex-specific timing of mate searching and territory prospecting in the nightingale: Nocturnal life of females. *Proceedings of the Royal Society B* 276:2045–2050.

Royama, T. 1992. *Analytical population dynamics.* Chapman & Hall, London.

Rubolini, D., F. Spina, and N. Saino. 2004. Protandry and sexual dimorphism in trans-Saharan migratory birds. *Behavioral Ecology* 15:592–601.

Rudebeck, G. 1950. Studies on bird migration based on field studies in southern Sweden. *Vår Fågelväldsupplement* 1:1–148.

Ruegg, K. C. 2008. Genetic, morphological, and ecological characterization of a hybrid zone that spans a migratory divide. *Evolution* 62:452–466.

Ruegg, K. C., R. J. Hijmans, and C. Moritz. 2006. Climate change and the origin of migratory pathways in the Swainson's Thrush, *Catharus ustulatus. Journal of Biogeography* 33:1172–1182.

Ruegg, K. C., H. Slabbekorn, S. Clegg, and T. B. Smith. 2006. Divergence in mating signals correlates with ecological variation in the migratory songbird, Swainson's Thrush (*Catharus ustulatus*). *Molecular Ecology* 15:3147–3156.

Ruegg, K. C., and T. B. Smith. 2002. Not as the crow flies: A historical explanation for circuitous migration in Swainson's thrush (*Catharus ustulatus*). *Proceedings of the Royal Society B* 269:1375–1381.

Runge, M. C., and P. P. Marra. 2005. Modeling seasonal interactions in the population dynamics of migratory birds. Pages 375–389 in *Birds of two worlds* (R. Greenberg and P. Marra, eds.). Johns Hopkins University Press, Baltimore.

Ruth, J. M. 2006. Partners in flight. http://www.partnersinflight.org/. Accessed June 27, 2011.

Salomonsen, F. 1955. The evolutionary significance of bird-migration. *Det Kongelige Danske Videnskabernes Selskab Biologiske Meddelelser* 22(6):1–62.

Salomonsen, F. 1972. Zoogeographical and ecological problems in arctic birds. *Proceedings of the International Ornithological Congress* 15:25–77.

Sandberg, R., and B. Holmquist. 1998. Orientation and long-distance migration routes: An attempt to evaluate compass cue limitations and required precision. *Journal of Avian Biology* 29:626–636.

Sanderson, F. J., P. F. Donald, D. J. Pain, I. J. Burfield, and F. P. J Bommel. 2006. Long-term population declines in Afro-Palearctic migrant birds. *Biological Conservation* 131:93–105.

Saracco, J. F., D. F. DeSante, M. P. Nott, and D. R. Kaschube. 2009. Using the MAPS and MOSI programs to monitor landbirds and inform conservation. *Proceedings of the 4th International Partners in Flight Conference,* McAllen, TX.

Sauer, J. R., J. E. Hines, and J. E. Fallon. 2003. *The North American Breeding Bird Survey, results and analysis 1966–2002, version 2003.1.* Patuxent Wildlife Research Center, Laurel, MD.

Sauer, J. R., J. E. Hines, J. E. Fallon, K. L. Pardieck, D. J. Ziolkowski, Jr., and W. A. Link. 2011. The North American Breeding Bird Survey, results and analysis 1966–2009. Version 3.23.2011. http://www.mbr-pwrc.usgs.gov/bbs/bbs.html. Accessed June 7, 2011.

Sauer, J. R., B. G. Peterjohn, and W. A. Link. 1994. Observer differences in the North American Breeding Bird Survey. *Auk* 111:50–62.

Sauer, J. R., S. Schwartz, and B. Hoover. 1996. The Christmas bird count home page. Patuxent Wildlife Research Center, Laurel, MD.

Schlichting, C., and M. Pigliucci. 1998. *Phenotypic evolution.* Sinauer Associates, Sunderland, MA.

Schmidt-Nielson, K. 1983. *Animal physiology: Adaptation and environment.* 3rd ed. Cambridge University Press, Cambridge.

Schmutz, J. K. 1987. Habitat occupancy of disturbed grasslands in relation to models of habitat selection. *Journal of Range Management* 40:438–440.

Schuchmann, K.-L. 1996. Hummingbirds—ecophysiological and behavioral adaptations to extreme environmental conditions and to limited food resources: A review. *Proceedings of the Brazil Ornithological Congress* 5:67–76.

Schuchmann, K.-L., ed. 1999. Family Trochilidae. Pages 468–535 in *Handbook of the birds of the world*, Volume 5 (J. del Hoyo, A. Elliott, and J. Sargatal, eds.). Lynx Edicions, Barcelona.

Schüz, E. 1952. *Vom Vogelzug*. Schops, Frankfurt.

Schüz, E. 1951. überblick über die Orientierungsversuche der Vogelwarte Rossitten. *Proceedings of the International Ornithological Congress* 10:249–268.

Schwabl, H., and B. Silverin. 1990. Control of partial migration and autumnal behaviour. Pages 144–155 in *Bird migration: Physiology and ecophysiology* (E. Gwinner, ed.). Springer, Berlin.

Schwabl, H., J. C. Wingfield, and D. S. Farner. 1984. Endocrine correlates of autumnal behaviour in sedentary and migratory individuals of a partial migratory population of the European Blackbird (*Turdus merula*). *Auk* 101:499–507.

Schwartz, P. 1964. The Northern Waterthrush in Venezuela. *Living Bird* 3:169–184.

Seagle, S. W., and B. R. Sturtevant. 2005. Forest productivity predicts invertebrate biomass and Ovenbird (*Seiurus aurocapillus*) reproduction in Appalachian landscapes. *Ecology* 86:1531–1539.

Seastedt, T. R., and S. F. Maclean. 1979. Territory size and composition in relation to resource abundance in Lapland Longspurs breeding in arctic Alaska. *Auk* 96:131–142.

Sekercioglu, C. H., S. H. Schneider, J. P. Fay, and S. R. Loarie. 2008. Climate change, elevational range shifts, and bird extinctions. *Conservation Biology* 22:140–150.

Sherry, T. W., and R. T. Holmes. 1992. Population fluctuations in a long-distance neotropical migrant: Demographic evidence for the importance of breeding season events in the American Redstart. Pages 431–442 in *Ecology and conservation of neotropical migrant landbirds* (J. M. Hagan III and D. W. Johnston, eds.). Smithsonian Institution Press, Washington, DC.

Sherry, T. W., and R. T. Holmes. 1995. Summer versus winter limitation of populations: What are the issues and what is the evidence? Pages 85–120 in *Ecology and management of neotropical migratory birds* (T. E. Martin, and D. M. Finch, eds.). Oxford University Press, Oxford.

Shields, W. M. 1982. *Philopatry, inbreeding, and the evolution of sex*. State University of New York Press, Albany.

Siegel, R. B., R. L. Wilkerson, K. J. Jenkins, R. C. Kuntz II, J. R. Boetsch, J. P. Schaberl, and P. J. Happe. 2007. Landbird monitoring protocol for national parks in the North Coast and Cascades network. In Techniques and methods, Book 2—Collection of environmental data, section A6. U.S. Department of the Interior, U.S. Geological Survey. http://www.birdpop.org/DownloadDocuments/nccn_protocol.pdf. Accessed September 8, 2012.

Sillet, T. S., and R. T. Holmes. 2002. Variation in survivorship of a migratory songbird throughout its annual cycle. *Journal of Animal Ecology* 71:296–308.

Simmons, R. C. 1981. *The American colonies: From settlement to independence*. W. W. Norton, New York.

Skellam, J. G. 1966. Seasonal periodicity in theoretical population ecology. Pages 179–205 in *Proceedings of the 5th Berkeley Symposium on Mathematical Statistics and Probability*, Volume 4. (L. M. Le Cam, and J. Neyman, eds.), University of California Press, Berkeley, 1967.

Skutch, A. 1950. The nesting season of Central American birds in relation to climate and food supply. *Ibis* 92:185–222.

Skutch, A. F. 1954. Life histories of Central American birds. *Pacific Coast Avifauna* 31:1–449.

Skutch, A. F. 1960. Life histories of Central American birds. *Pacific Coast Avifauna* 34:1–593.

Skutch, A. F. 1961. Helpers among birds. *Condor* 63:198–226.

Skutch, A. F. 1967. Life histories of Central American highland birds. *Publications of the Nuttall Ornithological Club* 7.

Skutch, A. F. 1969. Life histories of Central American birds. *Pacific Coast Avifauna* 35:1–580.

Skutch, A. F. 1972. Studies of tropical American birds. *Publications of the Nuttall Ornithological Club* 10.

Skutch, A. F. 1976. *Parent birds and their young.* University of Texas Press, Austin.

Skutch, A. 1980. Arils as a food of tropical American birds. *Condor* 82:31–42.

Skutch, A. F. 1985. Clutch size, nesting success, and predation on nests of neotropical birds, reviewed. Pages 575–594 in *Neotropical ornithology* (P. A. Buckley, M. S. Foster, E. S. Morton, R. S. Ridgely, and F. C. Buckley, eds.). Ornithological Monographs 36. American Ornithologists' Union, Washington, DC.

Slatkin, M. 1984. Ecological causes of sexual dimorphism. *Evolution* 38:622–630.

Smith, A. D., and A. M. Dufty, Jr. 2005. Variation in the stable-hydrogen isotope composition of Northern Goshawk feathers: Relevance to the study of migratory origins. *Condor* 107:547–558.

Smith, H. G., and J. A. Nilsson. 1987. Intraspecific variation in migratory patterns of a partial migrant, the Blue Tit (*Parus caeruleus*): An evaluation of different hypotheses. *Auk* 104:109–115.

Smith, T. B., S. M. Clegg, M. Kimura, K. C. Ruegg, B. Mila, and I. J. Lovette. 2005. Molecular genetic approaches to linking breeding and overwintering areas in five Neotropical migrant passerines. Pages 222–234 in *Birds of two worlds* (R. Greenberg and P. Marra, eds.). Johns Hopkins University Press, Baltimore.

Smithburn, K. C., T. P. Hughes, A. W. Burke, and J. H. Paul. 1940. A neurotropic virus isolated from the blood of a native of Uganda. *American Journal of Tropical Medicine and Hygiene* 20:471–492.

Smythies, B. E. 1953. *The birds of Burma.* Oliver and Boyd, Edinburgh.

Snell-Rood, E. C., and D. A. Cristol. 2005. Prior residence influences contest outcome in flocks on non-breeding birds. *Ethology* 111:441–454.

Snow, D. W. 1967. *A guide to moult in British birds.* British Trust for Ornithology, Hertfordshire, UK.

Snyder, N. F. R., and J. W. Wiley. 1976. *Sexual size dimorphism in hawks and owls of North America.* Ornithological Monographs 20. American Ornithologists' Union, Washington, DC.

Solokov, L. V., M. Y. Markovets, A. P. Shapoval, and Y. G. Morozov. 1998. Long-term trends in the timing of spring migration of passerines on the Courish Spit of the Baltic Sea. *Avian Ecology and Behaviour* 1:1–21.

Sorenson, L. G., and S. R. Derrickson. 1994. Sexual selection in the Northern Pintail: The importance of female choice versus male-male competition in the evolution of sexually-selected traits. *Behavioral Ecology and Sociobiology* 35:389–400.

Sowls, L. K. 1955. *Prairie ducks: A study of their behavior, ecology, and management.* Stackpole Books, Harrisburg, PA.

Spackman E, D. S., R. Slemons, K. Winker, D. Suarez, M. Scott, and D. Swayne. 2005. Phylogenetic analyses of type A influenza genes in natural reservoir species in North America reveals genetic variation. *Virus Research* 114:89–100.

Spottiswoode, C. N., and A. P. Møller. 2004. Extrapair paternity, migration, and breeding synchrony in birds. *Behavioral Ecology and Sociobiology* 15:41–57.

Spottiswoode, C., A. P. Tøttrup, and T. C. Coppack. 2006. Sexual selection predicts advancement of avian spring migration in response to climate change. *Proceedings of the Royal Society B* 273:3023–3029.

Staicer, C. A. 1992. Social behavior of the Northern Parula, Cape May Warbler, and Prairie Warbler wintering in second-growth forest in southwestern Puerto Rico. Pages 308–320 in *Ecology and conservation of Neotropical migrant landbirds* (J. M. Hagan III and D. W. Johnston, eds.). Smithsonian Institution Press, Washington, DC.

Steele, K. E., M. J. Linn, R. J. Schoepp, N. Komar, T. W. Geisbert, and R. M. Manduca. 2000. Pathology of fatal West Nile virus infections in native and exotic birds during the 1999 outbreak in New York City. *Veterinary Pathology* 37:208–224.

Steiof, K. 2005. Do migratory birds spread poultry flu? *Berichte zum Vogelschutz* 42:15–32.

Stenger, J., and J. G. Falls. 1959. The utilized territory of the Ovenbird. *Wilson Bulletin* 71: 125–140.

Stevenson, H. M. 1957. The relative magnitude of the trans-Gulf and circum-Gulf spring migrations. *Wilson Bulletin* 69:39–77.

Stewart, R. E. 1952. A life history study of the yellow-throat. *Wilson Bulletin* 65:99–115.

Stewart, R. E., and J. W. Aldrich. 1951. Removal and repopulation of breeding birds in a spruce-fir forest community. *Auk* 68:471–482.

Stiles, F. G., and S. M. Smith. 1977. New information on Costa Rican waterbirds. *Condor* 79:91–97.

Stinchcomb, G. E., T. C. Messner, S. G. Driese, L. C. Nordt, and R. M. Stewart. 2011. Pre-colonial (A. D. 1100–1600) sedimentation related to prehistoric maize agriculture and climate change in eastern North America. *Geology* 39:363–366.

Stone, W. 1937. *Birds studies at old Cape May: An ornithology of coastal New Jersey.* Volumes 1 and 2. Delaware Valley Ornithological Club, Philadelphia.

Storm, R. M., ed. 1966. *Animal orientation and navigation.* Oregon State University Press, Corvallis.

Streby, H., and D. Andersen. 2011. Seasonal productivity in a population of migratory songbirds: why nest data are not enough. *Ecosphere* 2:art78. http://dx.doi.org/10.1890/ES10-00187.1 Accessed September 8, 2012.

Stresemann, E. 1963. The nomenclature of plumages and molts. *Auk* 80:1–8.

Stresemann, E., and V. Stresemann. 1966. Die Mauser der Vögel. *Journal of Ornithology* 107 (Sonderheft) Striges:357–375.

Studds, C. E., and P. P. Marra. 2005. Nonbreeding habitat occupancy and population processes: An upgrade experiment with a migratory bird. *Ecology* 86:2380–2385.

Stutchbury, B. J. M., E. A. Gow, T. Done, M. MacPherson, J. W. Fox, and V. Afanasyev. 2011. Effects of post-breeding moult and energetic condition on timing of songbird migration into the tropics. *Proceedings of the Royal Society B* 278:131–137.

Stutchbury, B. J. M., and E. S. Morton. 2001. *Behavioral ecology of tropical birds.* Academic Press, New York.

Stutchbury, B. J. M., E. S. Morton, and B. Woolfenden. 2007. Comparison of the mating systems and breeding behavior of a resident and a migratory tropical flycatcher. *Journal of Field Ornithology* 78:40–49.

Stutchbury, B. J. M., S. A. Tarof, T. Done, E. A. Gow, P. M. Kramer, J. Tautin, J. W. Fox, and V. Afanasyev. 2009a. Tracking long-distance songbird migration by using geolocators. *Science* 323:896.

Stutchbury, B. J. M., S. A. Tarof, T. Done, E. A. Gow, P. M. Kramer, J. Tautin, J. W. Fox, and V. Afanasyev. 2009b. Tracking long-distance songbird migration by using geolocators: Supporting online material (Materials and Methods, Fig S1, Table S1, References). *Science* 323.

Sullivan, K. A. 1989. Predation and starvation: age-specific mortality in juvenile juncos (*Junco phaeonotus*). *Journal of Animal Ecology* 58:275–286.

Sutherland, W. J. 1996. Predicting the consequences of habitat loss for migratory populations. *Proceedings of the Royal Society B* 263:1325–1327.

Sutherland, W. J. 1998. The effect of local change in habitat quality on populations of migratory species. *Journal of Applied Ecology* 35:418–421.

Swanson, D. 1989. *Breeding biology of the White-winged Dove (Zenaida asiatica) in south Texas.* MSc diss., Texas A&I University, Kingsville.

Sykes, P. W., Jr., C. B. Kepler, D. A. Jett, and M. E. DeCapita. 1989. Kirtland's Warblers on the nesting grounds during the post-breeding period. *Wilson Bulletin* 101:545–558.

Tabb, E. C. 1979. Winter recoveries in Guatemala and southern Mexico of Broad-winged Hawks banded in south Florida. *North American Bird Bander* 4:60.

Tacha, T. C. 1988. Social organization of Sandhill Cranes from mid-continental North America. *Wildlife Monographs* 99.

Tacha, T. C., S. A. Nesbitt, and P. A. Vohs. 1992. Sandhill Crane (*Grus canadensis*). In The birds of North America online (A. Poole, ed.). Cornell Laboratory of Ornithology, Ithaca, NY. http://bna.birds.cornell.edu.bnaproxy.birds.cornell.edu/bna/species/031/articles/introduction. Accessed September 8, 2012.

Takahashi, A., D. Ochi, Y. Watanuki, T. Deguchi, N. Oka, V. Afanasyev, J. W. Fox, and P. N. Trathan. 2008. Post-breeding movement and activities of two Streaked Shearwaters in the north-western Pacific. *Ornithological Science* 7:29–35.

Tauber, M. J., C. A. Tauber, and S. Masaki. 1986. *Seasonal adaptations of insects.* Oxford University Press, Oxford.

Taylor, P. B. 2006b. Spotted Flycatcher *Muscicapa striata*. Pages 114 in *Handbook of the birds of the world.* Volume 11 (J. del Hoyo, A. Elliott, and D. Christie, eds.). Lynx Edicions, Barcelona.

Taylor, P. B. 2006a. Gambaga Flycatcher *Muscicapa gambagae*. Pages 114–115 in *Handbook of the birds of the world.* Volume 11 (J. del Hoyo, A. Elliott, and D. Christie, eds.). Lynx Edicions, Barcelona.

Taylor, R. J. 1984. *Predation.* Chapman and Hall, New York.

Taylor, R. M., T. H. Work, H. S. Hurlbut, and F. Rizk. 1956. A study of the ecology of West Nile virus in Egypt. *American Journal of Tropical Medicine and Hygiene* 5:579–620.

Telfair, R. C., II. 2006. Cattle Egret *Bubulcus ibis*. In The birds of North America online (A. Poole, ed.). Cornell Laboratory of Ornithology, Ithaca, NY. http://bna.birds.cornell. edu.bnaproxy.birds.cornell.edu/bna/species/113/articles/introduction. Accessed September 8, 2012.

Temeles, E. J. 1985. Sexual size dimorphism of bird-eating hawks: The effect of prey vulnerability. *American Naturalist* 125:485–499.

Terborgh, J. 1974. Preservation of diversity: The problem of extinction prone species. *BioScience* 24:715–722.

Terborgh, J. 1989. *Where have all the birds gone?* Princeton University Press, Princeton, NJ.

Terborgh, J. 1990. Mixed flocks and polyspecific associations: Costs and benefits of mixed groups to birds and monkeys. *American Journal of Primatology* 21:87–100.

Terborgh, J. W., S. K. Robinson, T. A. Parker III, C. A. Munn, and N. Pierpoint. 1990. Structure and organization of an Amazonian forest bird community. *Ecological Monographs* 60:213–238.

Terrill, S. 1987. Social dominance and migratory restlessness in the Dark-eyed Junco (*Junco hyemalis*). *Behavioral Ecology and Sociobiology* 21:1–11.

Terrill, S. 1990. Ecophysiological aspects of movements by migrants in the wintering quarters. Pages 130–143 in *Bird Migration: Physiology and ecophysiology* (E. Gwinner, ed.). Springer, Berlin.

Terrill, S. B. 1991. Evolutionary aspects of orientation and migration in birds. Pages 180–201 in *Orientation in birds* (P. Berthold, ed.). Birkhauser Verlag, Basel, Switzerland.

Terrill, S. B., and K. B. Able, K. B. 1988. Bird migration terminology. *Auk* 105:205–206.

Terrill, S. B., and P. Berthold 1990. Ecophysiological aspects of rapid population growth in a novel Blackcap (*Sylvia atricapilla*) population: An experimental approach. *Oecologia* 85:266–270.

Thaxter, C. B., A. C. Joys, R. D. Gregory, S. R. Baillie, and D. G. Noble. 2010. Hypotheses to explain population change among breeding bird species in England. *Biological Conservation* 143:2006–2019.

The Nature Conservancy. 2007. NatureServe. NatureServe Explorer Species Database. The Nature Conservancy. http://www.natureserve.org/explorer/servlet/NatureServe?init=Species. Accessed February 9, 2011.

Thiollay, J. M. 1988. Comparative foraging success of insectivorous birds in tropical and temperate forests: Ecological implications. *Oikos* 53:17–30.

Thomas, A. L. R., and A. Hedenström. 1998. The optimum flight speeds of flying animals *Journal of Avian Biology* 29:469–477.

Thomas, C. D., and J. J. Lennon. 1999. Birds extend their ranges northward. *Nature* 399:213.

Thomas, W. R., M. J. Pomerantz, and M. E. Gilpin. 1980. Chaos, asymmetic growth and group selection for dynamic stability. *Ecology* 61:1312–1320.

Thomson, R. L., J. T. Forsman, and M. Mönkkönen. 2003. Positive interactions between migrant and resident birds: Testing the heterospecific attraction hypothesis. *Oecologia* 134:431–438.

Thorup, K., I.-A. Bisson, M. S. Bowlin, R. A. Holland, J. C. Wingfield, M. Ramenofsky, and M. Wikelski. 2007. Evidence for a navigational map stretching across the continental U.S. in a migratory songbird. *Proceedings of the National Academy of Sciences* 104:18115–18119.

Thorup, K., T. E. Ortvad, J. Rabøl, R. A. Holland, A. P. Tøttrup, and M. Wikelski. 2011. Juvenile songbirds compensate for displacement to oceanic islands during autumn migration. *PLoS ONE* 6(3):e17903.

Thorup, K., and J. Rabøl. 2001. The orientation system and migration pattern of long-distance migrants: Conflict between model predictions and observed patterns. *Journal of Avian Biology* 32:111–119.

Thorup, K., and J. Rabøl. 2007. Compensatory behaviour after displacement in migratory birds: A meta-analysis of cage experiments. *Behavioral Ecology and Sociobiology* 61:825–841.

Tinbergen, N. 1946. Sperber als Roovijand van Zangvogels. *Ardea* 34:1–123.

Todd, W. E. C. 1940. *The birds of western Pennsylvania*. University of Pittsburgh Press, Pittsburgh.

Tøttrup, A. P., K. Rainio, T. Coppack, E. Lehikoinen, C. Rahbek, and K. Thorup. 2010. Local temperature fine-tunes the timing of spring migration in birds. *Integrative Comparative Biology* 50:293–304.

Tøttrup, A. P., and K. Thorup. 2008. Sex-differentiated migration patterns, protandry and phenology in North European songbird populations. *Journal of Ornithology* 149:161–167.

Tramer, E. 1974. Proportions of wintering North American birds in disturbed and undisturbed dry tropical habitats. *Condor* 76:460–464.

Tramer, E., and T. A. Kemp. 1980. Foraging ecology of migrant and resident warblers and vireos in the highlands of Costa Rica. Pages 285–297 in *Migrant birds in the Neotropics: Ecology, behavior, distribution, and conservation* (A. Keast, and E. S. Morton, eds.). Smithsonian Institution Press, Washington, DC.

Trivers, R. L. 1972. Parental investment and sexual selection. Pages 136–179 in *Sexual selection and the descent of man, 1871–1971* (B. Campbell, ed.). Aldine, Chicago, IL.

Tryjanowski, P., S. Kuzniak, and T. Sparks. 2002. Earlier arrival of some farmland migrants in western Poland. *Ibis* 144:62–68.

Tryjanowski, P., S. Kuzniak, and T. H. Sparks. 2005. What affects the magnitude of change in first arrival dates of migrant birds. *Journal of Ornithology* 146:200–205.

Turell, M. J., M. L. O'Guinn, D. J. Dohm, and J. W. Jones. 2001. Vector competence of North American mosquitoes (Diptera: Culicidae) for West Nile virus. *Journal of Medical Entomology* 38:130–134.

Turney, C. S. M., and H. Brown. 2007. Catastrophic early Holocene sea level rise, human migration and the Neolithic transition in Europe. *Quaternary Science Reviews* 26:2036–2041.

Udvardy, M. D. F. 1969. *Dynamic zoogeography*. Van Nostrand Reinhold, New York.

United Nations Environmental Program. 2004. Convention on migratory species. http://www.cms.int/. Accessed February 8, 2012.

U.S. Bureau of the Census. 1990. *1990 Census*. U.S. Bureau of the Census, Washington, DC.

U.S. Centers for Disease Control and Prevention. 2007. Key facts about avian influenza (bird flu) and avian influenza A (H5N1) virus. http://www.cdc.gov/flu/avian/gen-info/facts.htm. Accessed January 3, 2008.

U.S. Congress. 1900. Lacey Act. Congressional Record, 56th Congress, 1st Session, April 30, 1900, pages 4871–4872.

U.S. Congress. 1973. Endangered Species Act. 7 U.S.C. § 136, 16 U.S.C. § 1531 et seq.

U.S. Department of Agriculture. 2012. Jack pine interactive native range distribution map. http://www.plantmaps.com/nrm/pinus-banksiana-jack-pine-range-map.php. Accessed April 12, 2012.

U.S. Geological Survey. 2003. West Nile virus maps: Birds. http://westnilemaps.usgs.gov/2002/usa_avian_apr_22.html. Accessed June 7, 2004.

U.S. Geological Survey. 2011. National Atlas of the United States: breeding bird survey. U.S. Geological Survey, Biological Resources Division. http://www.nationalatlas.gov/mld/bbsrtsl.html. Accessed June 7, 2011.

U.S. Fish and Wildlife Service, Region 2. 2009. *Spotlight species action plan: The Golden-cheeked Warbler*. U.S. Fish and Wildlife Service, Washington, DC.

U.S. Fish and Wildlife Service. 1990. Endangered and threatened wildlife and plants; final rule to list the Golden-cheeked Warbler as endangered. *Federal Register* 55:53153–53160.

U.S. Fish and Wildlife Service. 2006. History of the U.S. Fish and Wildlife Service National Wildlife Refuge System. *Encyclopedia of Earth*. http://www.eoearth.org/article/History_of_the_U.S._Fish_and_Wildlife_Service_National_Wildlife_Refuge_System. Accessed June 11, 2011.

U.S. Fish and Wildlife Service. 2011a. Species profile: Whooping Crane *Grus americana*. http://ecos.fws.gov/speciesProfile/profile/speciesProfile.action?spcode=B003. Accessed June 27, 2011.

U.S. Fish and Wildlife Service. 2011b. Species profile for the endangered Golden-cheeked Warbler. http://ecos.fws.gov/speciesProfile/profile/speciesProfile.action?spcode=B07WactionPlans. Accessed June 8, 2011.

U.S. Fish and Wildlife Service. 2011c. Endangered species: Kirtland's Warbler *Dendroica kirtlandii*. http://www.fws.gov/midwest/endangered/birds/Kirtland/kiwafctsht.html. Accessed June 27, 2011.

U.S. Fish and Wildlife Service. 2011d. Neotropical Migratory Bird Conservation Act. http://www.fws.gov/birdhabitat/Grants/NMBCA/index.shtm. Accessed September 8, 2012.

U.S. National Oceanic and Atmospheric Administration. 2008. Freeze/frost maps. http://www.ncdc.noaa.gov/oa/climate/freezefrost/frostfreemaps.html. Accessed November 19, 2011.

U.S. National Park Service. 2011. Birds of Padre Island National Seashore, U.S. National Park Service. http://www.nps.gov/pais/planyourvisit/upload/Birdguide.pdf. Accessed May 19, 2011.

Urban, E. K., C. H. Fry, and S. Keith. 1986. *The birds of Africa*, Volume 2. Academic Press, New York.

Urban, E. K., C. H. Fry, and S. Keith. 1997. *The birds of Africa*, Volume 5. Academic Press, New York.

Uyeki, T. M. 2009. Human infection with highly pathogenic avian influenza (H5N1) virus: Review of clinical issues. *Clinical Infectious Diseases* 49:279–290.

Van de Casteele, T., and E. Matthysen. 2006. Natal dispersal and parental escorting predict relatedness between mates in a passerine bird. *Molecular Ecology* 15:2557–2565.

van der Graaf, A. J., J. Stahl, A. Klimkowska, J. P. Bakker, and R. H. Drent. 2006. Surfing on a green wave—how plant growth drives spring migration in the Barnacle Goose *Branta leucopsis*. *Ardea* 94:567–577.

Van Horne, B. 1983. Density as a misleading indicator of habitat quality. *Journal of Wildlife Management* 47:893–901.

van Noordwijk, A. J. 1989. Reaction norms in genetical ecology. *BioScience* 39:453–458.

van Noordwijk, A. J. 2003. Climate change: the early bird. *Nature* 422:29.

van Noordwijk, A., F. Pulido, B. Helm, T. Coppack, J. Delingat, H. Dingle, A. Hedenström, H. van der Jeugd, C. Marchetti, A. Nilsson, and J. Pérez-Tris. 2006. A framework for the study of genetic variation in migratory behavior. *Journal of Ornithology* 147:221–233.

Vega-Rivera, J. H. 1997. *Premigratory movements of a long distance migratory species: The Wood Thrush (Hylocichla mustelina)*. Ph.D. diss., Virginia Polytechnic Institute and State University, Blacksburg.

Vega-Rivera, J. H., C. A. Haas, J. H. Rappole, and W. J. McShea. 2000. Parental care of fledgling Wood Thrushes. *Wilson Bulletin* 112:233–237.

Vega-Rivera, J. H., W. J. McShea, and J. H. Rappole. 2003. Comparison of breeding and postbreeding movements and habitat requirements for the Scarlet Tanager (*Piranga olivacea*) in Virginia. *Auk* 120:632–644.

Vega-Rivera, J. H., W. J. McShea, J. H. Rappole, and C. A. Haas. 1999. Postbreeding movements and habitat use of adult Wood Thrushes in northern Virginia. *Auk* 116:458–466.

Vega-Rivera, J. H., M. A. Ortega-Juerta, S. Sarkar, and J. H. Rappole. 2011. Modelling the potential winter distribution of the endangered Black-capped Vireo (*Vireo atricapilla*). *Bird Conservation International* 21:92–106.

Vega-Rivera, J. H., J. H. Rappole, W. J. McShea, and C. A. Haas. 1998a. Pattern and chronology of prebasic molt for the Wood Thrush and its relation to reproduction and migration departure. *Wilson Bulletin* 110:384–392.

Vega-Rivera, J. H., J. H. Rappole, W. J. McShea, and C. A. Haas. 1998b. Wood Thrush post-fledging movements and habitat use in northern Virginia. *Condor* 100:69–78.

Verhulst, P.-F. 1845. Recherches mathématiques sur la loi d'accroissement de la population. *Nouv. mém. de l'Academie Royale des Sci. et Belles-Lettres de Bruxelles* 18:1–41.

Verhulst, P.-F. 1847. Deuxième mémoire sur la loi d'accroissement de la population. *Mém. de l'Academie Royale des Sci., des Lettres et des Beaux–Arts de Belgique* 20:1–32.

Verner, J. 1977. On the adaptive significance of territoriality. *American Naturalist* 111:769–775.

Vickery, P. D., M. L. Hunter, Jr., and J. V. Wells. 1992. Use of a new reproductive index to evaluate relationship between habitat quality and breeding success. *Auk* 109:697–705.

Visser, M. E., and C. M. Lessells. 2001. The costs of egg production and incubation in Great Tits (*Parus major*). *Proceedings of the Royal Society B* 268:1271–1277.

Visser, M. E., A. J. van Noordwijk, J. M. Tinbergen, and C. M. Lessells. 1998. Warmer springs lead to mistimed reproduction in great tits (*Parus major*). *Proceedings of the Royal Society B* 265:1867–1870.

Vitz, A. C., and A. D. Rodewald. 2006. Can regenerating clearcuts benefit mature-forest song-birds? An examination of post-breeding ecology. *Biological Conservation* 127:477–486.

von Haartman, L. 1971. Population dynamics. Pages 391–459 in *Avian biology*. Volume 1 (D. S. Farner, and J. R. King, eds.). Academic Press, New York.

Wagner, R. H., M. D. Schug, and E. S. Morton. 1996. Condition-dependent control of paternity by female purple martins: Implications for coloniality. *Behavioral Ecology and Sociobiology* 38:379–389.

Wallenstein, A., V. J. Munster, N. Latorre-Margalef, M. Brytting, J. Elmberg, R. A. M. Fouchier, T. Fransson, P. D. Haemig, M. Karlsson, A. Lunkkvist, A. D. M. E. Osterhaus, M. Stervander, J. Waldenström, and B. Olsen. 2007. Surveillance of influenza A virus in migratory waterfowl in northern Europe. *Emerging Infectious Diseases* 13:404–411.

Wallraff, H. G. 2004. Avian olfactory navigation: Its empirical foundation and conceptual state. *Animal Behaviour* 672:189–204.

Walsberg, G. E. 1983. Avian ecological energetics. Pages 120–161 in *Avian biology*. Volume 7 (D. S. Farner and J. R. King, eds.). Academic Press, New York.

Walter, H. 1979. *Adaptations to prey and habitat in a social raptor*. University of Chicago Press, Chicago.

Ward, D. H., E A. Rexstad, J. S. Sedinger, M. S. Lindberg, and N. K. Dawe. 1997. Seasonal and annual survival of adult Pacific brant. *Journal of Wildlife Managment* 61:773–781.

Ward, P., and P. Jones. 1977. Pre-migratory fattening in three races of the red-billed Quelea *Quelea quelea* (Aves: Ploceidae), an intra-tropical migrant. *Journal of Zoology London* 181:43–56.

Ward, P., and A. Zahavi. 1973. The importance of certain assemblages of birds as information centres for food finding. *Ibis* 115:517–534.

Waser, P. M., and W. T. Jones. 1983. Natal philopatry among solitary mammals. *Quarterly Review of Biology* 58:355–390.

Watte, B. C., and R. L. Phillips. 1994. An approach to controlling Golden Eagle predation on lambs in South Dakota. Pages 28–30 in *Proceedings of the 16th Vertebrate Pest Conference* (W. S. Halverson and A. C. Crabb, eds.). University of California, Davis.

Weatherhead, P. J. 1998. Natal philopatry and local resource competition in the Common Goldeneye. *Journal of Avian Biology* 29:321–322.

Weatherhead, P. J., and M. R. L. Forbes. 1994. Natal philopatry in passerine birds: Genetic or ecological influences. *Behavioural Ecology* 5:426–433.

Weber, J.-M. 2009. The physiology of long-distance migration: Extending the limits of endurance metabolism. *Journal of Experimental Biology* 212:593–597.

Webster, M. S., P. P. Marra, S. M. Haig, S. Bensch, and R. T. Holmes. 2002. Links between worlds: Unraveling migratory connectivity. *Trends in Ecology and Evolution* 17:76–83.

Webster R. G., W. B., O. Gorman, T. Chambers, and Y. Kawaoka. 1992. Evolution and ecology of influenza A viruses. *Microbiological Reviews* 56:152–179.

Webster, R., M. Peiris, H. Chen, and Y. Guan. 2006. H5N1 outbreaks and enzootic influenza. *Emerging Infectious Diseases* 12:3–8.

Weeks, H. P. Jr., 1994. Eastern Phoebe *Sayornis phoebe*. In The birds of North America online (A. Poole, ed.). Cornell Laboratory of Ornithology, Ithaca, NY. http://bna.birds.cornell.edu.bnaproxy.birds.cornell.edu/bna/species/094/articles/introduction. Accessed September 8, 2012.

Wehner, R. 1998. Navigation in context: grand theories and basic mechanisms. *Journal of Avian Biology* 29:370–386.

Weir, J. T., and D. Schluter. 2004. Ice sheets promote speciation in boreal birds. *Proceedings of the Royal Society B* 271:1881–1887.

Weise, C. M. 1963. Annual physiological cycles in captive birds of differing migratory habits. *Proceedings of the International Ornithological Congress* 13:983–993.

Weisstein, E. W. 1999. Logistic equation. MathWorld—A Wolfram Web Resource. http://mathworld.wolfram.com/LogisticEquation.html. Accessed February 10, 2011.

Welsh, D. A. 1975. Savannah Sparrow breeding and territoriality on a Nova Scotia dune beach. *Auk* 92:235–251.

Wenink, P. W., and J. Baker 1996. Mitochondrial DNA linkages in composite flocks of migratory and wintering Dunlins (*Calidris alpina*). *Auk* 113:744–756.

Wennerberg, L. 2001. Breeding origin and migration pattern of Dunlin (*Calidris alpina*) revealed by mitochondrial DNA analysis. *Molecular Ecology* 10:1111–1120.

Westneat, D. F., and P. W. Sherman. 1997. Density and extra-pair fertilizations in birds: A comparative analysis. *Behavioral Ecology and Sociobiology* 41:205–215.

Westneat, D. F., P. W. Sherman, and M. L. Morton. 1990. The ecology and evolution of extra-pair copulations in birds. *Current Ornithology* 7:331–369.

Wetmore, A. 1926. *The migrations of birds.* Oxford University Press, London.

Wetmore, A., R. F. Pasquier, and S. L. Olson. 1984. *The birds of the Republic of Panama,* Volume 4. Smithsonian Institution Press, Washington, DC.

Whitaker, D. M., and I. C. Warkentin. 2010. Spatial ecology of migratory passerines on temperate and boreal forest breeding grounds. *Auk* 127:471–484.

Whitcomb, R. F. 1977. Island biogeography and habitat islands of eastern forest. *American Birds* 31:3–5.

White, C. M., N. J. Clum, T. J. Cade, and W. G. Hunt. 2002. Peregrine Falcon *Falco peregrinus.* In The birds of North America online (A. Poole, ed.). Cornell Laboratory of Ornithology, Ithaca, NY. http://bna.birds.cornell.edu.bnaproxy.birds.cornell.edu/bna/species/660/articles/introduction. Accessed September 8, 2012.

Whooping Crane Eastern Partnership. 2011. Whooping Crane Eastern Partnership. http://www.bringbackthecranes.org/. Accessed December 1, 2011.

Wiens, J. A. 1977. On competition and variable environments. *American Scientist* 65:590–597.

Wiens, J. A. 1983. Avian community ecology: An iconoclastic point of view. Pages 353–403 in *Perspectives in ornithology* (A. Brush and G. Clark, eds.). Cambridge University Press, London.

Wiens, J. A., and M. I. Dyer. 1975. Simulation of Red-winged Blackbird impact on grain crops. *Journal of Applied Ecology* 12:63–82.

Wiggins, D. A., T. Pärt, and L. Gustaffson. 1994. Seasonal decline in Collard Flycatcher *Ficedula albicollis* reproductive success: An experimental approach. *Oikos* 70:359–364.

Wikelski, M., E. M. Tarlow, A. Raim, R. H. Diehl, R. P. Larkin, and G. H. Visser. 2003. Costs of migration in free-flying songbirds. *Nature* 423:704.

Wilcove, D. 1990. A quiet exit. *Living Bird Quarterly* 9:10–11.

Wiley, R. W., and E. G. Bolen. 1971. Eagle-livestock relationships: Livestock carcass census and wound characteristics. *Southwestern Naturalist* 16:151–169.

Williams, G. G. 1945. Do birds cross the Gulf of Mexico in spring? *Auk* 62:98–111.

Williams, G. G. 1947. Lowery on trans-Gulf migration. *Auk* 64:217–238.

Williams, G. G. 1950. The nature and causes of the coastal hiatus. *Wilson Bulletin* 62:175–182.

Williams, G. G. 1951. Letter to the editor. *Wilson Bulletin* 63:52–54.

Williams, G. G. 1958. Evolutionary aspects of bird migration. Pages 53–85 in *Lida Scott Brown Lectures in Ornithology*. University of California at Los Angeles, Los Angeles.

Williams, T. C., and J. M. Williams. 1990. The orientation of transoceanic migrants. Pages 7–20 in *Bird migration: Physiology and ecophysiology* (E. Gwinner, ed.). Springer-Verlag, Berlin.

Williamson, M. H. 1972. *The analysis of biological populations*. Edward Arnold, London.

Williamson, P. 1971. Feeding ecology of the Red-eyed Vireo (*Vireo olivaceus*) and associated foliage gleaning birds. *Ecological Monographs* 41:129–152.

Willis, E. O. 1966. The role of migrant birds at swarms of army ants. *Living Bird* 5:187–231.

Willis, E. O. 1967. The behavior of Bicolored Antbirds. *University of California Publications in Zoology* 79:1–127.

Willis, E. O. 1972. *The behavior of Spotted Antbirds*. Ornithological Monographs 10. American Ornithologists' Union, Washington, DC.

Willis, E. O. 1973. *The behavior of Oscellated Antbirds*. Smithsonian Institution Press, Washington, DC.

Willis, E. O. 1988. Land-bird migration in Sao Paulo, southeastern Brazil. *Ornithological Monographs* 19:756–764.

Wilson, A., and C. L. Bonaparte. 1808–1814 [reprinted 1858]. *American ornithology of the natural history of the birds of the United States*. Porter and Coates, Philadelphia.

Wiltschko, R. 1992. Das Verhalten verfrachteter Vogel. *Vogelwarte* 36:249–310.

Wiltschko, W., and R. Wiltschko. 1978. A theoretical model for migratory orientation and homing in birds. *Oikos* 30:177–187.

Wiltschko, R., and W. Wiltschko. 1999. The orientation system of birds: Pt. 4. Evolution. *Journal of Ornithology* 140:393–417.

Wiltschko, R., and W. Wiltschko. 2003. Mechanisms of orientation and navigation in birds. Pages 433–456 in *Avian migration* (P. Berthold, E. Gwinner, and E. Sonnenschein, eds.). Springer, Heidelberg.

Wingfield, J. C., and B. Silverin. 1986. Effects of corticosterone on free-living male Song Sparrows *Melospiza melodia*. *Hormones and Behavior* 20:405–417.

Wingfield, J. C., and D. S. Farner. 1993. Endocrinology of reproduction in wild species. Pages 164–328 in *Avian biology*. Volume 9 (D. S. Farner and J. R. King, eds.). Academic Press, New York.

Wingfield, J. C., M. C. Moore, and D. S. Farner. 1983. Endocrine responses to inclement weather in naturally breeding populations of White-crowned Sparrows. *Auk* 100:56–62.

Winker, K. 2000. Migration and speciation. *Nature* 404:36.

Winker, K. 2009. Reuniting phenotype and genotype in biodiversity research. *BioScience* 59:57–665.

Winker, K. 2010a. *On the origin of species through heteropatric differentiation: A review and a model of speciation in migratory animals*. Ornithological Monographs 69. American Ornithologists' Union, Washington, DC.

Winker, K. 2010b. Subspecies represent geographically partitioned variation, a gold mine of evolutionary biology, and a challenge for conservation. Pages 6–23 in *Avian subspecies* (K. Winker and S. M. Haig, eds.). Ornithological Monographs 67. American Ornithologists' Union, Washington, DC.

Winker, K., P. Escalante, J. H. Rappole, M. A. Ramos, R. Oehlenschlager, and D. W. Warner. 1997. Periodic migration and lowland forest refugia in a sedentary Neotropical bird, Wetmore's Bush-Tanager. *Conservation Biology* 11:692–697.

Winker, K., K. G. McCracken, D. D. Gibson, C. L. Pruett, R. Meier, F. Huettmann, M. Wege, I. V. Kulikova, Y. N. Shuravlev, M. L. Perdue, E. Spackman, D. L. Suarez, and D. E. Swayne. 2007. Movements of birds and avian influenza from Asia into Alaska. *Emerging Infectious Diseases* 13:547–552.

Winker, K., and C. L. Pruett. 2006. Seasonal migration, speciation, and morphological convergence in the genus *Catharus* (Turdidae). *Auk* 123:1052–1068.

Winker, K., and J. H. Rappole. 1992. Timing of migration in the Yellow-bellied Flycatcher in south Texas. *Condor* 94:526–529.

Winker, K., J. H. Rappole, and M. A. Ramos. 1995. The use of movement data as an assay of habitat quality. *Oecologia* 101:211–216.

Winker, K., D. W. Warner, and A. R. Weisbrod. 1992. The Northern Waterthrush and Swainson's Thrush as transients at a temperate inland stopover site. Pages 384–402 in *Ecology and conservation of Neotropical migrant landbirds* (J. M. Hagan III and D. W. Johnston, eds.). Smithsonian Institution Press, Washington, DC.

Winkler, D. 2005. How do migration and dispersal interact? Pages 401–413 in *Birds of two worlds* (R. Greenberg, and P. Marra, eds.). Johns Hopkins University Press, Baltimore.

Winterbottom, J. M. 1943. On woodland bird parties in Northern Rhodesia. *Ibis* 85:437–442.

Wires, L. R., F. J. Cuthbert, D. R. Trexel, and A. R. Joshi. 2001. *Status of the Double-crested Cormorant (Phalacrocorax auritus) in North America.* University of Minnesota, Department of Fisheries and Wildlife, St. Paul, MN.

Witmer, M. C., D. J. Mountjoy, and L. Elliot. 1997. Cedar Waxwing *Bombycilla cedrorum.* In The birds of North America (A. Poole, ed.). Cornell Laboratory of Ornithology, Ithaca, NY. http://bna.birds.cornell.edu.bnaproxy.birds.cornell.edu/bna/species/309/articles/introduction. Accessed September 8, 2012.

Wolfson, A. 1948. Bird migration and the concept of continental drift. *Science* 108:23–30.

Wood, B. 1992. Yellow Wagtail *Motacilla flava* migration from West Africa to Europe: Pointers towards a conservation strategy for migrants on passage. *Ibis* (Supplement) 1:66–76.

Woog, F., M. Wink, E. Rastegar-Pouyani, J. Gonzalez, and B. Helm. 2008. Distinct taxonomic position of the Madagascar stonechat (*Saxicola torquatus sibilla*) revealed by nucleotide sequences of mitochondrial DNA. *Journal of Ornithology* 149:423–430.

World Health Organization. 2007. Epidemic and pandemic alert response: Avian influenza. http://gamapserver.who.int/mapLibrary/Files/Maps/Global_SubNat_H5N1inAnimal ConfirmedCUMULATIVE_20070524.png. Accessed May 24, 2007.

World Organization for Animal Health (Organisation Mondiale de la Santé Animale). 2006. Wild birds' role in HPAI crisis confirmed. http://www.oie.int/eng/press/en_060602. htm. Accessed May 24, 2007.

World Organization for Animal Health (Organisation Mondiale de la Santé Animale). 2011. Update on highly pathogenic avian influenza in animals (type H5 and H7). http://www.oie.int/animal-health-in-the-world/update-on-avian-influenza/2011/. Accessed December 1, 2011.

Wunderle, J. 1992. Sexual habitat segregation in wintering Black-throated Blue Warblers in Puerto Rico. Pages 299–307 in *Ecology and conservation of Neotropical migrant landbirds* (J. M. Hagan III and D. W. Johnston, eds.). Smithsonian Institution Press, Washington, DC.

Xu, X., K. Subbarao, N. Cox, and Y. Guo. 1999. Genetic characterization of the pathogenic influenza H5N1 virus: Similarity of its hemagglutinin gene to those of H5N1 viruses from the 1997 outbreaks in Hong Kong. *Virology* 261:15–19.

Xu, X., X. Zheng, and H. You. 2009. A new feather type in a nonavian theropod and the early evolution of feathers. *Proceedings of the National Academy of Sciences* 106:832–834.

Yaremych, S. A., R. E. Warner, P. C. Mankin, J. D. Brawn, A. Raim, and R. Novak. 2004. West Nile virus and high death rate in American Crows. *Emerging Infectious Diseases* 10(4):709–711.

Yasué, M., C. J. Feare, L. Bennun, and W. Fiedler. 2006. The epidemiology of H5N1 avian influenza in wild birds: Why we need better ecological data. *BioScience* 56:923–929.

Ydenberg, R. C., R. W. Butler, and D. B. Lank. 2007. Effects of predator landscapes on the evolutionary ecology of routing, timing and molt by long-distance migrants. *Journal of Avian Biology* 38:523–529.

Yeagley, H. L. 1947. A preliminary study of a physical basis of bird navigation. *Journal of Applied Physiology* 18:1035–1063.

Yohannes, E., H. Biebach, G. Nikolaus, and D. J. Pearson. 2009. Migration speeds among eleven species of long-distance migrating passerines across Europe, the desert and eastern Africa. *Journal of Avian Biology* 40:126–134.

Yohannes, E., K. A. Hobson, and D. J. Pearson. 2007. Feather stable-isotope profiles reveal stopover habitat selection and site fidelity in nine migratory species moving through sub-Saharan Africa. *Journal of Avian Biology* 38:347–355.

Yohannes, E., K. A. Hobson, D. J. Pearson, and L. I. Wassenaar. 2005. Stable isotope analyses of feathers help identify autumn stopover sites of three long-distance migrants in northeastern Africa. *Journal of Avian Biology* 36:235–241.

Yohannes, E., M. Valcu, R. W. Lee, and B. Kempenaers. 2010. Resource use for reproduction depends on spring arrival time and wintering area in an arctic breeding shorebird. *Journal of Avian Biology* 41:580–590.

Yosef, R. 1996. Loggerhead Shrike (*Lanius ludovicianus*). In The birds of North America online (A. Poole, ed.). Cornell Laboratory of Ornithology, Ithaca, NY. http://bna.birds.cornell.edu.bnaproxy.birds.cornell.edu/bna/species/231/articles/introduction. Accessed September 8, 2012.

Young, B. E. 1991. Annual molts and interruption of the fall migration for molting in Lazuli Buntings. *Condor* 93:236–250.

Young, B. E. 1996. An experimental analysis of small clutch size in tropical house wrens. *Ecology* 77:472–488.

Young, W. 2005. To the Editor. *Audubon Naturalist Society Newsletter*, August–September.

Zahavi, A. 1971. The social behavior of the White Wagtail *Motacilla alba alba* wintering in Israel. *Ibis* 113:203–212.

Zimmer, J. T. 1938. Notes on migrations of South American birds. *Auk* 55:405–410.

Zimmerman, J. L. 1966. Effects of extended tropical photoperiod and temperature on the Dickcissel. *Condor* 68:377–387.

Zink, R. M. 2002. Towards a framework for understanding the evolution of avian migration. *Journal of Avian Biology* 33:433–436.

Zink, R. M. 2011. The evolution of avian migration. *Biological Journal of the Linnean Society of London* 104:237–250.

Zink, R. M., and J. B. Slowinski. 1995. Evidence from molecular systematics for decreased avian diversification in the Pleistocene epoch. *Proceedings of the National Academy of Sciences* 92:5832–5835.

Zink, R. M., and J. Klicka. 1997. The importance of recent ice ages in speciation: A failed paradigm. *Science* 277:1666–1669.

Ziolkowski, D., Jr., K. Pardiek, and J. R. Sauer. 2010. On the road again: For a bird survey that counts. *Birding* 42(4):32–40.

TAXONOMIC INDEX

INDEX